T0269487

SUSTAINABLE POWER TECHNOLOGIES AND INFRASTRUCTURE

SUSTAINABLE POWER TECHNOLOGIES AND INFRASTRUCTURE

Energy Sustainability and Prosperity in a Time of Climate Change

GALEN J. SUPPES and TRUMAN S. STORVICK
Department of Chemical Engineering,
The University of Missouri Columbia,
Columbia, Missouri

AMSTERDAM • BOSTON • HEIDELBERG • LONDON
NEW YORK • OXFORD • PARIS • SAN DIEGO
SAN FRANCISCO • SINGAPORE • SYDNEY • TOKYO
Academic Press is an imprint of Elsevier

Academic Press is an imprint of Elsevier
125, London Wall, EC2Y 5AS.
525 B Street, Suite 1800, San Diego, CA 92101-4495, USA
225 Wyman Street, Waltham, MA 02451, USA
The Boulevard, Langford Lane, Kidlington, Oxford OX5 1GB, UK

ISBN: 978-0-12-803909-0

British Library Cataloguing-in-Publication Data

A catalogue record for this book is available from the British Library.

Library of Congress Cataloging-in-Publication Data

A catalog record for this book is available from the Library of Congress.

For Information on all Academic Press publications visit our website at http://store.elsevier.com/

Publisher: Joe Hayton
Acquisition Editor: Raquel Zanol
Editorial Project Manager: Mariana Kühl Leme
Editorial Project Manager Intern: Ana Claudia A. Garcia
Production Project Manager: Kiruthika Govindaraju
Designer: Matthew Limbert

CONTENTS

LIST OF FIGURES

PREFACE

A typical list of the top 10 of the world's largest corporations includes seven oil and gas corporations, one electrical utility company, Walmart, and one more that can vary from year to year. Energy and its application is the world's largest industry sector and impacts our lives on a daily basis. In view of the trillions of dollars of annual revenue, it should be no surprise that the many perspectives advertised to the public on energy alternatives are greatly influenced by the special interests of these corporations.

The primary purpose of this book is to provide perspectives on energy utilization, planning, and paths to sustainability that are free of special interests—perspectives that can be verified by facts and common sense.

One of the most misunderstood of the energy sources is nuclear energy. In fact, this book is a revised version of the book entitled *Sustainable Nuclear Power*, which was published in 2007. Between 2007 and 2015, a number of unforeseen events occurred; including development of tight oil reserves in the United States, increased use of wind power that is backed up by more plentiful natural gas reserves, and the tsunami in Japan that caused meltdowns at the Fukushima power plant.

The greatest mistakes that could be made in energy planning are mistakes due to a lack of understanding of the full potential and full implications of pursuing or not pursuing energy alternatives. The emphasis of this book revision remains nuclear power because of the extent to which it is misunderstood. However, the impact of tight oil production, wind power, new natural gas reserves, and even new modes of transportation have warranted an increase in the covering of these topics in this new edition entitled *Sustainable Power Infrastructure—Energy Sustainability and Prosperity in a Time of Climate Change.*

NUCLEAR POWER IN HISTORY

In 1939, scientists in Europe reported that when Uranium was radiated with neutrons the Uranium atom nucleus underwent "fission." The two pieces of the "split" Uranium nucleus formed the nuclei of two smaller atoms, two or three neurons and a huge pulse of energy. This experiment suggested that it would be possible to make a very powerful nuclear weapon using a Uranium fission chain reaction as the source of explosive energy. World War II was just underway and all nuclear fission experiments in the United States were classified "top secret!"

The public announcement of progress on the secret weapons program was a powerful explosion over Hiroshima Japan on August 6, 1945 that incinerated the city center. Three days later a second explosion over Nagasaki Japan produced similar results. This led the Japanese to surrender ending World War II with the documents signed on September 2,1945 on the USS *Missouri*. This was accomplished just six years after the announcement of nuclear fission in 1939!

Still under the vale of top secret, Admiral Hyman Rickover was assigned the task of "taming" the nuclear energy release to produce a reactor to provide power for submarines. A fully functional USS *Nautilus* was launched in September 1954. The contractors that built submarines also built fossil fuel (coal)-fired power plants—nuclear energy looked like an option for producing electricity for the domestic market.

President Dwight Eisenhower delivered the "Atoms for Peace" speech on December 8, 1953 essentially inviting Admiral Rickover to share the technology to build the first commercial nuclear power plant. Adapting the nuclear fuel for the submarine for a reactor to produce commercial electricity was done. The first commercial nuclear power plant was completed at Shippingsport, PA, on May 26, 1958. This is a remarkably short 19 years from the discovery of nuclear fission to the scientific and technology development that produced commercial nuclear power!

After 57 years of commercial US nuclear power production, the United States has 99 nuclear power plants producing 799 billion kWh of electricity (19.5%) of the total commercial electricity. These reactors require 18,692 metric tons of uranium metal per year to produce new fuel.

The global numbers of reactors (including the United States) are 437 producing 2411 billion kWh (11.5%) of the total global electricity production. This will require 66,888 metric tons of uranium to produce the enriched uranium fuel. The US Defense Department nuclear technology was essential for the rapid growth of domestic nuclear power.

While the required technologies have been demonstrated, the vision that defines paths to economically sustainable commercialization is lacking. It is the vision and discussions to select these new technologies that make this book unique.

Our purpose for writing this book is to help you, our reader, understand energy alternative sources, how energy is used to meet transportation and residential needs, and how nuclear power is one (maybe the only) means to provide abundant and sustainable power to the world.

The book contents are designed to incrementally build on the science and engineering foundations of those early engineers and scientists. Most of them did not have a background in nuclear science or engineering. Our plan is that this text will encourage young scientists and engineers to join the tasks developing the necessary technical advances. The text is written so that the curious citizens can be informed about this energy supply options. Citizen support is important to future expansion of this energy options.

The sustainability that nuclear fission energy can bring includes more than electrical power. Nuclear power can extend sustainability through applications to transportation and space heating using grid electricity. This has been the topic of multiple Reports to Congress and Congressional Research Service reports. Where possible, these reports are cited and directly quoted.

The text is designed to be understandable to interested citizens and legislators. They will ultimately lead the government to change policies and allow nuclear energy to be safer, more available, and free of nuclear waste handling issues.

ORGANIZATION OF THE BOOK

The chapters of this book are intended to be self-contained. This results in duplication of topics, but it should be easier to read where you have special interest.

Chapters 1–3 cover the history of energy, the reserves, and some of the renewable resources available to us. Energy history on Earth begins with the sunlight that forms wood and living organisms. Wood and living organisms are the raw materials stored by nature and transformed by geological processes over millions of years. This is the source of coal, petroleum, and natural gas. Wood warmed the cradle of civilization. Coal was probably first used 2000 years ago, and more recently, it powered the start of the industrial revolution. Liquid fuel is easy to use in engines and served to power the modern automobile. The first fixed wing aircraft flew using liquid fuel just after 1900. Liquid petroleum fuels made the twentieth century the petroleum age with the mid-century addition of pipeline distribution of natural gas.

Chapter 4 evaluates alternative energy sources and technologies. We depend exclusively on petroleum fuels for transportation, and any interruption of our supply of imported petroleum can quickly become an economic and social problem. Coal is the main source of fuel for electric power production; here, the competition from other energy sources keeps the price of electricity fairly stable. Known and emerging technologies can stabilize energy prices and create security from unemployment and military conflict.

The story of energy through the nineteenth and twentieth centuries depended on the work produced by hot gases expanding in engines, the machines designed to do work. The science and technology of the development of these new machines is summarized in Chapter 5. Chapters 6 and 7 describe the technologies that provide transportation, electricity, and the equipment we use to heat and cool our homes and workplaces.

Prior to radiation from the sun touching the earth, it was atomic energy that shaped the universe. The discovery of nuclear fission just prior to World War II drove the research into the "top secret" underground. The power of nuclear energy was announced to the public near the end of World War II with two thunderous explosions over Japan. Two city centers were leveled and thousands of people evaporated.

This unfriendly introduction to nuclear energy has produced an attitude among many that everything nuclear should be banned. Even the name "nuclear magnetic resonance imaging" (an important medical diagnostic tool) had to be changed to calm patients' anxiety using the diagnostic tool. Attitude not withstanding, nuclear energy touches us in the form of radiation from the sun and geothermal heat every day. As we confront the depleting oil and coal reserves, keep in mind there is a huge energy reserve available using nuclear energy.

Today, there are 100 nuclear power plants in the United States. They produce about 19% of the electricity we use every day. Fission a few pounds of uranium as "nuclear fuel" and it replaces thousands of tons of coal or diesel fuel. A nuclear powered submarine can cruise under water for months instead of hours. Nuclear reactors provide electrical power without the air pollution associated with burning coal, petroleum, and natural gas.

The source of this nuclear energy originated when atoms were formed, long before our solar system existed. All of the atoms that we now find in the gases, liquids, and solids we use were assembled in and among the stars from the particles and energy that. make up our sun and the rest of the Milky Way Galaxy. Remarkably, the history of energy starts, and ends, with nuclear energy.

Chapters 1 through 7 develop the case for nuclear energy to provide our electrical, transportation, and residential power needs. Nuclear processes are put in context as natural processes that are vital to our ecosystem. This includes the warmth of the sun's radiation and the role nuclear decay plays in maintaining the earth's molten core and creating the habitable surface where we live.

Chapter 8 is an extended discussion of the future of nuclear power. There is a section on nuclear science—a complex subject made remarkably simple to understand and use with the tools the scientists have provided. Safety is important with radiation exposure limits and safety presented. The international fleet of nuclear reactors is copies of the reactors designed for submarines with improvements in instrumentation and operating procedures. The spent fuel from these commercial power reactors has produced only 2% of the energy available in that fuel. Should that spent fuel be chemically reprocessed to recover fuel values? There are Generation IV reactors that have been designed and tested that can operate on reprocessed fuel. These developments expand the available nuclear energy by about 1000fold. Is there a role for "small" reactors? We include an introduction to potential for fusion reactors—a technology "in the making." Is it possible produce energy the way the sun does it? The Federal Government has had a huge impact on the development of nuclear science and technology. It's certainly worth a look at the reports to congress. This is our longest, multitopic chapter.

Chapter 9 discusses the use of the electrical power grid as the conduit through which nuclear energy can be delivered to serve energy needs. The grid provides the energy for transportation and space heating for homes and businesses. Technologies like plug-in hybrid electric vehicles and heat pumps for winter home heating can use nuclear energy from the grid to reduce (eliminate) the need to import both petroleum and natural gas.

Chapter 10 presents topics related to the future of the "nuclear energy" option. There will be alternatives to nuclear power since it has become "global." There are third-world societies that have domestic energy sources. The manufacturing infrastructure will "speak" to their sources of electricity. Is cost the only (or best) metric for selecting future energy options? An experiment to watch is the growth of the modular electric vehicle. It has been demonstrated that the electric vehicle has won the fuel to electricity to travel distance contest. The near term future will be much like today. The long-term vision of the energy future dims as decades are added to the estimates.

As this is written there is another international conference on climate change. Carbon dioxide in the atmosphere from combustion of fossil fuels has been identified as the culprit. A decision to "tax" carbon dioxide emissions or a call for a reduction in fossil fuel combustion will certainly impact the sources of energy in the future.

ACKNOWLEDGMENTS

Discussions with colleagues led us to write about the strong interaction between science and technology, economics, and legislation in our modern society. We thank the hundreds of students who have been patient with our presentations and taught us that the KISS (Keep It Simple Stupid) theory of teaching really works. Complex solutions are best formulated by answering a series of well-stated simple questions that leave a trail to that complex problem solution.

Thanks are due for the hospitality of the Chemical Engineering Department shown by a Professor in retirement (TSS). This was an inspiration that made partnering in this writing project possible.

CHAPTER 1

Energy and Civilization

Contents

The remarkable improvement in the standard of living in the United States during the twentieth century is unprecedented in world history. Travel that once took months can now occur in hours. Viruses that had the capability of genocide have all but been eradicated.

These advances are attributed to technology. Technology has proved to be decisive in winning wars, and good technology choices have determined the fate of countries and empires.

Technology is not science; it is the way that science is used to make those things we use such as automobiles, refrigerators, and computers. Science education begins as part of the K-12 education made available to all in the United States. Technology is usually covered in engineering and medical-related education programs. The average person receives an education on technology through conversations with friends and through articles written by journalists who are not educated on technology.

Sustainable Power Technologies and Infrastructure.
DOI: http://dx.doi.org/10.1016/B978-0-12-803909-0.00001-4

The average person is not well informed to provide informed influence on how nations should proceed to use and develop new technology. While it is not within the capabilities of a single book to provide a complete education on technology, a book is capable of presenting more complete descriptions of the advantages and of specific technologies such as the energy-related topics presented in this book.

A first step to providing more complete descriptions of energy technologies is to identify the source of energy and how all the forms of energy at our disposal share a common origin. We simply tap into energy sources at different stages of its passage through time.

ENERGY IN TODAY'S WORLD

Sunlight is yesterday's atomic energy. The energy stored in wood and vegetable oils was yesterday's sunlight. Yesterday's wood and vegetable oils are today's coal and crude oil. Yesterday's coal and crude oil are today's natural gas. A description of these natural energy stockpiles and their history sets the tone for subsequent discussion on technology using these energy reserves.

Nature used time to transform the sunlight to wood, oil, coal, petroleum, and natural gas. Today, man can transform these reserves in a matter of hours. Relatively simple processes for converting petroleum into gasoline have evolved into technologies that allow coal to be chemically taken apart and put back together at the molecular level. Fuel cells can convert chemical energy directly to electricity without combustion.

To understand the advantages and disadvantages of nature's various energy reserves requires an understanding of engines and power cycles. Studying the text on gasoline engines shows in a matter of minutes how these machines work. Likewise, processes for converting coal into electricity that took a couple centuries to develop can now be quickly explained.

At the start of the twentieth century, suitable liquid fuels were rare, and the proper match of a fuel with an engine was an art. Today, we can move vehicles or produce electricity from energy originating in petroleum, coal, natural gas, wood, corn, trash, sunlight, geothermal heat, wind, or atomic energy. Each of these can be used in different ways. Natural gas, for example, can be used directly in spark-ignition engines, converted to gasoline fuel, converted to diesel fuel, converted to hydrogen fuel, or used as fuel to produce electricity.

Today's world is one where technology can do much more than what might be cost-effective or sustainable. For example, it is possible to use an atomic accelerator to convert cheap metals into gold, and it is possible to separate from sea water many valuable metals including gold and uranium. These technologies are simply not cost-effective either from a "dollar" or "energy input" perspective.

So, which technologies are the right technologies to use today to provide us and our children the best possible futures?

The process for unlocking the potential of technology starts with asking the right questions. Both history and science are part of the story we tell.

Gasoline from Coal Technology

In 1940, Germany was converting coal into high-quality diesel and jet fuel, and they were able to sustain this industry (aside from allied bombing) using coal that was considerably more expensive to mine than the vast, rich reserves of today's Wyoming coal. Wyoming has vast supplies of coal in 40-foot thick seams just less than 100 feet below the surface—it can literally be harvested and loaded into trucks for a few dollars a ton.

Synthetic fuel production, as an alternative to crude oil that was not available, was sustainable in Germany in 1940. *Why is it not sustainable today with cheaper coal, 60 years of scientific and technological advances, and pipeline distribution that does not rely on costly petroleum tanker shipment from the middle east?* Originally, the German synthetic fuel process was designed to produce refinery feedstock. Can the synthetic fuel industry compete today by producing a fuel that can be directly used in engines? If the refinery could be bypassed, the cost advantages of synthetic fuels might advance it over petroleum alternatives but not at the low price of petroleum in January 2015.

South African synthetic fuel (the German Fischer–Tropsch process) facilities were able to sustain production of synthetic oil from coal in competition with world crude oil prices at $10 per barrel in the late 1990s. Canadian syncrude (synthetic crude oil) facilities are reported to be producing petroleum from oil sands at $10–$12 per barrel. The oil sand reserves are estimated to be about the same size as world reserves of petroleum. Today, Canadian oil sands are used instead of imported oil—the technology is sustainable and profitable.

Why have South Africa and Canada been able to incubate these industries during the past few decades while the United States *failed and, today, remains* *without a significant synthetic fuel industry to replace crude oil imports that exceed $200 billion per year?* Lack of competitive technology is not at fault.

Repeatedly, US voters have given the mandate to foster cost-competitive alternatives to imported petroleum. *Do US policies foster the development of replacements for petroleum, or do US policies lock in competitive advantages for petroleum over alternatives?* When you get past the hype of fuel cells, ethanol, and biodiesel, a comparison of US tax policies on imported crude oil relative to domestic fuel production reveals practices that favor crude oil imports. These and similar policies are the economic killers of a technology that might eliminate the need to import fuels and would create quality US jobs.

When considering alternative fuels understand that the liquid fuel distribution infrastructure and the refineries are controlled by corporations with vested interests in gasoline and diesel fuel. With this and other barriers to commercialization in the United States, the most likely options to succeed are those that do not rely on new fuels and distribution infrastructure. The two options are electrical power and natural gas, and of these, natural gas imports have recently been significantly reduced by production of natural gas from domestic shale oil deposits.

Natural gas provides limited advantage over petroleum, but recently the price of natural gas per unit of energy has become competitive with gasoline. Electrical power provides a domestic alternative that does not rely on a new fuel distribution infrastructure—a reliance on diverse indigenous energy supplies creates stability in prices and reliability in supply. Electricity is the one option that can substantially replace petroleum as *the* transportation fuel. Of the options to produce electrical power, nuclear stands out due to its abundance and its fuel supply provides electrical power without the generation of greenhouse gases.

The utility of electrical power is extended to automobiles with "plug-in" hybrid electric vehicles (PHEVs). PHEVs can use electrical power to replace all imported oil without producing air pollution. Use of PHEVs could reach cost parity with conventional gasoline vehicles in a matter of months if development and production of the technology was made a national priority. In a decade of evolution, the average consumer could save $1,000–$2,000 over the life of a vehicle using these technologies rather than the conventional gasoline engines.

Having missed the entry positions on technologies like Fischer–Tropsch fuels and Canadian tar sands is PHEV technology an opportunity? If PHEV technology is the right opportunity at the right time, is it also the last real opportunity before other nations challenge the economic might of the United States?

What about nuclear energy and nuclear waste? Can the fission products in spent nuclear fuel be separated from the bulk of the waste—the bulk being mostly uranium that when recovered can be valuable as fuel?

Sustainable Nuclear Energy

Figure 1.1 summarizes the legacy of 30 years of nuclear power production in the United States. While much attention has been paid to the radioactive spent fuel generated by commercial nuclear power, the fact is that 30 years of fission products from all the US facilities would occupy a volume less than the size of a small house. On the other hand, the inventory of

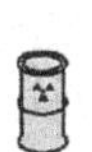

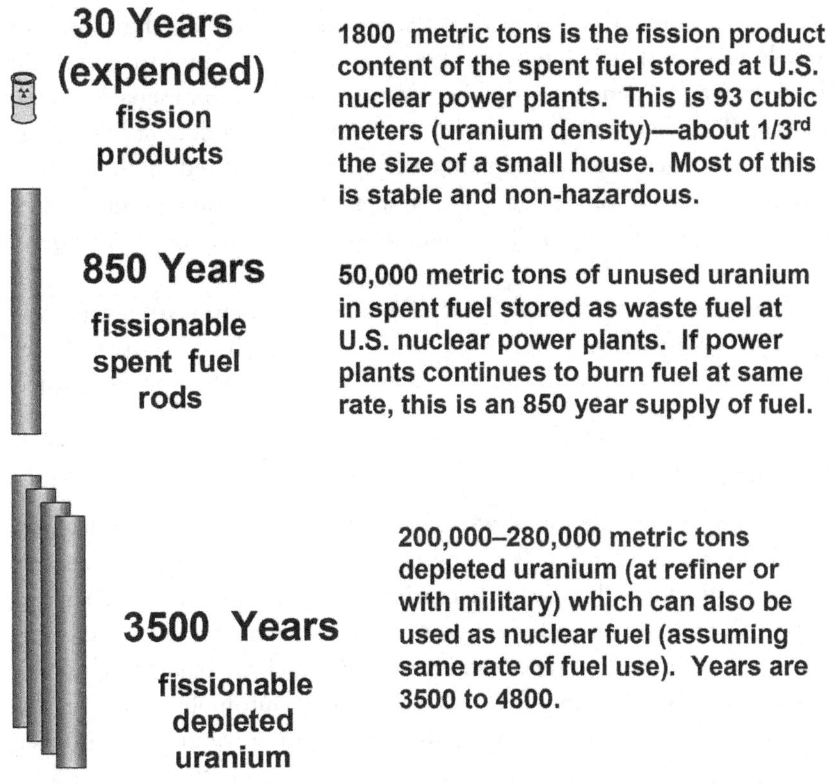

Figure 1.1 The legacy of 30 years of commercial nuclear power in the United States including 30 years of fission products that are of little value and sufficient stockpiled fissionable fuel to continue to produce electrical power at the same rate for another 4350 years.

stockpiled fissionable material in the form of spent fuel and depleted uranium could continue to supply 18% of the electrical power to the United States for the next 350 years without additional mining and while reducing any waste or hazard. This fuel inventory is a valuable resource and represents material that has already been mined, processed, and stored in the United States.

Reprocessing spent nuclear fuel emerges as the key to sustainable, abundant, and cheap electricity. Reprocessing is removing the small fraction; that is, "*the fission products" from spent nuclear fuel . . . recovering the bulk that is potentially valuable* nuclear fuel. The removal of the fission products is easier (its chemistry) than mechanically concentrating the fissionable U-235 isotope (isotope enrichment), used to convert natural uranium into reactor fuel grade uranium. The energy inventory illustrated in Figure 1.1 is available with chemical reprocessing of the spent nuclear fuel. Generation IV nuclear reactors should be developed to use the "whole ton" of natural uranium. These technologies can actually use the "nuclear waste" generated by the existing fleet of commercial nuclear reactors. This electrical power would be generated producing little to no greenhouse gases while eliminating the spent fuel stored at the nuclear power plant facilities. About 3.4% of spent fuel is fission products. Less than 0.5% of the fission products require long-term radioactive storage or burial.

Technologies are available that allow nuclear power to meet every aspect of sustainability. The inventory of uranium that has already been mined will produce energy longer than scientists can reasonably project new energy demands or sources.

The Critical Path

Within the past decade, there were serious concerns about the availability and high price of liquid fuels for automobiles. The past couple of decades have seen fracking (hydraulic fracturing of shale formations containing "tight oil" and natural gas) yield new and significant US reserves of fuels. At the same time, the nuclear industry is on track to build new nuclear power plants and several electrical and PHEV automobiles are on the market and on an evolutionary path that can improve and sustain their presence.

The past decade has witnessed a transformation from "urgency" in our energy plight to an era of real opportunity. If handled carefully this opportunity will yield sustainable and abundant energy for centuries, major improvements in quality of life (faster and lower cost transportation), and reduced greenhouse gas emissions.

SOURCES OF ENERGY

The past 100 years are like a blink of the eye in the history of the earth, yet, within the last century, scientists have unraveled the history of energy. This story goes hand in hand with the history of the universe. Following energy back in time takes you to the origin of the universe.

Your body is powered by the energy stored in the chemical bonds of the food you eat. The energy in this food is readily observed by taking a match to a dried loaf of bread and watching it burn. Both your body and the fire combine oxygen and the bread to form water and carbon dioxide. While the fire merely produces heat in this reaction, your body uses the energy in a very complex way to move muscles and produce the electrical energy of your nervous system to control motion and thought.

Both your body and the fire use the chemical energy stored in the starch molecules of the bread. This energy is released as chemical bonds of starch and oxygen and are converted to chemical bonds in water and carbon dioxide. Even the molecules your body retains will eventually return to carbon dioxide, water, and minerals.

The energy in the chemical bonds of the food came from photosynthesis that uses the energy from the sun to combined carbon dioxide and water to produce vegetation and oxygen. While the oxygen and carbon stay on earth and cycle between vegetation and the atmosphere, solar radiation has a one-way ticket into the process where it provides the energy to make life happen. Without this continuous flow of energy from the sun, our planet would be cold and lifeless.

The radiation that powers the photosynthesis is produced by the virtually endless nuclear reaction in the sun. In this process, hydrogen atoms combine to form helium. When hydrogen atoms join to form more stable helium, the total mass is slightly reduced. The lost mass is converted into energy according to Einstein's equation, $E = mc^2$. Enough mass (hydrogen) was formed during the birth of the universe to keep the stars shining during the past 13–14 billion years, the best estimate of the age of the universe.

The presence of different elements in our planet, solar system, and galaxy reveals energy's history. All forms of energy on earth originated at the birth of the universe. Our life and the machines we use depend on energy's journey, catching a ride as the energy passes by. We are literally surrounded with energy in hydrogen, uranium, and chemical bonds that limit our use of this energy largely determined by our choices and, in some cases, our pursuit of technology to better use available resources.

NATURE'S METHODS OF STORING ENERGY

All forms of energy, whether nuclear, chemical energy in coal, chemical energy in petroleum, wind, or solar, are part of energy's journey that started with the birth of the universe. In our tiny corner of the universe, the energy output of the sun dwarfs all other energy sources. Nuclear fusion in the sun releases massive amounts of energy. The only way this energy can escape from the sun is in the form of radiation. Radiation output increases as temperature increases. Somewhere along the journey, the sun came into a balance with its radiant energy loss tending to decrease the temperature at the same rate as the nuclear fusion worked to increase the temperature. In this process, the outward force of the constant nuclear explosions is balanced by the sun's gravitational force to form a nearly perfect sphere.

Before life evolved, solar energy shined on the earth. If you close your eyes and look at the sun, your face warms while the top of your head receives little of this warming energy. The radiation causes the earth's equator to be warmer than the poles. These temperature differences cause wind in the atmosphere and ocean currents.

Before life existed on the earth, the sun's radiation formed water vapor and caused it to rise from the oceans into the atmosphere. This water vapor catches the wind and is blown to the mountains where it is cooled to form rain. The high elevation of this water in the mountains is pulled by gravity giving it energy to flow downhill. Rocks and gravel dissipate this energy on its journey back to the oceans—the potential energy from water's height in the mountain is converted to the kinetic energy of the water moving through rapids and eventually to thermal energy as it reaches the ocean.

The first primitive organic life appeared on earth about 3 billion years ago. Progressively, more complex molecules were formed. The key to life on earth was the molecule chlorophyll. This is nature's "workshop tool" that takes water and carbon dioxide from the air (plus a "pinch" of minerals), powered by solar energy to give us photosynthesis first occurring about one billion years later [2]. Small variation in the chlorophyll molecule produced the spectacular variety of green plants—potential food and fiber that supported life for our early ancestors.

When the earth was formed, the geological record showed that the atmosphere was rich in carbon dioxide and some really toxic gases. The toxic gases were gradually removed by natural chemical reactions. Photosynthesis allowed the carbon dioxide in the air to be combined with water to form vegetation and oxygen. The removal of carbon dioxide and replacing it with oxygen was an essential step providing for human development from animals and the evolution to man.

The vast majority of the early vegetation fell to the ground and decomposed combining with oxygen returned it to carbon dioxide and water. Some fell to the bottom of swamps where oxygen could not reach it as fast as it piled up. At locations where fallen vegetation accumulated faster than oxygen and microbes could convert it back to water and carbon dioxide, the deposits were buried deeper and deeper making it even more difficult for oxygen to reach them. After a sufficiently long time, the vegetation rearranged into more stable deposits that we call coal. Different types of coal developed depending upon the depth, temperature, and moisture of the deposits. This preservation process was particularly effective in swamps where the water reduced the rate at which oxygen could reach the fallen vegetation.

The surfaces of oceans, lakes, and ponds were inhabited by bacteria called phytoplankton (small, floating, or weakly swimming animals or plant life in water). The cells of these phytoplankton contained oils that are in some ways similar to the corn oil used to cook French fries. When these phytoplanktons died, most of them were converted back to carbon dioxide and water by the oxygen dissolved in the water or by animal feeding. Some were swept to ocean depths where oxygen was absent. Here they accumulated. These dead bacteria were often buried by silt. The combination of time and pressure caused by the overburden of water and silt transformed these deposits, in the absence of oxygen, to petroleum oil.

In the turmoil of erosion, volcanoes, and general continental drift, large deposits of coal and oil made it back to the surface where, in contact with oxygen, they oxidized back to water and carbon dioxide. Other deposits persist for us to recover. Still other deposits were buried deeper, reaching higher pressures due to overburden and higher temperatures, due to the earth's geothermal heat. There, the coal and petroleum converted to a combination of natural gas and high-carbon deposits of hard coal or carbon in the form of graphite.

Over tens of millions of years, the solar energy working with life on earth continued to form the energy deposits of coal, petroleum, and natural gas. Currently, yesterday's radiation is available as vegetation such as wood, corn, palm oil, etc. Today's radiation is available as sunlight, wind, ocean currents, and the hydro energy of water in high-altitude rivers and lakes.

The legacy of the universe is all around us. Compared to our consumption of energy, the fusion energy available in the hydrogen of the waters of the ocean is almost endless. Uranium available in the soil and dissolved in the ocean can produce energy by fission. All atoms smaller than iron could be fused to form iron while all atoms larger than iron could undergo nuclear decay or fission (splitting) to form iron; both processes involve the nuclei of atoms and release vast amounts of energy as they naturally occur.

The geothermal energy (heat) of the earth originated at the birth of the universe. The cosmic forces at the beginning formed the atoms that collected, formed rocks that became earth with a molten center. If the heat were "turned off" the core of the earth would have long ago cooled and solidified. Adding to the heat of colliding masses (the drifting continents), uranium and other larger molecules are constantly undergoing nuclear rearrangements (including fission) from the surface to the center of the earth. The fission energy release occurs in one atom at a time, but the total energy released adds up. The released heat maintains molten magma from earth's core to near the surface. On the surface we see some of this energy released as volcanic eruptions and geysers.

It is important to recognize that nuclear conversions have always played a vital role in the evolution of life on earth. A natural nuclear reactor actually formed in Oklo, Gabon (Africa), about 2 billion years ago. This occurred when the higher concentration of U-235 existed in the ore at Oklo—the high concentrations actually produced a fission chain reaction of the U-235. When the Oklo uranium ore was recently mined, there was a lower-than-normal U-235 concentration and traces of fission products and plutonium were found in those uranium deposits.

We did not "invent" nuclear processes (or even nuclear reactors). We have learned to control nuclear processes and harness the energy released. We have options on where and how we can tap into available energy sources to power our modern machines and half way through the twentieth century, nuclear power became an option available to meet rapidly increasing energy demands.

MAN'S INTERACTION WITH NATURE'S STOCKPILES AND RENEWABLE ENERGIES

Primitive man was successful in tapping into the easily available and easily usable forms of energy. He lived in warmer climates where the solar energy protected him from the cold. Even the most primitive animals, including early man, responded to their need to nourish their bodies with food.

As the use of fire developed, man was able to move into colder climates where the energy in wood was released by burning campfires. Animal fat and olive oil were soon discovered to be useful sources of fuel to feed the fire for heat and light. These fats and oils were observed to burn longer and could be placed in containers or wrapped on the end of a stick to create a torch—hence, they are early endeavors into fuel processing. Whale blubber was later added to animal fat and olive oil for food and fuel.

The wheel and axle was another early step in developing energy technology. The wheel and axle assisted man to use his physical energy to move heavier loads. The cart was made more effective moving heavier loads using domesticated animals to pull it. For stationary applications, water wheels and windmills converted the hydraulic and wind energies into shaft work for many applications including pumping water and grinding grain. Wind energy powered ships to explore new lands, establish trade, and expanded the fishing industry.

Machines using wind and hydraulic energy made it possible for one person to do the work of many—freeing up time for them to do other tasks. An important task was educating the young. Time was also available for the important tasks of inventing newer and better machines. Each generation of new machines enhanced man's ability to educate, invent, discover, and add to leisure time.

Societies prospered when they used the freedom created by machines to educate their youth and to create new and better machines. Inventions/discoveries extended to medicines that conquered measles and polio. The benefits of modern society are available because of the effective use of energy and the way energy-consuming machines enhanced man's ability to perform routine tasks freeing time for education, discovery, and innovation. Civilization that emerged prospered.

History shows that civilization evolves based on technology. For man, the "survival of the fittest" is largely the survival of the culture most able to advance technology. In modern history, while Hitler's technology dominated the World War II battle field, Germany was winning the war. As the Allies developed new technology that surpassed German technology, the Allies began to dominate the battlefields leading to military victory.

If you drive through the Appalachian Mountains, you can see how coal seams (varying from an inch to over a foot thick) once buried a few hundred feet underground are now exposed on the open cliffs. The upheaval that created mountains also brought up deposits of coal and oil. At cliffs like these, man first discovered coal. Coal was considerably easier to gather at these locations than firewood, and gradually replaced firewood that was becoming scarce. Marco Polo observed "black rocks" being burned for heat in China during his 1275 travels [3]. Coal's utility caught on quickly. Between 1650 and 1700, the number of ships taking coal from Newcastle to London increased from 2 to about 600. In 1709, British coal production was estimated to be 3 million tons per year. Benjamin Franklin noted (1784) that the use of coal rather than wood had saved the remaining English forests and urged other counties to follow suit.

When oil was found seeping from the ground, it could be collected and used to replace an alcohol–turpentine blend called camphene (camphene being less expensive than whale oil) [4]. Eventually, mining and drilling techniques were developed to produce larger deposits of coal and oil found underground.

From a historic perspective, energy technology has tended to feed upon itself and make its utility increase at increasing rates. Large deposits of coal allowed a few men to gather as much fuel as everyone in a community gathered wood a few centuries earlier. Easy and efficient gathering of fuel freed up more time and resources to develop new and better machines that used the dependable fuel supply.

Prior to the nineteenth century, the decisions to proceed with newly demonstrated technology were easy because the benefits were obvious. The vast amounts of virgin wilderness dwarfed the small tracts of land being devastated by poor mining practices, and an energetic entrepreneur could simply go to the next town to build the next generation of machines as local markets were dominated by established businesses/corporations.

At the end of the nineteenth century, vast tracts of land and the ocean were no longer barriers to the ambitions of the people managing corporations. The telegraph allowed instant communication and steam engines on ships and locomotives allowed most places to be accessed in a matter of days. The time arrived when a budding entrepreneur could no longer go to the next town to get outside the influence of existing corporations. In energy technology, the time had arrived when companies became monopolistic energy empires.

The growth of local businesses into corporations with expanding range of influence made their products quickly available to more people. The benefits were real, but the problems were also real. One problem was that innovation was being replaced by business strategy determining those technologies that would be developed.

For energy options to be commercialized today, both technical and nontechnical barriers must be overcome. The nontechnical barriers generated by corporations and their far-reaching political influences are often greater than the technical barriers. These nontechnical barriers must be understood and addressed to adopt new energy options.

INDUSTRIAL REVOLUTION AND ESTABLISHMENT OF ENERGY EMPIRES

Standard Oil Monopoly

The Standard Oil monopoly of the early twentieth century demonstrated what happens when a corporation looses sight of providing consumers a product and focuses on strategies to produce a profit.

After the civil war, men swarmed to western Pennsylvania to lay their claims to land in a "black gold rush." John D. Rockefeller was among these pioneers. Within 1 year of discovering the economic potential of drilling for petroleum, overproduction saw the price drop from $20 per barrel to 10 cents per barrel. Rockefeller realized that the key to making money in oil was not getting the oil out of the ground but, rather, transporting the crude oil to refine it into products for distribution to customers [5]. Rockefeller crossed a line when he changed his corporate philosophy to one of profiting by stifling the competition through monopolistic control of refining and distribution of all petroleum products.

The Atlantic and Great Western Railway controlled the cheap rail transit in western Pennsylvania and this controlled the oil market. Rockefeller was able to prevent his competition from using that railroad to sell their oil.

This was a change in paradigm for the energy industry. One company controlled the market by controlling access to the commodity. While nations and shipping fleet owners had done this in the past, this time it was different. This was one company operating in a free country aggressively moving to eliminate all competition.

Artificially inflated oil prices were just like taxes, without representation. When previously faced with a similar situation the people united to bring on the American Revolutionary War. The adverse impacts of the oil monopoly were difficult to quantify, unlike a tax on tea, and there was no precedent to show the way to reasonable remedies.

Competitors of Standard Oil were stifled. Some of the competition sold out. High consumer prices, higher than the free market would bear, were the result of this monopoly. Technology and innovation were also stifled. Corporate success was determined by controlling access to the products, not by using the "best" technology.

In the past, improving technology benefited both the consumer and the company. When business "savvy" replaced innovation, the company was at odds with what was good for the consumer. The creative innovation and technology was now forced off the highway of ideas onto the back roads where progress was slow.

For 32 years, Standard Oil profited from its monopoly on oil refining and product distribution. In May 1911, the US Supreme Court Chief Justice White wrote the decision which mandated that Standard Oil divest of assets to break its monopoly within 6 months. This was accomplished by forming six independent "Standard Oil Companies" geographically separated. Five of these firms control the crude oil to market business in the United States today. In 1974, assets of the descendants of John D. Rockefeller were estimated to have the largest family fortune in the world estimated to be $2 billion.[1]

During the twentieth century, pioneers had reached the end of land in the habitable frontier. The steam engine and telegraph provided the means by which companies could extend their influence across the country and the globe. One can argue that the international nature of today's mega corporations elevates them to a status as great as the nations they claim to serve. One can further argue that a mega corporation can be a friend or enemy to a society in the same sense that a neighboring country can be a friend or an enemy.

In 1942, Senator Harry S. Truman led an investigating committee on treasonous prewar relationships between General Motors, Ethyl Corporation, Standard Oil, and DuPont in collaboration with German company IG Farben. Company documents show corporate agreements designed to preserve the corporations no matter who won World War II. Corporate technology exchanges compromised the competitive edge held by the United States entering World War II including leaded gasoline technology (critical for high octane aircraft fuels) and noncompetitive stances on synthetic rubber technology. *At the time, British intelligence called Standard Oil a hostile and dangerous element of the enemy.* (Stephenson, 1976; Borkin, 1978) [6–8]. Continuation of this behavior led to anticompetitive-related antitrust hearings on leaded gasoline technology against these American companies in 1952.

With technology and innovation taking a back seat to business interests, politics and energy technology became perpetually tangled. The larger companies were formally pursuing their agendas even when they were in conflict with public interest and involved collaboration with the enemies of our nation's closest allies.

A corporation that profits by providing consumer products more efficiently and at a lower cost is significantly different than a corporation that makes profit by controlling the supply/price of a consumer commodity or product. The Rockefeller Oil monopoly demonstrated that the profits were greater for the business strategy that is in conflict with national benefit. In the end the only punishment was the mandated divestment of the Rockefeller Oil properties. The Rockefeller family emerged as the wealthiest family in the world.

What was the real precedent set by the Standard Oil monopoly? Was it that monopolies will not be allowed, or was it that great fortunes can be made and kept even if you are caught? The consequences are that business practices and not technical merit tend to have increasing impact on which technologies become commercial and ultimately benefit the public. With the introduction of the corporate lobbyist, technical merit is debatably in at least third place in this hierarchy.

Innovation in a World of Corporate Giants

There is little doubt that obstructed commercialization stifles technology innovation. Companies and individuals have no incentive to build a nuclear powered automobile since the government would not allow this vehicle to be used on the highways (for good reason). Restricted commercialization can be good by redirecting efforts away from projects that endanger the public. Restricted commercialization can be bad when the motivation is to maintain a business monopoly and to stifle competition.

History has shown little evidence of the impact of unrestricted entrepreneurism because modern history has been dominated by business activity rather than technical innovation. The true potential of unrestricted entrepreneurism is rarely seen. World War II is the best example in recent history illustrating what happens when we focus on developing the best technology available and the machines to get jobs done. During the 10-year period from 1940, the following technologies were developed:

- Nuclear bomb
- Jet aircraft
- Radar
- Transistors
- Intercontinental rockets
- Guided missiles
- Synthetic oil produced from coal
- Mass production of military aircraft and tanks
- Swept wing and flying wing aircraft
- Stealth submarine technology
- A plastics industry including synthetic rubber, nylon, and synthetic fiber.

Many of the commercial advances between 1946 and 2000 occurred because defining technology was developed during World War II. Specific twentieth-century accomplishments that fall into this category are nuclear power, jet air travel, landing a man on the moon, guided missile technology, transistor-based electronics and communication, stealth aircraft including the B1 bomber, and the modern plastics industry.

The technological developments of World War II show what can be achieved when technology and commercialization become a national goal. If technology has inherent limits on the good it can provide, we haven't reached these limits. We are limited in this pursuit, by public opposition to some new technologies to become part of our societal infrastructure, and by the willingness of corporations or governments to invest in development and commercialization of some new technology ideas.

At the beginning of the twenty-first century, politics and energy technology are hopelessly entangled. In 2012, seven of the top ten Fortune Global 500 companies were in energy or energy use technology (e.g., the automotive industry) with a total revenue of $2.4 trillion. Management and business savvy usually trumps innovation in these companies; for example, mandated mileage increases for automobile and trucks imposed with industry opposition.

The United States rose to superpower status in the 1940s when the national focus was on developing and commercializing strategic technologies. History has shown powerful countries fall when the national focus switches from advancing technology to maintaining the steady flow of cash to corporations and well-connected individuals (maintaining corporate status quo). The Czars of Russia or aristocrats of Rome are two of many examples where common people were driven to revolt against a system dominated by the well-positioned elite.

Germany's synthetic oil production from coal is one strategic technology that did not become commercial in the United States following World War II. This technology is currently commercial in South Africa because they have no crude oil and there was a different attitude toward investing in this infrastructure. Today, in South Africa the industry is self-sustaining and the technology is being sold for use in other countries. These production facilities were designed and constructed by US energy engineering firms.

Canada started developing its oil sand resources in the 1960s even though the technology produced crude oil that cost more than the global price of crude oil. Because of continued, dedicated development, the oil sand now costs $10–$12 per barrel to produce as compared to crude oil at over $50 per barrel (year 2015) on the world market.

The Oil Economy Through 2009

The twenty-first-century civilization as we know it in the United States would be impossible without crude oil. We get over 90% of all automotive, truck, train, and air transport fuels from crude oil, as well as the majority of our plastics. The plastics are used to make everything from trash bags and paints to children's toys. If it is a solid device that is not paper/wood, ceramic/glass or metal, it is probably plastic.

Crude oil is a good fuel. Figure 1.2 illustrates how this natural product can be separated by boiling point range to provide gasoline, a middle fraction (kerosene, jet fuel, heating oil, and diesel fuel), and fuel oil (heavy oil used for boilers or large diesel engines for ships and electrical power). The natural distribution of crude oil into these three product classifications varies depending on the source of the crude oil.

Modern refining processes convert the crude oil into these three product categories and also provide chemical feedstocks. The modern refining process breaks apart and rearranges molecules to produce just the right

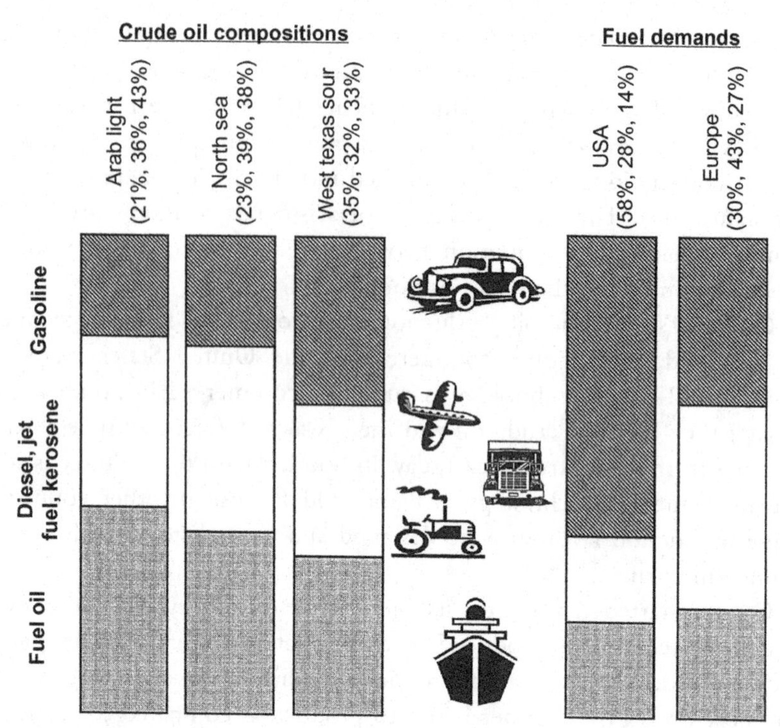

Figure 1.2 Typical composition of oil [9].

ratio of gasoline, diesel, and fuel oil. In the United States, this typically means chemically converting most of the "natural" fuel oil fraction and part of the middle fraction to increase the amount of gasoline.

Figure 1.3 is a detailed description of crude oil processing in the United States It includes imported and domestic production with other commercial applications. The diagram also illustrates how complex crude oil processing has become to provide the commercial product demands.

Figures 1.4 and 1.5 show the extent of US oil imports and how prices fluctuate. Since the year 2000, the United States was spending over $100 billion per year to import crude oil (estimated at over $200 billon per year in 2005). Cheap oil in 1997 and 1998 brought a prosperous US economy, more expensive oil in 2001 and 2002 added to the economic slump. During the next decade, prices continued to increase both due to limited supply and by increasing demand in countries like China. Since mid-2014, the market price of crude oil has dropped from about $105 to less than $60 per barrel. The economic impact of this slump will depend on how long it lasts and where the price "settles" during the next decade.

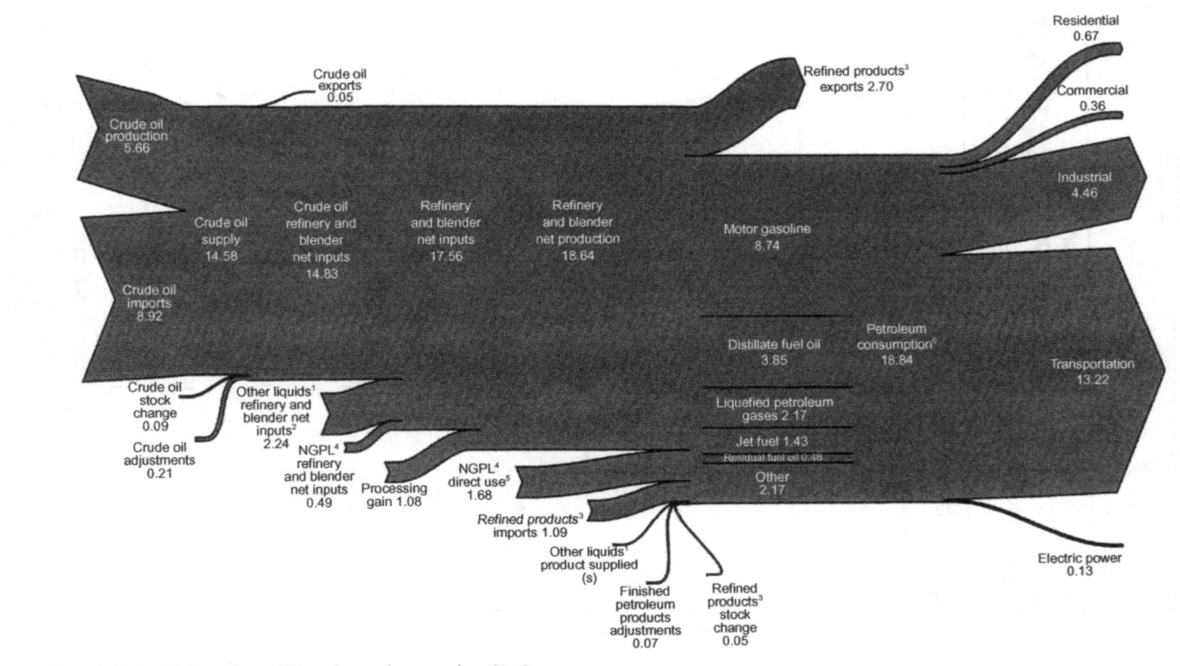

Figure 1.3 Year 2000 oil flow in million barrels per day [10].

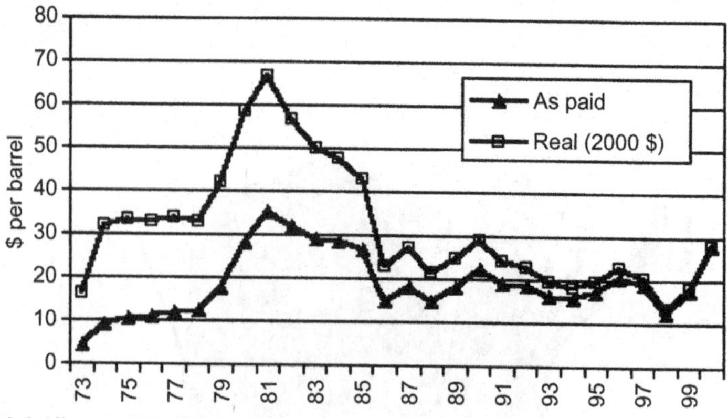

Figure 1.4 Historic US oil prices [11].

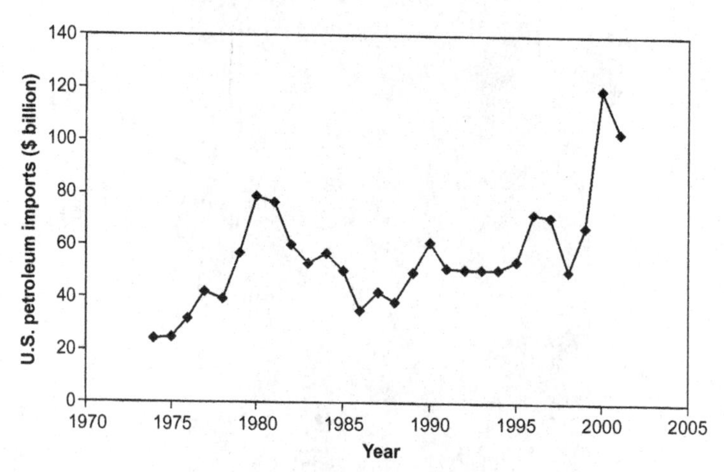

Figure 1.5 US imports of petroleum [12].

As illustrated in Figure 1.3, the United States imports well over half of the crude oil we process. One can argue whether the world will have enough oil for the next 100 years or maybe only the next 25 years. One thing is certain; the useful and significant domestic oil production in the United States has improved with the production of the "new" shale oil. It is certain that oil reserves are finite, that more oil will be available if the price of crude oil can sustain higher cost of production. Over half the world's known oil reserves are in the Middle East. Figure 1.6 shows this breakdown with Saudi Arabia, Iran, and Iraq each with greater reserves than any other country.

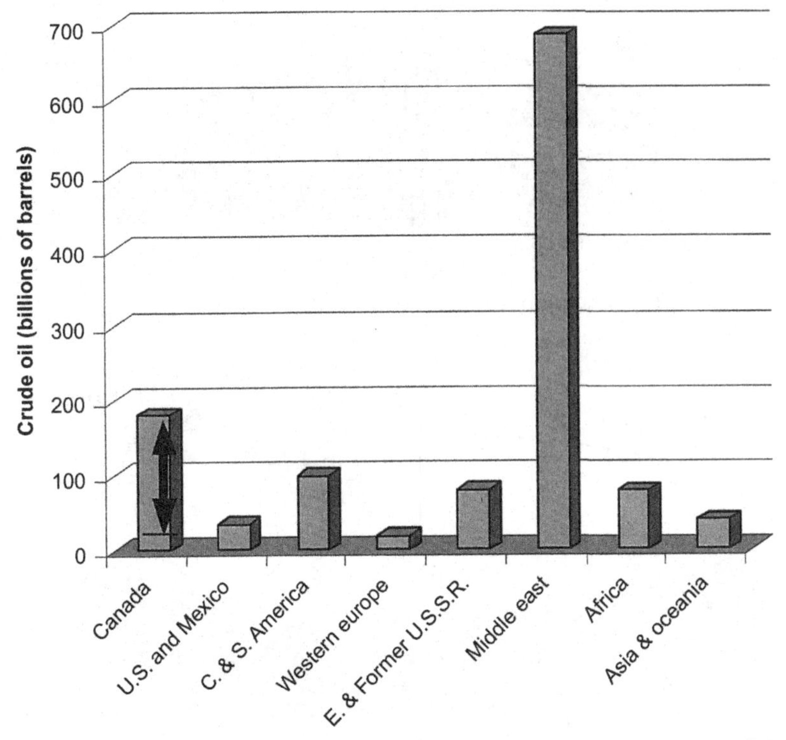

Figure 1.6 World oil reserves by region. Estimates of Canadian reserves by *Oil & Gas Journal* in 2003 are much higher than previous years—they likely include easily recovered oil sands [13].

Energy Sources

Petroleum provides more than 90% of vehicular fuels in the United States; but in addition, petroleum represents 53% of all energy consumed in the United States as summarized by Figure 1.7. The energy stocks used in the United States are in sharp contrast to US reserves (see Figure 1.8), and this will ultimately lead to energy crises.

To put things into perspective, with world energy consumption at the present rate, the world has approximately 3.6[1] years of petroleum (to supply all energy needs), 17 years of coal, 46 years of natural gas, and thousands years of uranium (assuming full use of uranium and ocean recovery) [14]. If the United States were the sole consumer of world energy

[1] For world oil reserves of 5.3e18 Btu and total world total energy consumption of 1.5e18 Btu/yr.

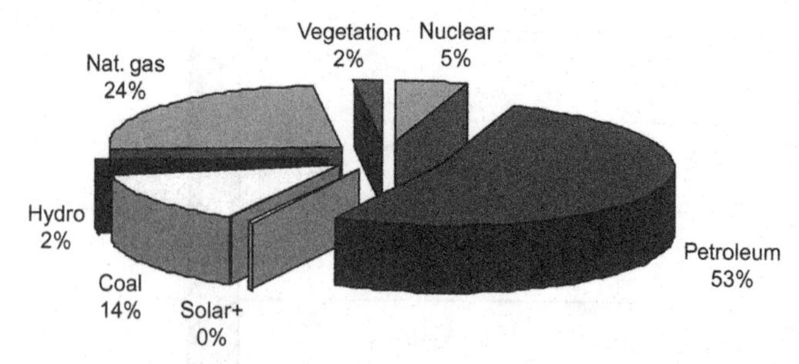

Figure 1.7 US Energy consumption by source.

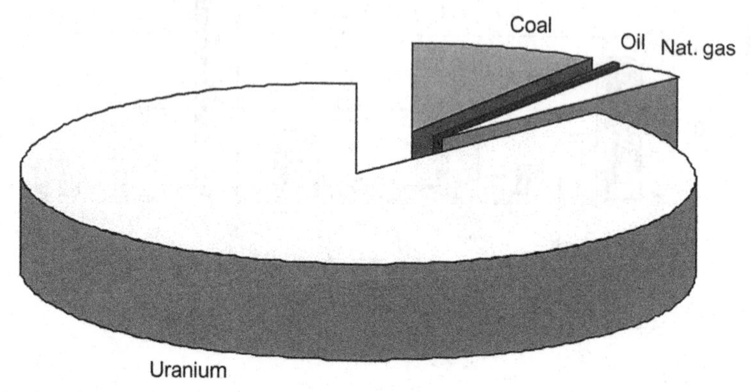

Figure 1.8 Estimate of US energy reserves.

reserves, world petroleum would last 75 years toward meeting all the US energy needs, coal 500 years, natural gas 1000 years, and uranium tens of thousands of years.

Presently, there is a mismatch between reserves and what is being consumed.

It is where consumption exceeds availability that technology makes the difference. There is little doubt—the world's demand for petroleum is rapidly exceeding the availability of petroleum. From the position of available energy reserves (see Figure 2.7), coal and nuclear are the obvious choices to replace petroleum. The right combination of technologies can make the difference between security and vulnerability; between cheap energy or economic recession due to restricted oil supply.

Oil Fracking and Horizontal Drilling

Technology on drilling for crude oil has advanced to the point where less than one of four oil wells drilled produced oil in the United States in the mid-1970s where essentially 100% of the wells now show producible oil. The improved technology starts with improved seismic mapping that identifies reserves and a process of horizontal drilling that bores into the geological formation for over a mile. At oil prices above about $70 per barrel this industry is prosperous.

The fracking (hydraulic fracturing the shale/rock formation) part of this technology occurs when high-pressure water is used to crack (fracture) the rocks in the formations to release oil. Contrary to much perception, the fracking occurs in geological formations typically a mile deeper than those formations yielding drinking water or otherwise impacting lifestyle on the surface. Problems do occur from other aspects of the drilling and contaminated water waste disposal; all of which must be controlled with high reliability.

International Energy Agency (IEA)'s 2014 World Energy Outlook indicates that the US domestic supplies will start to decline by 2020 "As tight oil output in the United States levels off, and non-OPEC supply falls back in the 2020s." The report states, "the Middle East becomes the major source of supply growth." This is consistent with reports from the US Energy Information Agency that forecast a plateau in US oil production after 2020.

The production numbers, reserve numbers, and oil prices have a consistency that indicates the United States has a window of opportunity of about 10 years (2015–2025) where it has the economy and time to make investments to secure the future generations. This statement is in agreement with the IEA article *United States must grasp opportunity to build sustainable energy system*, "The United States is in a strong position to deliver a reliable, affordable, and environmentally sustainable energy system, the IEA said today as it released a review of US energy policy. To do so, however, the country must establish a more stable and coordinated strategic approach for the energy sector than has been the case in the past."

This strategy includes sustainable electrical power production and new transportation technologies that can lead to major reductions in greenhouse gas emissions. These investments and other factors would improve the care with which we use reserves and ultimately extend the availability of affordable petroleum fuels unto the twenty-second century. The plan is to have lower cost and better performance alternatives—this is on the table and an agenda must be developed for it to happen.

COORDINATED STRATEGIC APPROACHES

A coordinated and strategic approach includes the following for the United States:

- Allow the US oil and natural gas industries to *take their course* and to nurture alliances with Canada since their oil sand and oil shale reserves can have a major impact.
- Allow the wind and solar energy industries to *take their courses.*
- Allow the biofuel industries to *continue to grow.*
- *Keep* the battery industries for transit and grid storage on course.
- Extend the nuclear power industry with new reactor technology.
- Answer the "calls" of the alarmists.
- *Attain* energy options previously considered unattainable.

Characterized as allowing to *take their course* are the three industry categories of (i) oil and gas, (ii) wind and solar, and (iii) biofuel industries. Great political favor has been shown to these industries with either favorable regulations or subsidies. Allowing these industries to "take their course" means to put a cap on the extent each is favored. Each has major flaws that limit their potentials as part of sustainable strategic solutions, but each also has advantages and niches to contribute sustainable industries. These industries each need to develop on paths that do not interfere with alternative technologies that might provide sustainable energy solutions.

Keeping the battery and nuclear industries on course is critical, because they will not only provide sustainable energy, but also provide major greenhouse gas emission reductions. The time may be at hand for a "fifth mode of transportation" that can be more efficient from both energy and transit time perspectives. Each of these deserves further discussion here and extended discussion in later chapters.

"Beware of the alarmists" falls into the same category as putting a cap on the amount of subsidies provided to the oil, gas, wind, solar, and biofuel industries. The common feature of these is they need the general public to ignore the rules of economic supply and demand, give each of these industries special competitive advantage with direct subsidies, or selectively provide an advantage over other industries by introducing new regulations. The best measure of whether a technology is worth the cost relative to its benefit is an open supply versus demand market.

If a technology costs more than alternatives, select the alternatives and save the dollars. There is any number of worthy projects that might be funded with public approval. The cost of a technology is the single most comprehensive metric for evaluating a technology that implicitly takes into account how worthy a technology is versus alternatives.

An exception to the use of cost as a primary factor for selecting technologies is for the fostering of new technologies that require a period of time for markets to develop and economies of scale to be realized. Taxpayers accept this approach as effective, but taxpayers are less accepting of recurrent subsidies to the point where "inferior" technologies interfere with the ability of more worthy technologies to obtain sustainability. The solution is the capping the total of subsidies applied toward any single industry; capping both in total dollars per year and the number of years.

Taxpayers are also accepting of "sin" taxes that are most prevalent on alcohol and tobacco. These taxes can be equally applied to industries that cause detriment to the environment. Carbon dioxide emissions may fall in this category. The use of the sin tax to account for hidden costs is far superior to subsidies since the tax can be defined to treat all industries the same.

The worst of all worlds is one of compounding and competing subsidies where any technology that is not backed by a strong lobbying group is left to flounder; this being the most accurate representation of current practices by many governments including the United States. From the engineering perspective, this approach of competing subsidies is characterized as being a system out of control that is bound to fail, possibly fail catastrophically.

When subsidies persist past the point of fostering a new industry and toward favoring specific industries; the impact is synonymous with the stifling of competing industries. Assuming this practice can be brought under control, it would be possible to sustainably achieve what was previously unattainable.

The energy and transportation infrastructures in 2015 are incrementally improved versions of the technologies developed in the 1940s—this history is a reflection of the impact of politics and lobbying on technology evolution. The topic of what is attainable warrants an introduction here and extended discussion in later chapters.

Nuclear Power

The processes that occur in nuclear fission reactors were not invented. Nuclear decay has been occurring on earth since its formation. What is

Table 1.1 Impacts as summarized in 1986 of coal versus nuclear for a 1 GW power plant operating for 1 year

	Coal	Nuclear
Occupation health deaths	0.5–5	0.1–1
Occupational health injuries	50	9
Total public and worker fatalities	2–100	0.1–1
Air emissions (tons)	380,000	6200
Radioactive emissions (curie)	1	28,000

new (within the past century) is scientific understanding of nuclear processes and the controlled use of nuclear fission to produce electrical power. There is much to be gained by the responsible use of nuclear fission to produce electricity.

Compared to the thousands who have died in the smog produced from burning coal or the thousands who have died in military conflicts to control the flow of petroleum, commercial nuclear power in the United States has demonstrated to be the safest and most environmentally friendly energy source. Table 1.1 statistics reported by Hinrichs [1] shows that the environmental impact of nuclear power is a small fraction of the impact of coal power. These lessons of history are clear. The extended statistics show that nuclear power in the United States is superior to all alternatives relative to total fatalities.

George Santana is credited with writing, "Those who fail to learn the lessons of history are destined to repeat them." Many great people have cited various forms of his statement. In energy technology, history's lessons are that more people will die due to coal and petroleum—from occupational accidents, military confrontations, and pollution generation—than with US nuclear energy. The environment will suffer more damage with coal and petroleum—from oil spills to the increase of carbon dioxide in the atmosphere. Responsible nuclear power generation should continue to be safe and more friendly to the environment provided engineers and scientists continue to apply the lessons they have learned to safely design and operate nuclear power facilities.

On the topic of nuclear waste (spent nuclear fuel, depleted uranium from nuclear fuel enrichment, excess weapons grade uranium, etc.), history shows that this "waste" should not be buried. Rather, processes should be developed to recover nuclear fuel values from it. Isolate the radioactive fission products for safe disposal. This topic will be treated in more detail in Chapter 8.

Tesla

Tesla Motors achieved what many considered unachievable in 2013. Tesla achieved profitability in a market of electric cars that were very expensive, had limited range, and did not have a network of refueling stations. Today, the company is expanding to include lower cost and extended range vehicles. They have added large-scale manufacturing of batteries to the plan to achieve economies of scale for lithium ion batteries. Several other automobile manufacturers are now offering electric vehicles in what appears to be a sustainable, competitive market that will expand.

The CEO of Tesla, Elon Reeve Musk, pursued the *electric automobile market in an approach previously thought to be* unattainable. He has clearly demonstrated the value of marketing and perseverance. He identified untapped markets to sustain a new industry.

Elen Musk was instrumental in a number of endeavors that made his fortune and that include in addition to Tesla Motors: SpaceX, SolarCity, Paypal, Zip2, and Hyperloop. The Hyperloop high-speed transportation system is an alternative to jet travel with the following advantages:

- The Hyperloop is faster than jet travel with proposed velocities up to 900 mph.
- It is based on railroad-type vehicles powered by batteries and operated in low-pressure tunnels to reduce friction.

The fastest trains today operate up to 200 mph by the Japanese Shinkansen line but with the record demonstration speeds of near 360 mph set by French TGV (French: *Train* à Grande Vitesse, "high-speed *train*") for rail travel and the Japanese Yamanashi test vehicle for maglev travel.

High speed land-based systems were proposed as early as the 1970s by Rand Corporation. Typical maximum travel speed for passenger jets is around 575 mph above land with air traffic control restrictions over cities and approaching airports. The Concord offered sustained supersonic transit service over the Atlantic Ocean for 27 years ending in October of 2003.

A system that would offer travel speeds in excess of 600 mph on land routes using electrical energy would be a significant advance in both quality of transportation and sustainability of transportation. The historic problem with such a system has been the high cost of the tunnel infrastructure. However, there is reason to believe that this cost barrier has been significantly reduced and that such system may become reality in the next couple of decades. Details are provided in Chapter 7 on the electric grid power.

Major simultaneous advances in sustained low-cost electricity, faster and lower cost travel, and substantially reduced greenhouse gas emissions are possible. The path there is through good, informed choices on technology and a lack of favoritism to any particular industry that would interfere the free market.

REFERENCES

[1] Hinrichs RA, Kleinbach M. Energy its use and the environment. 3rd ed. New York, NY: Brooks/Cole; 2002, Table 14.6.
[2] Tissot BP, Welte DH. Petroleum formation and occurrence. New York, NY: Springer-Verlag; 1978. p. 5.
[3] See <http://www.runet.edu/~wkovarik/envhist/2middle.html>.
[4] See <http://www.runet.edu/~wkovarik/envhist/4industrial.html>.
[5] Sampson A. The seven sisters. New York, NY: Bantam Books Inc.; 1981. p. 27.
[6] See <http://www.runet.edu/~wkovarik/envhist/7forties.html>; May 20, 2002.
[7] Stevenson W. A man called intrepid. New York, NY: Harcourt Brace Jovanovich; 1976.
[8] Borkin J. The crime and punishment of I.G. Farben. New York, NY: Free Press; 1978.
[9] Owen K, Coley T. Automotive fuels reference book. 2nd ed. Warrendale, PA: Society of Automotive Engineers; 1995.
[10] U.S. DOE Public Domain Image. See <http://www.eia.gov/totalenergy/data/annual/pdf/sec5_3.pdf>; 2011.
[11] See <http://www.eia.doe.gov/pub/oil_gas/petroleum/analysis_publications/oil_market_basics/default.htm>; May 17, 2002.
[12] See <http://www.eia.doe.gov/emeu/mer/>; May 17, 2002.
[13] PennWell Corporation. Oil Gas J December 23, 2002;100(52). Also see <http://www.eia.doe.gov/emeu/international/reserves.xls>.
[14] Consumption at 7.1e16Btu/yr in U.S. and oil, coal, and gas reserves of 5.3e18, 2.5e19, and 6.9e19. World consumption of 1.46e17, 1.18e17, and 1.4e18 for total of 1.4e18; use 1.5e18 for world. From <http://www.eia.doe.gov/pub/international/iealf/table11.xls>.

CHAPTER 2

Sources of Energy

Contents

Prior to the full-scale shale oil fracking boom (~2000–present), the evidence indicated that proven, recoverable oil reserves in the United States would only power our thirst for oil short term if cut off from oil imports. In fact, 2008 brought with it a doubling of gasoline prices to over $4 per gallon in the United States and sustained an economic recession due to limited crude oil supplies. Shale oil fracking brought jobs, money, and gas prices to less than $2 per gallon by the end of 2014.

Sustainable Power Technologies and Infrastructure.
DOI: http://dx.doi.org/10.1016/B978-0-12-803909-0.00002-6

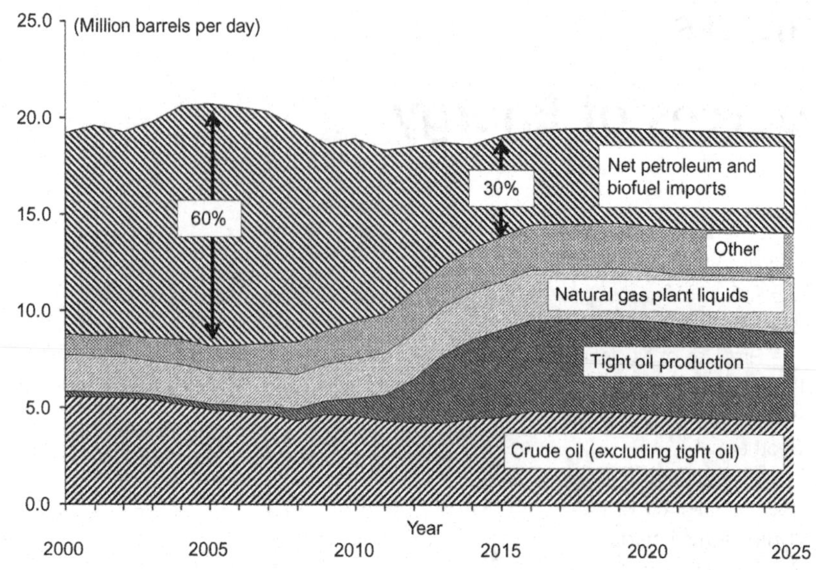

Figure 2.1 Past and projected source of liquid fuels in the United States including imports [1].

Today, because of fracking technology, the United States is a stronger country where the risk for rapid increases in fuel prices is reduced. The International Energy Agency (IEA) was correct when they reported:

> *The United States is in a strong position to deliver a reliable, affordable, and environmentally sustainable energy system; the IEA said today as it released a review of US energy policy. To do so, however, the country must establish a more stable and coordinated strategic approach for the energy sector than has been the case in the past.*

Figure 2.1 illustrates how from 2006 to 2015 oil fracking of tight oil in the United States halted a trend of increasing oil imports peaking at more than 60% of oil liquid fuel imports. Within these 9 years, imports decreased to 40% in 2012 and an estimated less than 30% by the end of 2014.

In parallel with the oil fracking boom has been the emergence of a sustainable electric car (hybrid and battery powered) battery industry including but not limited to the Tesla Motors. It is the fracking industry that has provided the United States with this window of opportunity. It is the battery industry that will bring the United States into an era of energy sustainability and prosperity while decreasing greenhouse gas emissions.

The battery industry is important because it allows the large-scale crossover of energy from the electrical grid directly to the transportation market. Through this, the diversity and stability of the grid electricity market can bring diversity and stability to the vehicular energy/fuel market. The diversity of the electricity grid includes the use of wind, solar, biomass, and nuclear energies; all of which can be used while reducing greenhouse gas emissions.

The big question in the immediate future is whether the United States will have a coordinated strategic approach that would be best for the United States, or whether the US economy will be dominated by the policies of Saudi Arabia and the OPEC nations. How they decide to open or close their "oil valve" to the global market during any particular month. The answer may be with the United States building a strong and healthy economy with greater than $55/barrel oil; a price that provides incentive for a sustained fracking industry and allows the battery industries to come into their own. Canada has successfully, historically implemented policies that protected its domestic oil industry; the United States has not.

In the discussion of sources of energy, some are easily used to produce liquid fuels and some are primarily useful for producing electricity. However, this distinction becomes less important as the US electric car battery industry grows.

COSMIC HISTORY OF ENERGY

The trajectory of the prehistory of human survival moving from "taming fire," winning metals from ore, and extending to the Industrial Revolution brought us to the 1800s, the chemical age. By 1900, it was observed that certain chemical elements slowly released penetrating radiation that they called x-rays (they didn't know what the radiation was!). By the 1920s, experiments demonstrated that the nucleus of an atom was positively charged and a magnetic field held negatively charged electrons in place around the nucleus forming the atom. Chemical compounds formed when two or more atoms shared electrons yielding a vast array of naturally occurring compounds and provided paths to synthesize lots of new synthetic materials.

This neat progression of understanding chemistry was interrupted in the 1930s. It was discovered there was an electronically neutral particle in the nucleus of an atom. They called it a "neutron." In 1939, scientists radiated uranium with neutrons and observed a huge release of energy. When a target uranium atom nucleus split (they called it "nuclear fission") there was formed a pair of atom nuclei with slightly less total mass than the target uranium atom. Two or three neutrons were also released when uranium fission occurred. These new atom nuclei quickly picked up electrons and they fit right into the period table chemists had developed. "Wow!" This suggested a way to quickly release huge amounts of energy (a powerful explosion!) by providing a uranium target atom for each neutron released leading to a "chain reaction." Historically, Adolf Hitler's Germany had conquered most small European countries and World War II was underway. The development of the technology to produce this "nuclear weapon" was classified "Top secret" and all further development of a nuclear bomb remained secret until well after the United States exploded two nuclear weapons over Japan, ending World War II.

The Source of Atoms

The source of the energy released by atomic fission had to come from the very small nucleus of the atom. Nearly all the mass of an atom is located in the nucleus because the negatively charged electrons surrounding the atom nucleus contain very little mass. Experiments showed that an atom nucleus was composed of just two particles—positively charged protons and neutrons. Classical physics could not explain how the huge repulsive force between the positively charged protons could be "packed" and held together in the "tiny" volume of an atom nucleus.

About 1663, Sir Isaac Newton demonstrated that "white" sunlight when passed through a glass prism would separate into a "spectrum" of colors from red to blue. Following a huge volume of experimental and theoretical research it was demonstrated that all atoms are characterized by the frequency of radiation they absorbed from the spectrum of a radiation beam or by the radiation they emit when they "cool" from excited states produced by a high-energy "pulse" of radiation. These observations are part of the demonstration that led to completing the quantum theory. These experiments involving radiation indicated that the radiation spectra—that includes sunlight—must be characterized as electromagnetic radiation. Long wavelengths are low energy carriers while short wavelengths are high-energy carriers.

Hence the model of an atom is composed of a positive charged nucleus surrounded by electrons—an electron for each proton in the nucleus—and each electron occupies a unique orbit a small distance from the atom nucleus. Classical physics would expect all the negatively charged electrons would be pulled into the positively charged atom nucleus. Quantum physics (remember, these are very small particles) says an electron can only change orbit when it absorbs a specific wavelength of "light" (spectroscopic analysis suggested this must be true). Since the quantum model of an atom is based on particles (protons, neutrons, and electrons) a "massless" particle was added to the model of the atom they called a "photon." Each photon carries the exact energy of a single wavelength in the spectrum of light. When a beam of light passes through a "sample" containing lots of atoms, an electron will move to a higher energy orbit and this wavelength will be missing in the light beam that passed through the sample. The addition of the photon "particle" completed the theoretical model that is the powerful nuclear theory of atoms used today.

The spectral energy of adsorption or emission is unique for every atom. This theoretical model is the basis for spectroscopic analysis. Astronomers use spectra from radiation sources in outer space that yields composition information of the atoms in outer space. The atmosphere covering the earth diminished the energy of the incoming radiation just like clouds decrease the energy of sunlight. The spectrometers mounted on the orbiting satellites of the earth removed the atmospheric layer and "sharpened" the spectral measurements to more clearly identify atoms in outer space.

Observation of the spectrum from cosmic explosions verifies that the intense temperature and pressures produced during these cosmic events are the source of the nuclei of the atoms represented in the periodic table. When neutrons and protons are "mashed" very close together, there is a very short-range "strong force" (that's its name because the nature of the force remains unknown!) that holds the protons and neutrons together to form that "tiny" atom nucleus.

What is Permanence?

The permanence scale in Figure 2.2 shows the binding energy in million electron volts (MeV), on the energy scale used for these subatomic particles) for each nucleon (a proton or a neutron) in the nucleus of an atom. The most stable atoms are those with the highest binding energy per nucleon. The nucleon-binding energy is plotted against the atomic mass number.

The maximum binding energy per nucleon in Figure 2.2 occurs at mass number 54, the mass number for iron. As the atomic mass number increases from zero, the binding energy per nucleon increases because the number of protons (with positive charge) increases, and protons strongly repel each other but that "strong force" holds them together to form a atom nucleus.

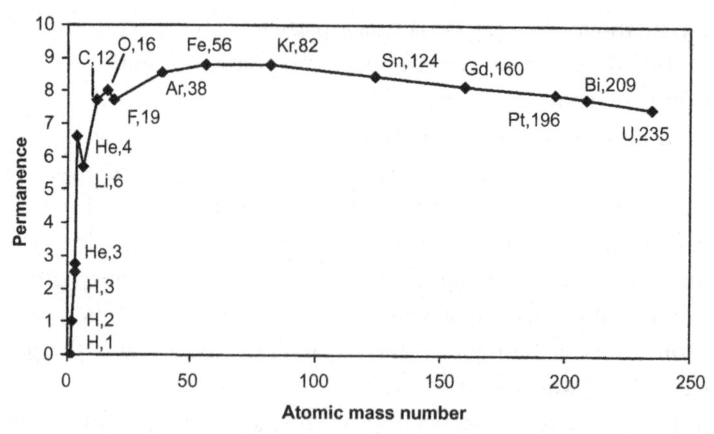

Figure 2.2 Impact of atomic mass number on permanence of atoms. H, hydrogen; He, helium; Li, lithium; C, carbon; O, oxygen; F, fluorine; Ar, argon; Fe, iron; Kr, krypton; Sn, tin; Gd, Gadolinium; Pu, plutonium; Bi, bismuth; and U, uranium.

The binding energy of the neutron (no charge) serves to assist holding the protons in the nucleus of the atom together. The atomic mass number for each atom is approximately the total mass of the protons and neutrons in the nucleus since the electrons (equal to the number of protons) have a small mass, about 1/1837 that of a proton.

Above a mass number of about 60, the binding energy per nucleon decreases. Visualize the large atom nucleus as many protons try to get away from each other packed with a larger number of neutrons all held very close together by that "strong force." Iron-56 is the most common iron isotope (about 92% of the mass of natural iron) and it has 26 protons and 30 neutrons. Natural uranium is composed of two isotopes: uranium-235 (about 0.073% of the mass) and uranium-238 (about 99.27%). U-238 nucleus holds together 92 protons and 146 neutrons. U-238 atoms decay very slowly. It takes about 4.5 billion years for one-half of a lump of pure U-238 to decay, so it is "almost" a stable isotope.

An additional comment: Notice that the nucleon-binding energy decreases as the atomic mass increases from 60 to 260. This is the source of the energy release in the nuclear fission process. Since the forces in the nucleus of large atoms are so carefully balanced, a small energy addition (a low-energy neutron entering the nucleus) will cause one or more of the nucleons to move just beyond the very short range of the "strong force" holding the nucleus together. This unleashes the strong repulsive forces acting between the many protons and the nucleus comes apart (flies apart) into two pieces. This releases lots of energy and two or three neutrons. The strong nuclear binding energies of the two pieces of the uranium atom form the nuclei of two smaller atoms.

These "new" atom nuclei quickly collect electrons and fall exactly into the chemical periodic table. The new small nuclei are usually "radioactive" spontaneously releasing subnuclear particles or radiation and thermal energy. There is only a small fraction of these "fission products" that continue to decay slowly for more than 100 years.

The arrays of different elements in our planet, the solar system, and the galaxy reveal their history. Hydrogen is the smallest of the atoms assigned an atomic number of one. Physicists tell us that during the birth of the universe it consisted mostly of hydrogen. Stars converted hydrogen to helium, and supernovas (see box "Supernovas") generated the larger atoms through atomic fusion.

Helium has 2 protons, lithium has 3 protons, carbon has 6 protons, and oxygen has 8 protons. The number of protons in an atom is referred to as the "atomic number" and identifies that atom. Atoms are named and classified by their atomic number. Atoms having between 1 and 118 protons have been detected and named (see box "Making New Molecules in the Lab"). The atomic mass is the sum of the mass of neutrons, protons, and electrons in an atom—the atomic mass and the atomic spacing determine the density (mass per unit volume) of natural bulk materials.

Protons are packed together with neutrons (subatomic particles without a charge) to form an atom nucleus. There are more stable and less stable combinations of these protons and neutrons. Figure 2.2 illustrates the permanence of nuclei as a function of the atomic mass number (the atomic mass is the sum of the protons and neutrons). Helium 3 (He,3) is shown to have a lower permanence than He,4. Two neutrons simply hold the two protons in He,4 together more firmly than the one neutron in He,3. In general, the number of neutrons in an atom must be equal to or greater than the number of protons, or that atom will be unstable and "decay" (release subnuclear particles or energy) moving to a more stable combination of protons and neutrons.

The wealth of information in Figure 2.2 demonstrates much about the composition of our planet. For example, how hydrogen atoms combine to form helium. It also illustrates that the "permanence" is greater for He,4 than for hydrogen (H,1). In these nuclear reaction processes, atoms tend to move "uphill" on the curve of Figure 2.2 toward more stable states. In Chapter 8, the concept of atomic stability will be discussed in greater detail, and the term "binding energy" will be defined and used in place of "permanence."

Supernovas—The Atomic Factories of the Universe

Astronomers have observed "lead stars" that produced heavier metals like lead and tungsten. Three have been observed about 1600 light years from Earth. To paraphrase a description of the process:

Stars are nuclear "factories" where new elements are made by smashing atomic particles together. Hydrogen atoms fuse to form helium. As stars age and use up their nuclear fuel (hydrogen), helium fuses into carbon.

Carbon, in turn, "grows" to form oxygen, and the process continues to make heavier elements until the natural limit at iron. To make elements heavier than iron, a different process is needed that adds neutrons to the atomic nuclei. Neutrons can be considered atomic "ballast" that carry no electric charge but serve to stabilize the strong repulsive forces of the protons.

Scientists believe there are two places where this "atom nucleus building" can occur: inside very massive stars when they explode as supernovas and more often, in normal stars at the end of their fuel supply when the gravitational forces cause them to collapse into a small, very dense object as they burn out.

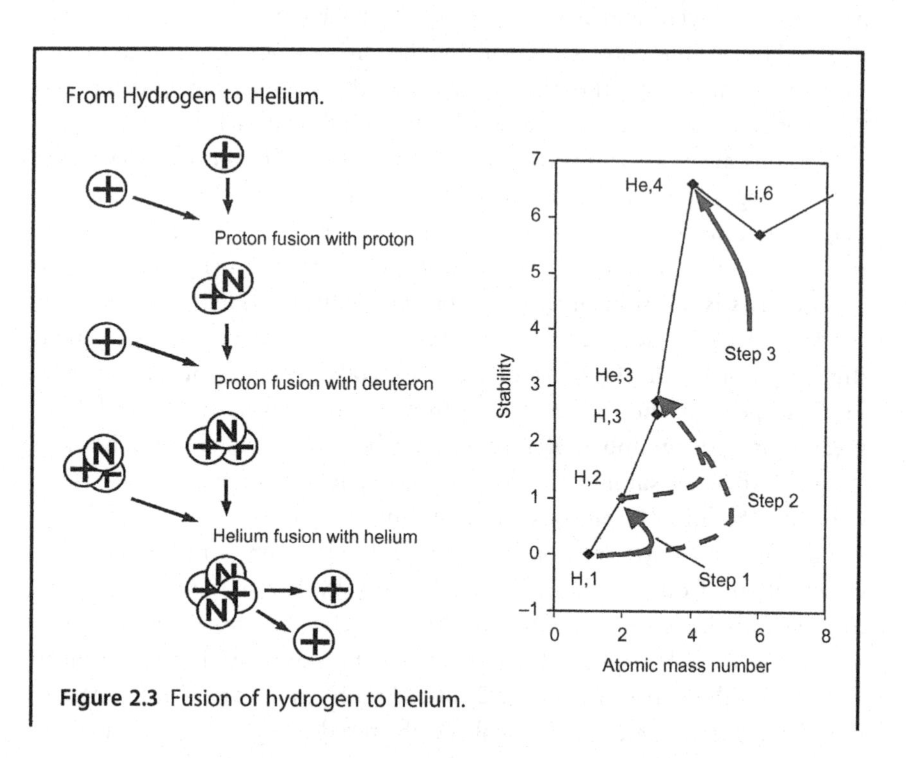

Figure 2.3 Fusion of hydrogen to helium.

> The most abundant atom in the universe is hydrogen. Hydrogen is the fuel of the stars. This diagram [2] illustrates how two protons join with two neutrons to produce a new stable molecule, helium. Considerable energy is released in the process, about 25 MeV of energy for every helium atom formed (570 million BTUs per gram helium formed).

To make atomic transitions to more permanent/stable atoms, extreme conditions are necessary (see box "Supernovas"). On the sun, conditions are sufficiently extreme to allow hydrogen to fuse to more stable, larger molecules. In nuclear fission reactions in a nuclear power plant or in Earth's natural uranium deposits, large molecules break apart (fission) to form more stable smaller atoms.

The most abundant atom in the universe is hydrogen. Hydrogen is the fuel of the stars. The diagram illustrates how four protons interact to produce a new stable molecule: helium. Considerable energy is released in this process—about 25 MeV of energy for every helium atom formed (that is 570 million BTUs per gram helium formed).

Figure 2.4 is a starting point for qualitative understanding of the history of energy. Nuclear reactions are the text providing the history of energy. The nontechnical history of energy goes something like this: Once upon a time—about 14 billion years ago—there was a "big bang." From essentially nothingness, in a very small corner of space, protons and helium were formed. Carbon, iron, copper, gold, and the majority of other atoms did not exist.

Unimaginably, large quantities of hydrogen and helium cluster together to form stars. The most massive of these stars formed supernovas. Fusion conditions were so intense in these supernovas that atoms of essentially all atomic numbers were formed. Hence, carbon, oxygen, iron, copper, gold, and the whole array of atoms that form solid objects on Earth were formed in these atomic factories.

The largest atoms formed were unstable and they quickly decayed or split by fission to form the nuclei of more stable smaller atoms. Uranium has intermediate stability. It can be induced to fission but it is stable enough to last for billions of years, very slowly undergoing spontaneous decay.

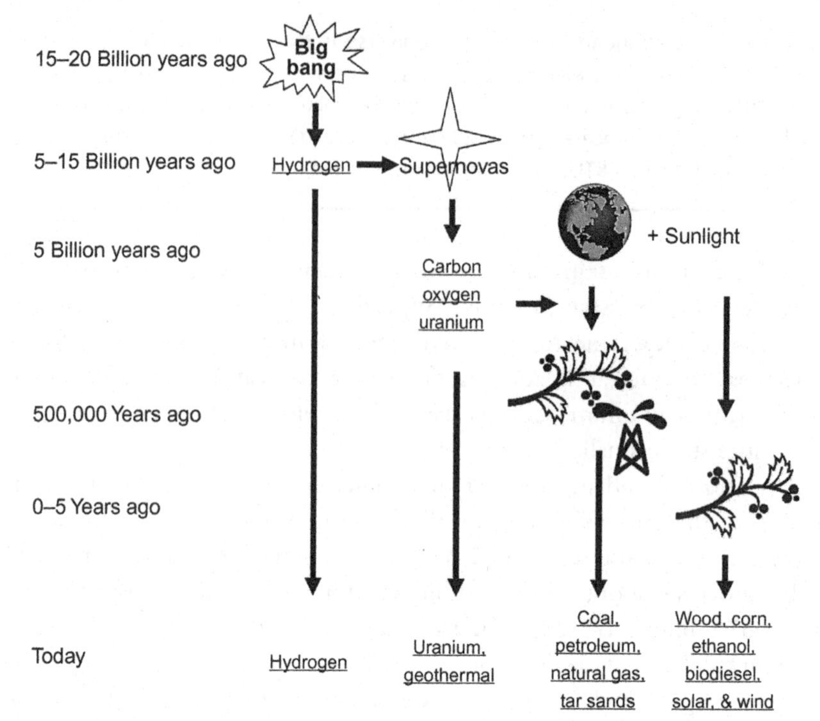

Figure 2.4 History of energy.

The spinning masses continued to fly outward from the big bang. As time passed, localized masses collected to form galaxies; and within these galaxies, stars, solar systems, planets, asteroids, and comets were formed.

Making New Molecules in the Lab

On a very small scale, scientists have been able to make new "heavy" atoms confirming the way a supernova could combine two smaller atoms to form a larger one. The heaviest atom that has been produced in the laboratory is Element 118 [3]. This was the work of an extended collaborative effort between scientists at the Joint Institute for Nuclear Research in Dubna, Russia and Lawrence Livermore National Laboratory in Berkeley, California. This very exacting experimental program was conducted over several years [4].

This is an example of one successful experiment. They used a cyclotron accelerator to produce a high-energy beam of a calcium isotope (atoms) to irradiate an isotope of californium (atoms). Six months of irradiation produced three identifiable Element 118 atoms! A statistical analysis of many months of measurements established that Element 118 decayed by emitting an alpha particle (a helium atom nucleus) in about 0.89 ms, establishing the half-life of Element 118. This decay reduces Element 118 to an Element 116 isotope that decays almost as quickly emitting one neutron followed by a second neutron. This decay destabilized Element 116 nucleus which fissions (splits apart) ending that experimental sequence. Studying the remains of these experiments (the fission products) made it possible to confirm Element 118 existed for an instant, a truly remarkable experimental accomplishment.

It is here where energy and the universe, as we know, took form. We are just beginning to understand the processes of the stars and supernovas to tap into the vast amounts of binding energy in an atom. The atomic binding energy available in one pound of uranium is equivalent to the chemical binding energy present in eight million pounds of coal!.[1]

NUCLEAR ENERGY

In theory, nuclear energy is available from all elements smaller (lighter) than iron through nuclear fusion and all elements larger (heavier) than iron through nuclear fission. Iron is on the peak of the permanence curve so it is also one of the most abundant elements. While all other "heavy" atoms can decay by spontaneous emission of subnuclear particles to gradually form iron. Iron does not spontaneously decay. When most of the nuclear energy of the universe is spent, iron and elements with similar atomic number (all being stable) will remain.

In general, the largest atoms are the most likely to quickly undergo nuclear decay such as the release of an alpha particle (a helium atom nucleus). Atoms that were heavier than uranium (see[22]) have undergone decay and fission and are no longer found on Earth. The amount of U-238 on Earth today is slightly less than half of what was present at Earth's formation. The amount of U-235 on Earth today is less than 1% of what was present at Earth's formation.

Of interest to us is the ability to use nuclear processes in a controlled and safe manner, because "controlled release" of the nuclear binding energy can be used to produce electricity. The energy released is the

[1] 12,000 tons per day (Chap. 4, Steam Turbine Section) * 365 days/year × 40%/30%/[750 kg × 1 ton/~1000 kg).

energy required to hold protons and neutrons, in atom nuclei as they combine and rearrange in the progression to higher "binding energy." We have been able to use nuclear fission on a practical/commercial scale with one naturally occurring element: Uranium. We could perform fission on elements heavier than uranium, but these are not available in nature. The hydrogen bomb is an example where the energy of fusion was used for massive destruction, but use of fusion for domestic energy production is much more difficult. Practical nuclear fusion methods continue to be an area of active research.

The only practical nuclear energy sources today are nuclear fission of uranium in nuclear reactors and the recovery of geothermal energy (heat) produced by nuclear decay under the surface of Earth. This occurs continuously with uranium, the primary fuel for both of these processes.

At 18.7 times the density of water, uranium is the heaviest of all the naturally occurring elements (the lightest is hydrogen; iron is 7.7 times the density of water). All elements (identified by the number of protons in the nucleus) occur in slightly differing forms known as isotopes. Each isotope forms by changing the number of neutrons packed into the nucleus of that atom.

Uranium has 16 isotopes of which only two are stable. Natural uranium consists of just two of those isotopes: 99.3% by weight U-238 and 0.71% U-235. (The atomic number of uranium is 72. The number of protons plus the number of neutrons is the mass number. U-238 has 72 protons + 166 neutrons = 238. U-235 has 72 protons + 163 neutrons = 235).

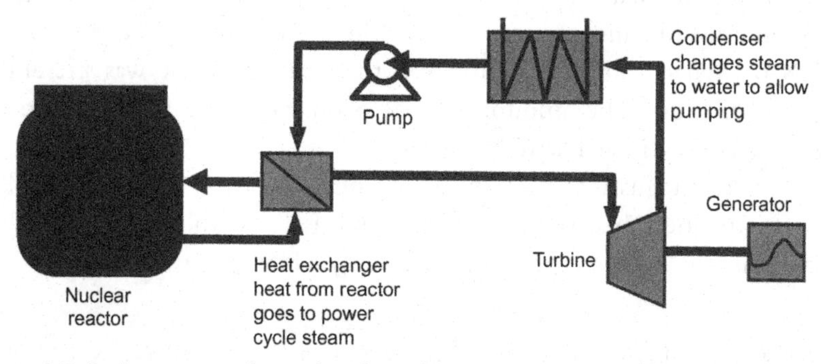

Figure 2.5 Basic steam cycle used with nuclear reactor source of heat.

Modern Nuclear Reactors in the United States

Modern nuclear power plants[24] use a pressurized [5] water reactor to produce thermal energy to produce the steam to drive a turbine and generate electricity. The fuel is 3%–4% U-235-enriched uranium oxide pellets sealed in tubes that are held in racks in the reactor pressure vessel. This maintains the geometry of the reactor core. The water that removes the heat from the core leaves the reactor at about 320°C, and it would boil at a pressure of about 70 atmospheres (850 psi). The pressure in the reactor vessel is held at 150 atmospheres (2250 psi), so it never boils. This hot water is pumped to a heat exchanger, where steam is produced to drive the turbines. The high-pressure reactor cooling water will always contain small amounts of radioactive chemicals produced by the neutrons in the reactor. This radioactivity never gets to the steam turbine where it would make it difficult to perform maintenance on the turbine and steam-handling equipment.

Large pressurized water reactors produce about 3900 MW of thermal energy to produce about 1000 MW of electric power. The reactor core contains about 100 tons of nuclear fuel. Each of the nuclear fuel racks has places where control rods can be inserted. The control rods are made of an alloy that contains boron. Boron metal absorbs neutrons, so with these rods in position, there will not be enough neutrons to initiate the chain reaction. When all of the fuel bundles are in position and the lid of the pressure vessel sealed, the water containing boric acid fills the pressure vessel. The control rods are withdrawn, and the boron water solution still absorbs the neutrons from U-235 fission. As the water circulates, boric acid is slowly removed from the water and the neutron production rate increases; the water temperature and pressure are closely monitored. When the neutron produces the rated thermal power of the reactor, the boron concentration in the water is held constant. As the fuel ages through its life cycle, the boron in the water is reduced to maintain constant power output.

If there is an emergency that requires a power shutdown, the control rods drop into the reactor core by gravity. The control rods quickly absorb neutrons, and fission power generation stops. The radioactive fission products in the fuel still generate lots of heat, as these isotopes spontaneously decay when fission stops. Water circulation must continue for several hours to remove this radioactive decay heat before the reactor lid can be removed to refuel or maintain the reactor vessel.

U-235 is slightly less stable than U-238 and when enriched to 3%–8% it can be made to release heat continuously in a nuclear reactor. Enriched to 90% U-235 a critical mass will undergo a chain reaction producing the sudden release of huge amounts of energy becoming a nuclear bomb. We have mastered the technology to perform both of these processes. U-238 decays slowly. About half of the U-238 present when Earth was formed about 4.5 billion years ago (and >99% of U-235) has decayed, keeping the Earth's interior a molten core.[23]

U-235 decays faster than U-238. We are able to induce fission of U-235 by bombarding it with neutrons. When one neutron enters a U-235 nucleus to form U-236, it breaks apart almost instantly because it is unstable. It breaks apart to form the nucleus of smaller atoms plus two or three neutrons. These two or three neutrons can collide with U-235 to produce another fission in a sustained chain reaction.

The First Use of Nuclear Power

Our use of nuclear fission was to make a bomb which was based on an uncontrolled chain reaction. A neutron chain reaction results when, for example, two of the neutrons produced by U-235 fission produce two new fission events. This will occur when nearly pure U-235 is formed into a sphere that contains a critical mass: about 60 kg of metal. Then in each interval of 10 billionth of a second, the number of fission events grows from 1, 2, 4, ... 64, ... 1024, 2048, ... as illustrated by Figure 2.6. The transition from very few fission events to an uncountable number occurs in less than a microsecond. The enormous energy released in this microsecond is the source of the incredible explosive power of a nuclear fission bomb.

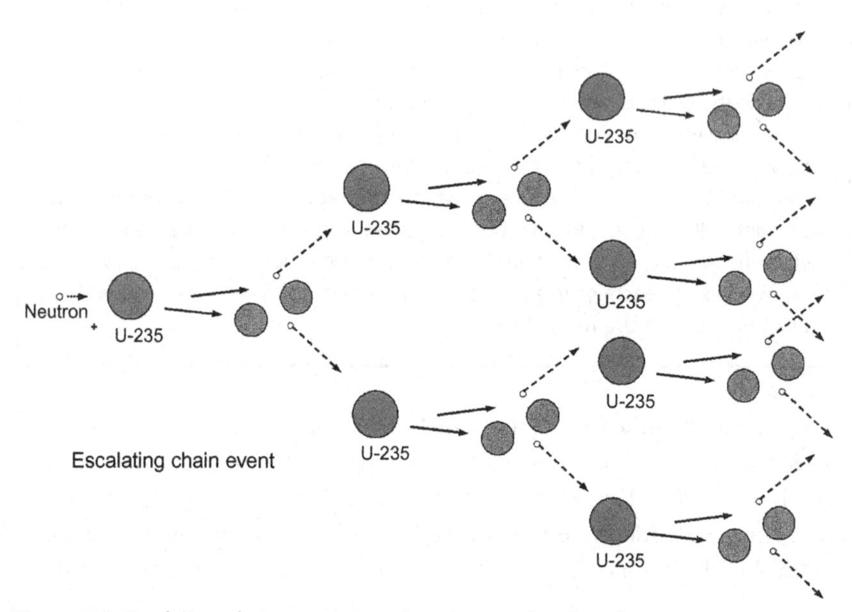

Figure 2.6 Escalating chain reaction such as in a nuclear bomb.

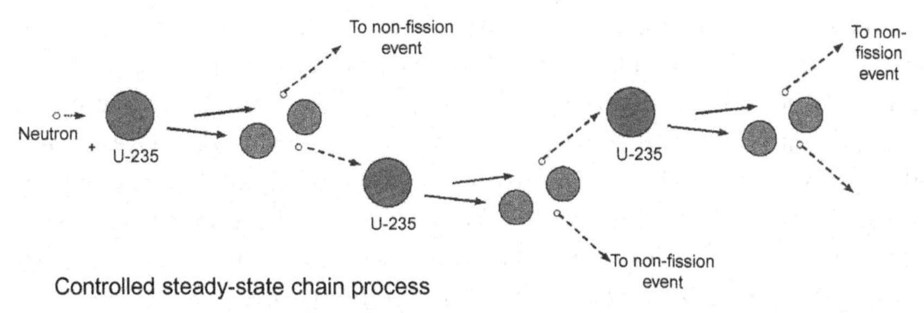

Figure 2.7 Controlled steady-state chain nuclear fission such as in a nuclear reactor.

This escalating chain reaction is to be distinguished from the controlled steady-state process as depicted by Figure 2.7. In a controlled steady-state process, a nearly constant rate of fission occurs (rather than a rapidly increasing rate) with a resulting constant release of energy.

The first nuclear bomb used in war exploded over Hiroshima, Japan, was a U-235 fueled bomb. Two hemispheres containing half of the critical mass are slammed together in a "gun barrel" with conventional explosive charges. In the resulting nuclear explosion, about 2% of the U-235 mass underwent fission. Everything else in the bomb was instantly vaporized. The fireball and the explosion shock wave incinerated and leveled a vast area of Hiroshima. This is the legacy of nuclear energy that indelibly etched fear into the minds of world citizens. The second explosion at Nagasaki was a plutonium bomb, followed by the development and testing of even more powerful and fearsome nuclear weapons during the Cold War period, adding to this legacy of fear.

For a nuclear bomb, the rapid chain reaction depicted in Figure 2.6 is competing with the tendency for the melting/vaporizing *u*ranium to rapidly "splat" over the surroundings. This "splatting" action tends to stop the chain reaction by separation of small pieces of uranium. Weapon's grade U-235 is typically at least 80% U-235; higher purities give increased release of the nuclear energy (more fission and less splatting).

The enormous energy available from U-235 in a very small space led US naval technologists to consider using nuclear energy to power submarines. The task is to configure the nuclear fuel (U-235 and U-238) so that exactly one of the neutrons produced by a U-235 fission produces only one new fission event. The shapes of the reactor core and control rods (that absorb neutrons) combine to serve as a "throttle" to match the energy release to load. The thermal energy produces steam that propels the vessel and provides electric power. All of this technology development was done with private industrial firms under contract by the military and was classified "top secret."

The industrial firms that built the nuclear reactors for the military also built steam turbines and generators for electric power stations. The first nuclear reactor built to produce electric power for domestic consumption was put into service in Shipingsport, Ohio, in 1957, just 15 years after the "Top Secret" Manhattan Project was assembled to build a nuclear weapon. This represents a remarkable technological achievement. Today, modern nuclear reactors produce electricity based on technology that closely mimics the reactors first used on the submarines.

Reserves

Uranium reserves are difficult to estimate; however, an accurate estimate can be made on the energy in the spent rods from US nuclear power generation. Current nuclear technology uses 3.4% of *u*ranium in fuel, leaving 96.6% of the uranium unused. The amount of nuclear fuel in spent nuclear fuel rods in US nuclear facilities contains as much energy as the entire recoverable US coal reserve. The depleted uranium left behind during the fabrication of the initial nuclear fuel rods has about four times as much energy as that remaining in the spent nuclear fuel rods. Combined, this stockpiled uranium in the United States has the capacity to meet all of the US energy needs with very little greenhouse gas emissions for the next 250 years.

Reprocessing Technology

Recovering fuel values from uranium that has already been mined and used will require reprocessing. A typical spent nuclear fuel rod in the United States contains about 3.4% fission products, 0.75%–1% unused U-235, 0.9% Pu-239 (Plutonium), 94.5% U-238, and trace amounts of atoms with atomic masses greater than U-235 and U-238 (referred to as transuranic elements).

Reprocessing spent nuclear fuel would make available the major fraction, the unspent uranium there most of the uranium that remains there. This would add to the stockpiled uranium fuel and in addition, separate the fission products. The fission products then become the "nuclear waste" with a very small quantity requiring storage of more than 100 years.

Reprocessing involves removing the 3.4% that is fission products and enriching U-235 and/or Pu-239 to meet the "specifications" of nuclear reactor fuel. The "specifications" depend on the nuclear reactor design. Nuclear reactors and fuel-handling procedures can be designed that allow nuclear fuel specifications to be met at lower costs than current reprocessing practice in France. For comparative purposes, the costs of coal, US nuclear fuel from mined uranium, and French reprocessed fuel are about 1.05, 0.68, and 0.90, respectively, cents per kWh or electricity produced.

Around 33% to 40% of the energy produced in nuclear power plants today originates comes from U-238 irradiated by neutrons in the reactor core. When a U-238 nucleus receives a neutron, there follows two nuclear decay events arriving at Pu-239 in about 90 days. That plutonium becomes nuclear fuel. For every three parts of U-235 fuel entering the reactor, about two parts of U-235 plus Pu-239 leave the reactor as "spent fuel." To date, these fuel values remain stored at the power plant site. All of the uranium, the two parts U-235 and Pu-239, are the target of reprocessing technology. To tap this stockpile of fuel, mixed oxide fuel (MOX—uranium and plutonium metal oxide) or new fast-neutron reactors could be put in place that produce more ("breed" Pu-239) Pu-239 than the combined U-235 and Pu-239 in the original spent fuel.

Decades of commercial nuclear power provide stockpiles of spent fuel rods. Billions of dollars have been collected on a 0.1 cent per kWh tax levied to decommission an "old" reactor. Decommissioning a "retired" reactor must include disposing the on-site spent fuel or reprocessing it. The remarkable safety history for US designed reactors is set against a costly history of regulations that limit slow technology development. These circumstances provide opportunity or perpetual problems, depending on the decisions made to use (or not use) advanced nuclear power options.

Figure 2.8 summarizes the accumulation of spent fuel currently being stored on-site at the nuclear power plants in the United States.

The United States uses about 100 GW of electrical power generating capacity from nuclear energy. The construction and startup of most of these plants occurred between 1973 and 1989. In 2007, the inventory of spent nuclear fuel corresponded to about 30 years of operation at 100 GW. Figure 2.8 approximates the total spent fuel inventories and cumulative inventories of U-235 and Pu-239 scenarios of reprocessing versus continued operation without reprocessing. Reprocessing is the key to decreasing Pu-239 and U-235 inventories and ending the accumulation of spent fuel at the nuclear reactor sites.

If reprocessing had started in 2015 to serve all current nuclear power plants, the current inventories, along with the Pu-239 that is generated as part of PWR operation, would provide sufficient Pu-239 to operate at

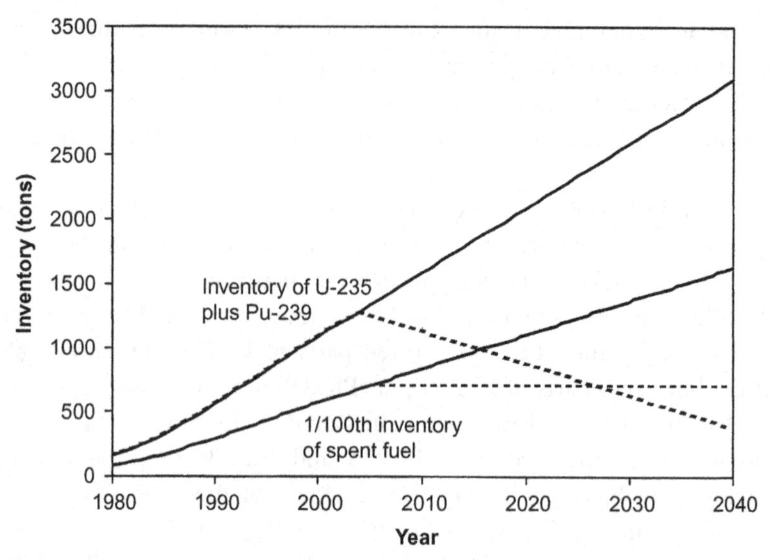

Figure 2.8 Approximate inventory of commercially spent nuclear fuel and fissionable isotopes having weapon potential (Pu-239 and U-235). The solid lines are for continued operation without reprocessing and the dashed lines are for reprocessing (starting in 2005) to meet the needs of current nuclear capacity.

existing capacity through 2065. If in 2005 the demand for Pu-239 and U-235 increased threefold (~300 GW capacity), the current inventories would last until about 2035. This does not depend on fast-neutron reactor technology to convert the much greater inventories of U-238 into Pu-239 fuel. This partly explains why breeder reactor research and operation were discontinued. Breeder reactors will not be needed for some time.

Fast-neutron reactor technology could allow nuclear reactors to meet all energy needs for the next 200–300 years without generating additional radioactive materials and without mining additional uranium. The potential of this technology should not be ignored.

Recoverable uranium on Earth could provide all energy needs for thousands of years (Actual estimates are millions of years, depending upon electrical energy consumption.). Reprocessing and fast-neutron reactors can use *u*ranium that is already mined and stored in various forms to provide all energy needs for hundreds of years while simultaneously eliminating nuclear waste. For this reason, an option to use thorium as a nuclear fuel can be delayed for several decades. It does remain a long-term nuclear option.

The fact is that nuclear power offers many outstanding options. The key missing aspect of the nuclear solution is the commitment to do it well and sustainably as opposed to allowing it to be dominated by knee-jerk reactions.

In the pursuit of sustainable energy, nuclear power emerges favorable for five reasons:

1. On a BTU basis, nuclear fuel is the least expensive and it is economically sustainable. Nuclear fuel has the potential to be 10 times less expensive than any alternative (less than $0.10 per MBTU).
2. Nuclear fuel is the most readily available fuel—it is stockpiled at commercial reactors in the form of spent fuel.
3. Nuclear fuel is the most abundant; enough has already been mined, processed, and stored in the United States to supply all US energy needs for centuries.
4. There is no technically available alternative to give sustainable energy supply for current and projected energy demand.
5. Nuclear energy generates a small fraction of greenhouse gas emissions relative to the best alternatives.

It is impractical to replace transportation fuels with biomass because there simply is not enough available biomass to meet these needs—let alone nontransportation energy demands. The long- term availability of petroleum has already led to military conflict to keep the oil flowing. The contribution to the trade deficit is also a drag on the US economy.

Natural gas power plants provide a short-term solution because they are cheaper to build and they convert about 50% of the combustion energy into electricity. Natural gas facilities also provide quick startup to compensate for unreliable wind. It is clear that the diverse mix of energy used for electrical power is better when both natural gas and nuclear are used to meet long-term goals on sustainable and inexpensive electrical power.

Coal will be an important global energy source for decades to produce electricity. Coal can remain for centuries as a feedstock to the chemical industry. However, coal is already used for about 50% of electric power production (see Table 2.1) even though nuclear energy is less expensive on a fuel basis.

Geothermal

Geothermal energy is the heat that is released from continuous nuclear decay occurring with uranium that is distributed throughout Earth on land and sea. Two factors lead to the accumulation of this heat: (i) thousands of

Table 2.1 US electricity power production in billions of kilowatt hours [6]

	1999		2012	
Coal	1881	50.1%	1514	36.9%
Natural gas	571	15.2%	1237	30.2%
Nuclear	728	19.4%	769	18.8%
Hydroelectric	319	8.5%	276	6.8%
Petroleum	118	3.1%	23.2	0.6%
Wood	37	1.0%	37.8	0.9%
Geothermal	14.8	0.4%	15.6	0.4%
Wind	4.5	0.1%	140.8	3.4%
Solar	0.5	0.0%	4.3	0.1%
Other bio	22.6	0.6%	19.8	0.5%
Misc	55	1.5%	56.1	1.4%
Total	3752		4095	

feet of the Earth's crust provide good insulation that reduces the rate of heat lost to outer space; and (ii) heavier elements (like uranium) are pulled by gravity toward the center of Earth, where these elements undergo natural radioactive decay releasing heat.

The warmer the geothermal heat source, the more useful the energy. For most locations, higher temperatures are located several thousand feet under the surface, and the cost of accessing them is great compared to alternatives. At the center of Earth, some 3700 miles below the surface, temperatures reach 9000°F and metals and rocks are liquid.[25]

At Yellowstone Park and Iceland, useful geothermal energy is available a few hundred feet under the surface or at the surface (hot springs and geysers). Even at these locations the cost of the underground pipe network necessary to create an electrical power plant is high. On a case-by-case basis, geothermal heating has been economical. Much of Iceland's residential and commercial heating is provided by geothermal energy (see box).

Geothermal Heating in Iceland

(from http://www.energy.rochester.edu/is/reyk/history.htm)

The first trial wells for hot water were sunk by two pioneers of the natural sciences in Iceland by Eggert Olafsson and Bjarni Palsson, at Thvottalaugar in Reykjavik and in Krisuvik on the southwest peninsula in 1755–1756. Additional wells were sunk by Thvottalaugar in 1928 through 1930 in search of hot water for space heating. They yielded 14 liters per second at a temperature of 87°C, which in November 1930 was piped 3 km to Austurbacjarskoli, a school in Reykjavik which was the first building to be heated by geothermal water. Soon after, more public buildings in that area of the city as well as about 60 private houses were connected to the geothermal pipeline from Thvottalaugar.

The results of this district-heating project were so encouraging that other geothermal fields began to be explored in the vicinity of Reykjavik. Wells were sunk at Reykir and Reykjahbd in Mosfellssveit, by Laugavegur (a main street in Reykjavik) and by Ellidaar, the salmon river flowing at that time outside the city but now well within its eastern limits. Results of this exploration were good. A total of 52 wells in these areas are now producing 2400 liters per second of water at a temperature of 62–132°C.

Hitaveita Reykjavikur (Reykjavik District Heating) supplies Reykjavik and several neighboring communities with geothermal water. There are about 150,000 inhabitants in that area, living in about 35,000 houses. This is way over half the population of Iceland. Total harnessed power of the utility's geothermal fields, including the Nesjavellir plant, amounts to 660 MW, and its distribution system carries an annual flow of 55 million cubic meters of water.

Some manufacturers refer to the use of groundwater or pipes buried in the ground used in combination with a heat pump as geothermal furnaces. These furnaces do not use geothermal heat. Rather, the large mass of Earth simply acts as energy storage to take in heat during the summer and give up heat during the winter.

RECENT SOLAR ENERGY

Use of Sunlight

Solar energy provides the most cost-effective means to reduce heating costs and can be used to directly produce electricity. Both can be cost-effective, depending on the local cost of conventional electrical energy alternatives.

Solar heating is the oldest, most commonly used and least expensive use of sunlight. Building location, orientation, and window location can be used to displace auxiliary heating such as a natural gas furnace. Windows located on the south side of a northern hemisphere building will bring in sunlight to heat during the winter. A strategically located tree or well-designed roof overhang can block the sunlight during the summer. The number and placement of windows will vary, based on design preference.

Aesthetics, solar functionality, and construction materials (siding on a building) are available for solar performance. New building designs are available that provide cost-effective combinations for solar systems.

Solar water heating systems are the next most popular use of solar energy. They use solar heat to reduce the consumption of natural gas or electricity to heat water. Clarence Kemp is known as the father of solar energy in the United States. He patented the first commercial Climax Solar Water Heater. This and competing systems sold about 15,000 units in Florida and California by 1937. In 1941, between 25,000 and 60,000 were in use in the United States, with 80% of the new homes in Miami having solar hot water heaters. Use outside the United States has developed, especially in regions where electricity costs are high and the climate is warm.

When confronted with the oil boycott and subsequent oil supply incentives, Israel proceeded with a major initiative to use solar water heaters. More than 90% of Israeli households owned solar water heaters at the start of the twenty-first century.[28] Solar water heaters are also quite popular in Australia. At sunny locations where electricity is expensive and where natural gas is not available, solar water heating is a good option. It is easy to store warm water so the solar energy collected during the day is available at night.

Considerable research has been conducted using mirrors to focus sunlight and generate the high temperatures required to produce steam for electrical power generation. To date, most of these systems are too costly. Alternatively, the direct conversion of sunlight to electricity is popular in niche markets, and new technology is poised to expand this application.

In the 1950s, Bell Laboratory scientists made the first practical photovoltaic solar cell. Today, that photovoltaic technology is widely used on flat screen computer monitors (provide electricity with the "picture code and it works) as well as for producing electrical power for electric devices in remote locations. These remote devices include highway signs that cannot be easily connected to grid electricity and small electrical devices like handheld calculators.

Solar energy is usually not used for power generation in competition with grid electricity. In some locations, photovoltaic cells on roofs provide an alternative for enthusiasts where consumer electrical prices are above $0.10 per kWh. Small solar roof units show a better payback to meet individual electrical needs than commercial units designed to sell bulk power to the electrical grid. While consumers will usually pay more than $0.08 per kWh for electricity, when selling excess electricity to the grid one typically receives less than $0.04 per kWh.

The south-facing walls and roof sections of every building in the United States are potentially useful locations for solar photovoltaic panels. Materials having both aesthetic and solar function are becoming available today. From this perspective, there is great potential for solar energy to replace a portion of grid electrical power. At 4.3 billion kWh in 2012, solar electrical power on the grid provided about 0.1% of the electrical energy production (see Table 2.1).

Based on Table 2.1, trends in electrical power generation are as follows: decreasing use of coal, increasing use of natural gas, near constant hydroelectric and nuclear, and increasing wind power.

Hydroelectric Energy

Water stored in high-elevation lakes or held by dams creates high-pressure water at the bottom of the dam. The energy stored in the high-pressure water can be converted to shaft work using hydroelectric turbines to produce electricity. Most of the good dam sites in the United States have been developed, so this is a mature industry. At 319.5 billion kWh in 1999, hydroelectric power on the grid provided 8.6% of the electrical energy production. Environmentalists oppose dam construction in the Northwest United States, and there is active lobbying (especially among native Americans living on the Reservation on the river) to remove dams on the Columbia River.

Wind Energy

Wind energy is one of the oldest and most widely used forms of power. Traditionally, the shaft work was used to pump water or mill grain. In the 1930s, an infant electrical power industry was driven out of business by policies favoring the distribution of fossil fuel electricity.

Between 1976 and 1985, over 5500 small electric utility units (1–25 kW) were installed on homes and remote locations in response to high oil prices. The installation and use dwindled when government subsidies ended in 1985. By the end of 2014, wind power increased to 4.5% of generated electricity in the United States. Goals are to increase this to as much as 20% [7].

The price of electrical power from wind has decreased from more than $1.00 per kWh in 1978 to about $0.066 per kWh in 2014 [8]. This does not include the cost of backup power when wind is not blowing.

The primary issue for more widespread use of wind power is not the cost of the wind turbines. The first and foremost issue is (i) wind is not dependable power on demand which means every MW of wind turbine power requires a MW of backup electrical power from an alternative source. One could argue on this basis that at best, wind power capital costs might be twice as much as other major power sources because of the constant need for backup power; other important issues are (ii) environmental impact (noise pollution, poor aesthetics, bird kills, etc.); and (iii) high maintenance costs because of the large number of wind turbines needed to generate as much power as a typical coal-fired power plant.

The dependability issue is the ability of wind power to supply continuous electrical power. The ability of a facility to provide electricity is characterized by its "capacity factor." This factor is the actual energy supplied by a wind turbine compared to the theoretical power supplied if it operated continuously at its design capacity. Wind power suffers from low-capacity factors because of the lack of wind at night and the lack of power demand when the wind is blowing. Table 2.2 provides capacity factors for 2012 for the most common sources of grid electrical power.

The capacity factor for wind is typically 0.2–0.35. Wind's low-capacity factor reflects the fact that wind is not always blowing while the low-capacity factor for natural gas is primarily a reflection of how natural gas is used to provide peak demand power. Nuclear power plants have long startup times, and so, they are the first sources of power used on the grid which results in high-capacity factors.

One of the less obvious opportunities for electrical power supply is energy storage. Storing wind energy as it is available for use during peak demand times will increase the value of the wind energy and would increase capacity factors. This could increase the value of wind energy from a wind farm by a factor of three or more. Such an increase in the

Table 2.2 Example capacity factors (2012)

Power source	Capacity factor
Coal	0.558
Natural gas	0.333
Nuclear	0.862
Hydro	0.400
Renewables	0.323
Petroleum	0.056

value of wind energy would change the economic outlook of wind power from marginal to profitable. Currently, it is more cost-effective to build a backup natural gas power plant than it is to use storage.

Solar Hydrogen

One of the hottest research topics of recent years (2010–2014) has been the use of sunlight to produce hydrogen. This can be done through engineering of bacteria or catalytic solar panels. The standard approach is to split water into hydrogen and oxygen.

Despite the publicity and tens of millions of dollars that the government is investing in this project, the technology has three fatal flaws; any one of which could be a roadblock to it becoming a sustainable technology.

The first flaw is the cost of collecting sunlight. Plants like corn or soybeans tend to self-assemble once the seed is put in the ground to cover the hundreds of square miles of land with plants. The sunlight is dispersed, and so large acreages of land are necessary. For all solar energy applications (except biomass), the cost of collecting the disperse sunlight translates to represents a huge investment for the infrastructure of covering the surface of hundreds of square miles of land with solar collectors or reactors.

The second flaw with solar hydrogen is the cost of collecting the hydrogen produced over hundreds of square miles of land. This is complicated because the hydrogen product must be kept separated from the oxygen that is also produced. Hydrogen–oxygen mixtures easily detonate yielding a powerful explosion risking the energy collector infrastructure. So, there is the additional capital cost to provide separators that collect the hydrogen and release the oxygen at each solar collector. This complicates the facility design that covers huge acreages.

The third flaw is that even if the hydrogen is collected, the large-scale storage and handling of hydrogen is still too costly and burdensome for widespread use in transportation. Hydrogen storage is an ongoing research topic across the globe.

If it ever becomes feasible to do this on a large scale, it is unlikely that conversion of water to hydrogen can compete with direct conversion of solar energy to electricity.

The United States is a rich country capable of devoting efforts to research that does not hold the prospect for immediate and sustainable utility. Research topics such as solar hydrogen deserve to be studied, but primarily in the context of advancing science and engineering rather than a realistic prospect of providing a new sustainable energy technology.

Biomass

Energy storage limits the utility for both wind power and solar energy. Nature's way for storing solar energy is to produce biomass. Biomass is biological material; grass, wood, corn, palm oil, and similar plant material. Time, the absence of oxygen, and compaction (promoted by Earth overburdens) convert biomass to coal, petroleum, or other geological variations of these materials.

Wood has been used through the ages to produce heat. Today, wood is burned to provide heat and very little is used to generate electricity, corn is converted to ethanol, and vegetable oils are converted to biodiesel. Unlike fossil fuels, biomass is not available in geological formations that have accumulated for years. Rather, biomass grows each year and must be harvested and used quickly to get quality fuel. The supply must be renewed every year.

The availability of biomass is reported as estimated amounts produced per year. A wide range of biomass types, as well as different terrain and climate, control biomass availability. The supply of biomass tends to increase with increasing prices. Table 2.3 summarizes example prices and availability of solid biomass (not including fats and oils). Updated prices on corn would need to be used for any estimates using that feedstock.

Table 2.3 US estimate of supplies and production potential of ethanol from biomass[a] [9]

	Price $/ dry ton	Quantity million dry tons/yr	Conversion gallons ethanol/ton	Ethanol equivalent millions of gallons/yr	Cost of feedstock/ gallon ethanol
$2.40 bu (56 lb) Corn, United States [10,11]	85.7	280	89	24920	$0.96
Refuse derived waste	15	80	80	6400	$0.19
Wood waste *cheap*	30	10	110	1100	$0.27
Wood waste *expensive*	45	80	110	8800	$0.41
Switchgrass *cheap*	25	5	95	475	$0.26
Switchgrass *expensive*	45	250	95	23750	$0.47

Except for reliable corn numbers, conversions are optimistic.

[a]Estimated per-ton yields of ethanol from corn, sorghum, refuse-derived fuel, wood waste, and switchgrass are 89, 86, 80, 110, and 95, respectively. Corn has 56 pounds per bushel with an assumed price of $2.40 per bushel ($85.71/ton) with an annual US production estimate of 10 billion bushels or 280 million tons.

Solid biomass is used for energy five different ways: (i) burning for heat or electrical power generation; (ii) conversion to ethanol; (iii) pyrolysis; (iv) gasification; and (v) anaerobic (without oxygen) methane production (landfill gas). The high cost of biomass makes conversion to liquid fuels and use as chemical feedstock the best applications for most of these renewable energy sources. Direct combustion and anaerobic methane are handled on a case-by-case basis, where they are generally profitable if the biomass has already been collected for other reasons (e.g., solid waste collection in cities). For quality liquid fuel production, two technologies stand out: ethanol production and coal gasification for Fischer–Tropsch liquid fuel production. When including oil seed crops (e.g., soybeans and rapeseeds) a third option for biodiesel is also becoming quite attractive.

Ethanol and Biodiesel

Biofuels like ethanol and biodiesel qualify as recent solar energy. Plants convert sunlight (radiation) to chemical bonds that store the energy for years if properly stored. The process of making ethanol and biodiesel is the process of changing nature's chemical bonds to those chemical bonds in liquid fuels useful for powering engines. Processes for making the liquid fuels include use of chemical reactions and processes to remove and purify the liquids from that part of the biomass that cannot be converted to either ethanol or biodiesel.

Table 2.3 shows the number of gallons of ethanol that can be produced from the most common forms of biomass. The corn data of Table 2.3 are important points of reference. Corn is the largest commodity crop in the United States and provides high yields of dense biomass. While the price per ton of corn is almost twice the price of large-volume switchgrass and wood, the processing and accumulation cost of corn is substantially less than the processing costs for the other feedstocks.

Dried distiller grain is the solid by-product sold as a high-protein, high-fat cattle feed when producing ethanol from corn. Over half of the corn costs are recovered by the sale of this by-product. These by-products make the use of corn, one of the most affordable relative to other biomass crops (e.g., sugarcane, not a temperate zone crop). Other biomass materials may actually have a higher yield of ethanol per acre, but they do not provide the valuable by-products.

Corn is the most commonly used biomass for producing ethanol. The production process consists of adding yeast, enzymes, and nutrients to the ground starchy part of corn to produce a beer. The beer contains from 4% to 18% ethanol, which is concentrated by distillation that is similar to the distillation used to produce whiskey. The final fuel ethanol must contain very little water for use as motor fuel. In gasoline, the water is an insoluble phase that extracts the ethanol and settles to the bottom of the tank. If this water gets to the engine, the engine will stall.

About 90 gallons of ethanol are produced from one ton of corn or about 2.5 gallons of ethanol can be obtained per bushel of ethanol. The cost to produce one gallon of ethanol from corn is equal to about $/bushel × 0.5 (by-product credit) ÷ 2.5 + $0.70 (processing costs). At $2.40/bu, this is $1.20/gallon. Ethanol has about two-thirds the energy content of gasoline, so these prices translate to $1.80 per equivalent gasoline gallon at $2.40/bu corn. Farmers have found that owning ethanol-producing facilities is a good investment because when corn prices are low, the profit margin for ethanol production can increase.

Estimates of gasoline used in US cars, vans, pickup trucks, and SUVs are about 130 billion gallons of gasoline per year. (These numbers agree with the motor gasoline consumption of 3.05 billion barrels reported elsewhere.) About 500 million prime, corn-producing acres would be required for ethanol to replace all of the gasoline. This is about one-quarter of the land in the lower 48 states. The lower 48 states have about 590 million acres of grassland, 650 million acres of forest, and 460 million acres of croplands (most is not prime acreage).

If all of the current corn crop were converted to ethanol, this would replace about 17 billion gallons of gasoline—less than 15% of our current consumption. Estimates of dedicating acreage for ethanol production equivalent (yield-based) to current gasoline consumption would require nine times the acreage used for the current US corn crop. This approach is not realistic. However, if hybrid vehicle technology doubles fuel economy and the electric power grid further reduces gasoline consumption to about 60 billion gallons, substantial ethanol replacement of gasoline is possible. Use of perennial crops would be a necessary component of large-volume ethanol replacement for gasoline.

Corn is an annual crop and must be planted each year. For this reason, costs for mass production of wood and grasses are potentially less than corn. In the late twentieth century, corn-to-ethanol production facilities dominated the biomass-to-ethanol industry. This was due to (i) less expensive conversion technologies for starch-to-ethanol production compared to cellulose-to-ethanol production required for wood or grasses;

and (ii) generally ambitious farmer-investors who viewed this technology as stabilizing their core farming business and providing a return on the ethanol production plant investment. State governments usually provide tax credits for investment dollars to build the ethanol plants, and there is a federal subsidy for fuel grade ethanol from biomass.

Because of lower feedstock costs (see Table 2.3), wood-to-ethanol and grass-to-ethanol technologies could provide lower ethanol costs—projections are as low as $0.90 per equivalent gasoline gallon. Research focus has recently been placed on cellulose-to-ethanol production. The cost of cellulose-to-ethanol production has decreased from more costly to about the same as corn-to-ethanol technology. Based on present trends, cellulose-to-ethanol technology cost could compete in the ethanol expansion in the twenty-first century. It would require large tracts of land dedicated to cellulose production.

The current world production of oils and fats is about 240 billion pounds per year (32.8 billion gallons, 0.78 billion barrels), with production capacity doubling about every 14 years. This compares to a total US consumption of crude oil of 7.1 billion barrels per year of which 1.35 billion barrels is distillate fuel oil (data for the year 2000).[41] With proper quality control, biodiesel can be used in place of fuel oil (including diesel) with little or no equipment modification. Untapped, large regions of Australia, Colombia, and Indonesia could produce more palm oil. This can be converted to biodiesel that has 92% of the energy per gallon as diesel fuel from petroleum. This biodiesel can be used in the diesel engine fleet without costly engine modifications.

In the United States, ethanol is the predominant fuel produced from farm commodities (mostly from corn and sorghum), while in Europe, biodiesel is the predominant fuel produced from farm commodities (mostly from rapeseed). In the United States, most biodiesel is produced from waste grease (mostly from restaurants and rendering facilities) and from soybeans.

In the United States, approximately 30% of crop area is planted to corn, 28% to soybeans, and 23% to wheat. For soybeans this translates to about 73 million acres (29.55 million hectares) or about 2.8 billion bushels (76.2 million metric tons). Soybeans are 18%–20% oil by weight, and if all of the US soybean oil production were converted to biodiesel, it would yield about 4.25 billion gallons of biodiesel per year. Typical high yields of soybeans are about 40 bushels per acre (2.7 tons per hectare), which translates to about 61 gallons of biodiesel per acre. By comparison, 200 bushels per acre of corn can be converted to 520 gallons of ethanol per acre.

Table 2.4 Comparison of annual US gasoline and diesel consumption versus ethanol and biodiesel production capabilities

Gasoline consumption (billions of gallons per year)	130
Distillate fuel oil (including diesel) consumption	57
Ethanol from corn [equivalent gasoline gallons]	25 [17]
Biodiesel from soybeans [equivalent diesel gallons]	4.25 [3.8]

Table 2.4 compares the consumption of gasoline and diesel to the potential to produce ethanol and biodiesel from US corn and soybeans. If all the starch in corn and all the oil in soybeans were converted to fuel, it would only displace the energy contained in 21 billion gallons of the 187 billion gallons of gasoline and diesel consumed in the United States. Thus, the combined soybean and corn production consumes 58% of the US crop area planted each year. It is clear that farm commodities alone cannot displace petroleum oil for transportations fuels. At best, ethanol and biodiesel production is part of the solution. The US biodiesel production in 2005 was about 30 million gallons per year compared to distillate fuel oil consumption of 57 billion gallons per year.

Converting corn and soybean oil to fuel is advantageous because the huge fuel market can absorb all excess crops and stabilize the price at a higher level. In addition, in times of crop failure, the corn and soybeans that normally would be used by the fuel market could be diverted to the food market. The benefits of using soybeans in the fuel market might be improved by plant science technology to develop higher oil content soybeans.

Soybeans sell for about $0.125 to $0.25 per pound, while soybean oil typically sells for about twice that ($0.25–$0.50 per lb). The meal sells for slightly less than the bean at about $0.11–$0.22 per pound. Genetic engineering that would double the oil content of soybeans (e.g., 36%–40%) would make the bean, on average, more valuable. In addition, the corresponding 25% reduction in the meal content would reduce the supply of the meal and increase the value of the meal. At a density of 0.879 g/cc, there are about 7.35 lbs of biodiesel per gallon. A price of $0.25 per lb corresponds to $1.84 per gallon; $0.125 per lb to $0.92 per gallon.

Fuel production from corn and soybean oil would preferably be sustainable without agricultural subsidies (none for ethanol use, biodiesel use, farming, or not farming). A strategy thus emerges that can increase the value of farm commodities, decrease crude oil imports, decrease the value of crude oil imports, and put US agriculture on a path of sustainability without government subsidies.

To be successful, this strategy would need the following components:

1. Develop better oil-producing crops.
   - Promote genetic engineering of soybeans to double oil content and reduce saturated fat content (saturated fats cause biodiesel to plug fuel filters at moderate low temperatures).
   - Promote the establishment of energy crops like the Chinese tallow tree in the South that can produce eight times as much vegetable oil per acre as soybeans.
2. Plan a future with widespread use of diesel engines and fuel cells.
   - Promote plug-in hybrid electric vehicle (PHEV) technology that uses electricity and higher fuel efficiency to replace 80% of gasoline consumption. Apply direct-use ethanol fuel cells for much of the remaining automobile transportation energy needs.
   - Continue to improve diesel engines and use of biodiesel and ethanol in diesel engines. Fuel cells will not be able to produce enough power to compete with diesel engines in trucking and farm applications for at least a couple of decades.
3. Pass antitrust laws that are enforced at the border. If the oil-exporting nations allow the price of petroleum to exceed $70 per barrel ($2.00 per gallon diesel, not including highway taxes), do not allow subsequent price decreases to bankrupt new alternative fuel facilities.
4. Fix the dysfunctional US tax structure.
   - Restructure federal and state taxes to substantially eliminate personal and corporate income taxes and replace the tax revenue with consumption taxes (e.g., 50%) on imports and domestic products. This would increase the price of diesel to $2.25 per gallon (red diesel, no highway tax).
   - Treat farm use of ethanol and biodiesel as an internal use of a farm product, and, therefore, no consumption tax would be applied. Increased use of oil crops would include use of rapeseed in drier northern climates (rapeseed contains about 35% oil) and use of Chinese tallow trees in the South. Chinese tallow trees can produce eight times as much oil per acre as soybeans. If Chinese tallow trees were planted in an acreage half that of soybeans and the oil content of soybeans were doubled, 17–20 billion gallons of diesel could be replaced by biodiesel allowing continued use of soybean oil in food applications. This volume of biodiesel production would cover all agricultural applications and allow the imports to be optional.

The PHEV technology would displace about 104 billion gallons per year of gasoline with electricity and increase energy efficiency. The electricity could be made available from the reprocessed spent nuclear fuel and adding advanced technology nuclear reactors. About half of the remaining 26 billion gallons of gasoline could be displaced with ethanol and half with continued use of gasoline.

In this strategy, up to 55 billion gallons of annual diesel and gasoline consumption would still need to come from fossil fuel sources. These could be met with petroleum, coal-derived liquid fuels (like Fischer–Tropsch fuels), and Canadian oil sand fuels. Increase of electric trains for freight could replace much of the 55 billion gallons.

It is in the farmer's interests to convert at least part of corn and oil seed (soybeans) to fuels since this creates an increased demand for their commodities and higher prices. When the resulting renewable fuel production capacity is combined with new shale oil fracking sources, the United States emerges with a much better fuel supply security in the year 2015 than in 2000.

Algal Biodiesel

Microorganisms like algae can be used to produce vegetable oils that can then be converted to biodiesel. Two categories of this technology are proposed: (i) use of bacteria to convert nutrients in a liquid to vegetable oils; and (ii) use of photosynthesis to produce the vegetable oil in algal or bacterial pools of water.

On the approach of converting nutrients, it is likely that sustainable approaches can be developed for bacteria (or other anaerobic microorganisms) to convert the sewage discharge of cities to a crop of microorganisms and methane. Those microorganisms could then be processed to produce a biodiesel product and solid biomass that could be burned as a fuel or sterilized and used as food in fish farms. This approach has two advantages over the photosynthesis approaches: (i) it uses a concentrated pool of nutrients that avoids the cost of collecting sunlight/radiation; and (ii) it receives the economic advantage of performing a waste treatment for which cities now provide a dedicated revenue stream to maintain. An advantage of this approach is that once successful, the technology can be expanded to allow other waste streams to be processed (other organic trash from cities) by the bacteria. The pools can incorporate a photosynthesis component to add to the waste conversion component. This photosynthetic component can be incrementally expanded as technology becomes available.

Algal biodiesel from photosynthesis can be achieved in two ways: (i) use concentrated light in reactors containing microorganisms; (ii) pools containing microorganisms dispersed over a large area to collect dispersed sunlight.

On the concentrated light photosynthesis, there is either the huge infrastructure cost of collecting and focusing sunlight or the huge cost of making electricity and converting it to light for the organisms. Collecting sunlight is expensive, and if collected, it would require more efficiency to convert the sunlight directly to electricity. An approach to produce electricity to generate light to yield "bugs" that can then be processed to produce liquid fuels does not survive an economic analysis—it is far better to use the electricity directly to charge batteries.

In a (2009) conference advocating algal biodiesel [15], even with high productivity of oil from algal biodiesel that a land mass the size of New Mexico [16] covered with water would be adequate to satisfy all our liquid fuel needs. There was an algal diesel test plot at Roswell, New Mexico [16] yielded disappointing results because the algae did not thrive in the cool weather that prevailed at night and much of the day. The fact is that huge expanses of the United States constantly seek more fresh water for cities and agriculture, and the assumption that water is not available for algal biodiesel. When climate, water, and relatively flat terrain needs are considered, possible locations for large-scale photo algal biodiesel production are limited to southern Louisiana and parts of Florida. In both instances, the first challenge will be to overcome environmentalists and environmental regulation to convert huge expanses of natural habitat to algal farms. Other technical and cost barriers include: (i) developing and maintaining the huge pond algal farms; (ii) preventing other microorganisms from overpowering the preferred algae in these ponds; (iii) the collection of algae for processing; and (iv) the huge volumes of algal fluids (intracellular water) that must be removed and processed to get to the oil.

To conclude, the only approach that has promise for sustainable and affordable production of biodiesel from microorganisms is the conversion of city sewage to oil by bacteria (or other organisms). When this industry is established and sustained, and if it overcomes all organism-processing hurdles, it would be realistic to believe that a sustainable photosynthesis-based industry could be established. Such an industry would first expand by supplementing the waste-based facilities. Algal-biodiesel technology is much like solar hydrogen where research can be justified because the United States can afford to perform research on energy that does not actually offer the prospect of sustainable industries in the next few decades.

FOSSIL FUELS

The "fossil" designation of certain fuels implies that the fuel energy originated from prehistoric vegetation or organisms. Sunlight is the only external source of energy received on earth. Sunlight supplied the energy for vegetation, the food for organisms. This suggests that fossil fuels are really "stored solar energy." They are the most commonly used energy source to provide heat and drive our machines. Fossil fuels tend to be concentrated at locations near the Earth's surface. This makes them easily accessible. Fossil fuel sources include the following:

- Coal
- Petroleum
- Heavy Oil
- Oil sands
- Oil shale
- Natural gas
- Methane hydrates

In Wyoming, there are vast coal seams 40-feet thick less than 100 feet underground. They can be "surface mined" rather than common underground "shaft coal mined." In the Middle East, hundreds of barrels per day of crude oil flow from a single well under the natural pressure in the deposit. Each source provides abundant energy.

Coal, petroleum, and natural gas are accessible fossil fuels and easy to use (see box "Petroleum and Gas"). They are by far today's most popular fuels. Table 2.5 summarizes the known accessible reserves of these fuels in

Table 2.5 Comparison of energy reserves and rates of consumption

Energy description	Amount (BTU)	Amount (barrels)
World recoverable coal reserves	2.5E 19	
World recoverable natural gas reserves	8.6E 19	
World recoverable petroleum reserves	7.0E 18	
US petroleum consumption (year 2000).	4.5E 16/yr	19.7E 9/day
US reserves of tight oil (fracking)	3.6E 17	58E 9 [13]

the world and in the United States. World recoverable reserves for coal, natural gas, and petroleum are 2.5E + 19 (25 billion billion), 8.6E + 19, and 7.0E + 18 BTUs, respectively [12]. For coal, the total estimated reserves are about a factor of 10 higher than the estimated recoverable reserves.

Petroleum Oil

In 2000, the United States consumed 19.7 million barrels of petroleum per day or 3.8E + 16 BTUs per year. This consumption would deplete known US petroleum reserves in about 3 years and estimated US petroleum reserves in about 7.6 years. The statistics is summarized in Table 2.5.

It is important to note that between 2005 and 2015 the estimates of world energy petroleum reserves increased by about 33% and the gas reserves increased by about 25%. By definition, the consumption of the reserves should result in decreases in reserves. These data indicate oil reserves are actually many times greater than reported in Table 2.5; it is more a matter of the cost of recovering the oil and the time to gear up the industry for that recovery.

From 2006 to 2008, rapid increases in oil demand, largely from China, outpaced the implementation of technology to access the increasing reserves that become available as the price increases. Ultimately, there is no energy crisis for the United States or the world; rather, there is simply a price to be paid for that energy. And there is the potentially very high price associated with poor planning as was indicated by the 2008 recession.

Heavy Oil and Oil Shale

Estimated energy reserves in heavy oil, oil sands, and oil shale dwarf known reserves in coal, natural gas, and petroleum. Figure 2.9 is an attempt to put the quantities of these reserves in perspective. One evolutionary route to form these three other oil reserves includes the advanced stages of petroleum decay. Petroleum deposits contain a wide range of hydrocarbons ranging from the very volatile methane to nonvolatile asphaltines/tars.

When petroleum is sealed securely in rock formations, the range of volatility is preserved for millions of years or converted to methane if buried deeper, where it is converted by geothermal heat.

U.S. Petroleum	1/10th X
World petroleum	1 X
World recoverable coal	5 X
World recoverable heavy oil and oil sands	10 X
World recoverable natural gas	>15 X
World recoverable (???) oil shale	>500 X
World methane hydrates	>5000 X
World recoverable uranium	>50,000 X

Figure 2.9 Comparison of estimated reserves of prominent fuels other than renewable fuels.

Petroleum and Gas—From the Ground to the Refinery

The inserted image [14] illustrates a typical petroleum reservoir and drilling used to recover petroleum. A rock cap has kept the reserve isolated from atmospheric oxygen for millions of years. Since the petroleum is lighter than water, it floats above any water in the petroleum deposit. Petroleum gases, including natural gas, are the least dense material in the formation and are

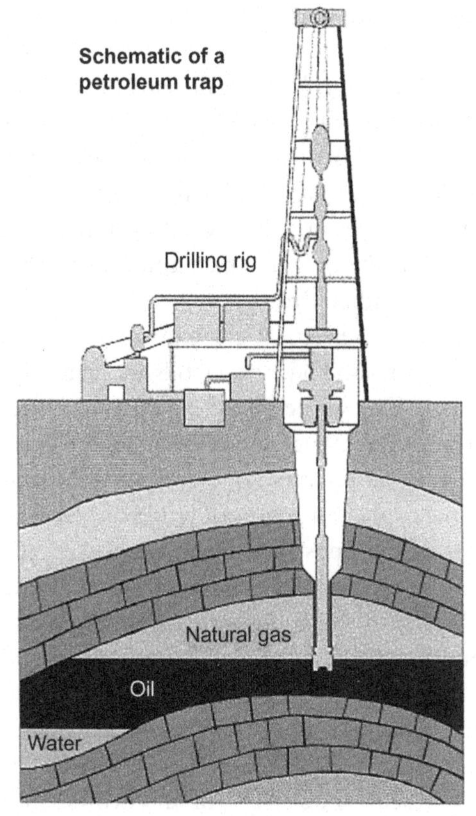

Figure 2.10 Illustration of petroleum drilling rig and reservoir.

located above the oil. Drilling and placing a pipe to the petroleum reserve allows recovery. Oil reserves recovered by conventional land drilling applications are typically several hundred feet to about a mile deep.

When the cap rock is fractured by an earthquake, by erosion of rocks above the formation or simply due to the porous nature of the rock, the more volatile components of petroleum escape. This leaves less-volatile petroleum residues in the forms of heavy oils, oil sand, or "tight oil" in natural oil shale formations.

Heavy oils are volatile-depleted deposits that will not flow to a producing well at reservoir conditions but need assistance for recovery. Oil sand liquid is a "heavy" oil—typically not mobile at reservoir conditions, but heat or solvents can make the oil flow through the reservoir porous rock. Oil in shale formations is usually immobile and present in "fine grained" rock formed when annual silt and clay settle out of the water in thin layers that does not allow oil flow. Unlike the oil in oil sands that can be extracted *in situ* or with small amounts of solvent and/or heat the oil in shale is more difficult to extract.

Heavy oil reserves in Venezuela are estimated to be from 100 billion barrels to one trillion barrels. These heavy oils are generally easier to recover than oil from oil sands. The United States, Canada, Russia, and Middle East have heavy oil reserves totaling about 300 billion barrels (conservative estimates). These heavy oil reserves are estimated to be slightly greater than all the more easily recovered conventional crude oil reserves.

Surface reserves of oil sands have been mined and converted to gasoline and diesel fuel since 1969 in Alberta, Canada. Production costs are about $20 per barrel of oil produced. This supplies about 12% of Canada's petroleum needs. The sands are strip-mined and extracted with hot water. Estimated reserves in Alberta are 1.2–1.7 trillion barrels with two open pit mines now operating. Other estimates put oil sand reserves in Canada, Venezuela, and Russia between 330 and 600 billion barrels. These estimates represent about 90% of the world's heavy oil (and oil sand) reserves in Western Canada and Venezuela. Total global oil sand reserves are 6–10 times the proven conventional crude oil reserves. (Conventional crude oil reserves are estimated at one trillion barrels.)

Oil Fracking Technology

During the past 20 years, horizontal drilling techniques have been developed and used to tap the huge quantities of natural gas and shale oil known to be in the vast shale formations. A vertical hole is drilled into the shale formation deep underground. A horizontal hole is then drilled from that vertical well into the shale formation. High-pressure water containing proprietary chemicals and fine sand is pumped into the shale, fracturing the shale rock. When the water is removed, the fine sand in the "fracking fluid" holds the thin layers of fractured shale open so that the oil and natural gas can flow to the vertical "production" well.

Shale oil is best characterized as relatively nonvolatile oil dispersed in shale. World shale oil reserves are estimated to be 600–3000 times the world crude oil reserves. Estimates specific to the western United States place reserves at 2–5 times the known world oil reserves. Tight oil reserves are the shale oil that can be recovered using 2015 fracking technology; the US tight oil reserves are approximately twice the US conventional petroleum reserves and are enough to last about 8 years without supplements from the other sources as illustrated in Figure 2.1.

Oil to be recovered from a drilled well must be liquid and flow through the geological formation. The oil in oil shale near the surface is not liquid; it is like a paste stuck in the shale. A shale oil formation deeper in the earth where temperatures are warm enough, the oil will be liquid rather than a paste. The shale is so fine grained that the oil and gas still won't flow and this is called "tight oil." "Fracking" fractures the shale and creates fractured rock paths for oil flow from the shale oil formation to the producing well.

Those highly productive geological formations in Texas and North Dakota extend into Canada. Shale oil fracking brings with it the following: (i) a currently realized increase in US oil and natural gas production; and (ii) improvement and lower cost approaches to fracking and horizontal drilling that have implications beyond shale oil reserves in the United States. The cumulative result is that stable oil productivity and prices should be realizable well into the 2030s.

While the industry has adopted terms like crude oil, heavy oil, oil sands, oil shale, tight oil, wet gas, and dry gas, the reality is that all combinations of volatility, ability to flow (viscosity and porosity), and accessibility (a few meters to 10,000 meters below the surface) exist. As a result technology like fracking can have an immediate impact on recovering oils from shale in North Dakota as well as extended impacts on accessing oils in different formations.

A reoccurring theme emerges, that energy reserves are important and immense (including energy in spent nuclear fuel rods) and the application of good technology is the determining factor. This was the case for the tight oil boom illustrated in Figure 2.1.

Natural Gas from Fracking

Shale gas fracking technology has introduced about 665 trillion cubic feet of recoverable natural gas reserves that is estimated to be about half of the total US recoverable reserves. Natural gas consumption in the United States is about 22 trillion cubic feet per year. This new natural gas brings with it a low-cost stability to electrical power production because natural gas electrical power plants are relatively inexpensive to build. Natural gas can also be used as a transportation fuel. These reserves further stabilize vehicular fuel prices and supplies, past the 2030s.

As an indicator of the extent to which natural gas can be used to replace gasoline, about 15% of Argentina's automobiles are natural gas powered. The top three nations based on the number of natural gas vehicles are Iran, Pakistan, and Argentina. These statistics natural gas can replace gasoline if needed or deemed a national priority.

Methane Hydrates

Methane gas often escapes from underground deposits (due to porous or cracked cap rock, erosion of overburden, etc.). If this gas is released deep in the ocean, where there is a combination of water pressure and low temperatures, methane hydrate is formed. Methane hydrate is ice that contains methane in the crystal structure of the ice. The hydrate is stable below the freezing point of water as well as at temperatures slightly warmer in deeper water where the pressure is high. Conditions are right for the formation of methane hydrates on the sea floor (below a few hundred feet of seawater) where the temperature is relatively constant near 4°C, the temperature of maximum water density methane hydrate conditions occur even in tropical oceans.

Methane may form from fossil fuels through reactions initiated by high temperatures from the decay of organic material. Methane is always present in crude oil wells, in coalmines, and surface swamps. The digestive systems of animals and people also contain anaerobic microbes in the digestive process that produces methane. If any of these sources of methane are released into water at cool temperatures and high pressures (due to depth of water), methane hydrates can form.

Also, very cold temperatures can cause methane hydrate formation without high pressures. The water in the Arctic permafrost is frozen and decaying organic material in the soil forms methane hydrate. As long as the permafrost is frozen it holds the methane in hydrate form. When permafrost thaws, the methane is released.

Methane from methane hydrate reserves is not currently recovered. The Japanese have great interest in technology to recover methane from deep ocean deposits. They lack natural fossil fuel reserves and there are deep coastal water areas that are potential sources of large methane hydrate deposits. In the United States, methane hydrates have received the attention of congressional hearings where reserves were estimated at

200,000 trillion cubic feet of natural gas. The hydrate reserves off the east and west coastal boundary waters are under the jurisdiction of the United States. Using conservative estimates, these hydrates contain enough methane to supply energy for hundreds of years.

Natural gas emissions from hydrate reserves can occur naturally. The methane greenhouse effect from released methane will probably occur whether or not we capture fuel energy. If we burn the natural gas the resulting carbon dioxide released would have about 10% the greenhouse effect of any released methane. Methane released from methane hydrates on the sea floor may have contributed to end some of the historic ancient ice ages. During an ice age, the sea levels drop as massive ice sheets form. The ocean water required to form the ice lowers the sea level that reduces the seawater pressure on the sea floor. The lower ocean bottom pressure melts the hydrate releasing methane that bubbles to the surface releasing it into the atmosphere.

Here, greenhouse gas emissions can be reduced while recovering usable natural methane by mining and recovering the methane reserves that are most likely to release naturally. Most hydrate mining research involves changing the temperature and pressure at the solid hydrate reserve location to cause the methane hydrates to melt and recover the released methane. Experts at a congressional hearing agreed that Alaska's North Slope was the most likely candidate for initial research on hydrate methane recovery because of relatively easy access (compared to the deepwater Gulf of Mexico) for the gas collection infrastructure and where the crude oil industry already exists.

Figure 2.9 summarizes the energies available in recoverable fuels. These fuels are all in concentrated deposits with the exception of uranium. The largest fraction of available uranium is dissolved in global seawater (oceans). The recovery process is known but costly so it's not competitive at today's low energy prices. These numbers approximate the size of the different energy reserves relative to available conventional crude oil.

Liquids from Coal and Natural Gas

Commercial processes for converting coal to diesel and gasoline date back to Germany before World War II. The processes for converting natural gas to a range of liquid fuels are actually less expensive than the processes that use coal.

Sasol Ltd., a corporation in South Africa, uses the Fischer–Tropsch process to convert coal to gasoline, diesel, and chemicals. This industry was developed since they have now crude oil deposits and as a result of embargoes against South Africa used to pressure South Africa to end racial discrimination.

Shell Corporation uses Fischer–Tropsch process to produce liquid fuels from natural gas from wells such as at Shell's Pearl GTL facility in Ras Laffan, Qatar. GTL stands for gas to liquids, and is a way to use stranded natural gas where converting the gas to liquid reduces transportation costs substantially.

As illustrated in Figure 2.1, the use of GTL technology is expected to continue to increase, slowly but steadily.

Lessons to Be Learned from History

History has shown that a good diversity in the electrical power infrastructure provides stable and reasonable prices as well as a reliable supply. This has been the case for decades in the United States. That diversity has increased in the past decade with coal reducing from over 50% to 36% of the source of US electrical power. Natural gas and wind power have increased; they have a synergy in that natural gas power plants are able to start or shut down on short notice to compensate for changing wind velocity. Nuclear has held its own at 19%–20%.

For liquid fuels, the stress of fuel imports and the high costs at the fuel pump have been resolved by a combination of technology and diversity as illustrated in Figure 2.1. Fracking technology took several years to develop but it rapidly became a main player and bringer of prosperity. Fuel prices of over $90 per barrel kick-started this industry and prices of over $50 per barrel should sustain the industry. Gas to liquid technology has also emerged as a contributor.

The real winner in the liquid fuel sector was the United States as illustrated in Figure 2.11 on the money flow out of the United States for oil imports. It took drastic conditions to bring the technology to bear to fix the problems, but it did happen. In 2008, 60% of US demands were met by oil imports at prices of $110 per barrel compared to 30% at $55 per barrel. The flow of money and opportunity from the US economy was decreased by 75% due to technology.

Three lessons stand out as most important from the past decade:

- Beware of alarmists
- Diversity creates stability

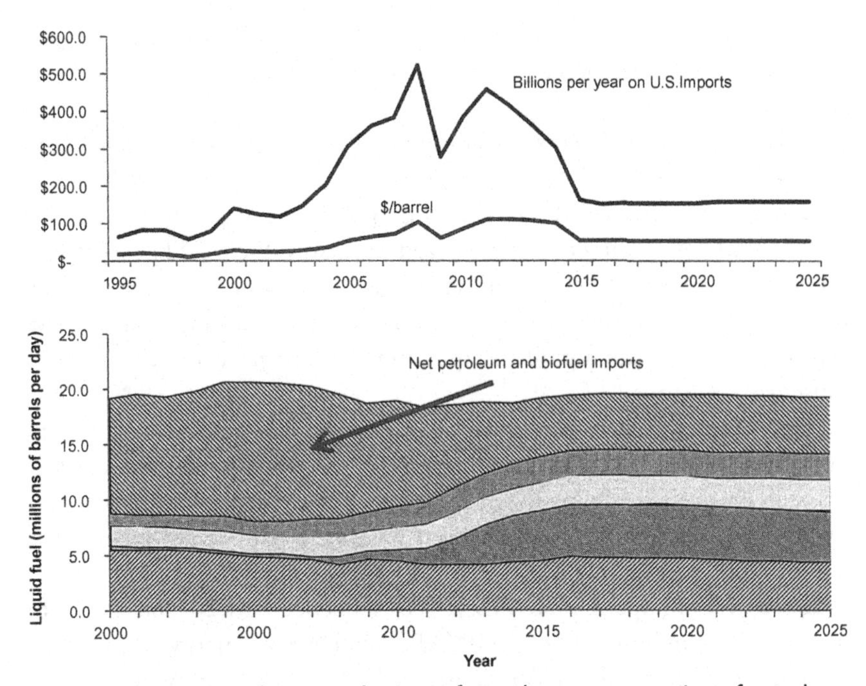

Figure 2.11 Historic and projected prices of petroleum, consumption of petroleum, and billions of dollars per year spent on oil imports by the United States.

- Technology provides diversity

  A fourth lesson is to:
- Create strategies and technologies for likely pending crises.

Alarmist agendas can come from corporations that seek profit as well as groups that focus on single issues. A decade ago the oil corporations used the alarmist approach in an attempt to gain access to oil reserves in Alaska's national parks. US leaders held strong to the position of protecting the parks, and a decade after the confrontation the United States has emerged with a better tight oil fracking industry and a reserve of oil in Alaska's national parks for a possible future rainy day.

On the other end of the spectrum, environmentalist groups or new industries bring alarmist agendas to either ban an industry or promote an industry. The alarmist approach is to ban nuclear power. A better approach is to recognize that the US nuclear is the safest of all the power industries. A solution is to make nuclear power even safer and to reprocess the spent nuclear fuel. A second alarmist approach is to point to a single solution such as wind power to reduce greenhouse gas emissions. This approach fosters sustainable industries that do not place a burden on society to subsidize the otherwise unsustainable industry.

The past decades have seen good trends in diversity for both liquid fuels and for electrical power generation. The greatest single game changing step on diversity is the widespread use of battery powered electric vehicles or direct-grid-powered electric vehicles. This great step would increase the diversity of both the liquid fuel and electric grid infrastructures (batteries can be used for peak load shifting). This single step would also create a path to rapid and substantial decreases in greenhouse gas emissions.

The United States would do well to increase nuclear power to 25%–30% and to adopt first reprocessing technology to use the decades of uranium and plutonium in spent fuel rods that have accumulated over 30 years of nuclear power production. The next step would include fast neutron flux reactor technology to tap the centuries of power in the U-238 that is in spent fuel rods. It costs between $1.20 and $1.40 to produce one million BTU (MBTU) of thermal energy from coal. Uranium fuel costs about $0.62 per MBTU when enriched to 3.4% U-235 isotope in the fuel. Higher U-235 enrichment reduces the cost of energy. Next generation IV reactors would improve the energy produced per ton of uranium fuel.

As with the past decades, technology enables diversity. These technologies include: Better batteries, technologies for better direct-grid-powered vehicles (fifth mode of transportation), and technologies for using the U-235, plutonium, and U-238 in spent nuclear fuel.

The changes that fracking and horizontal drilling technology have brought to the US economy and the energy sector have been remarkable. The resources and time needed to enact a *coordinated strategic approach for the energy sector* are clearly available. Much is attainable, and it is a matter of options US people and leadership choose. The following strategies have merit for setting policy:

- Allow key technologies to *take their course* without excessive federal subsidy that assists a particular technology. These subsidies put potential competitive technologies at a disadvantage.
- *Keep* battery technologies and electric vehicle technologies on course to create stability in transportation fuel prices, supply, and

sustainability (including reduced greenhouse gas emissions) by tapping that stability of the electrical grid infrastructure.

- *Target win–win technology approaches* such as converting waste into fuel (e.g., reprocessing spent nuclear fuel, converting sewage into liquid fuels) and providing faster and more convenient public transportation.
- *Biofuel technologies* will not replace petroleum, but they have advantages beyond the fossil fuels they replace. A range of technologies should be studied to advance both fundamental and applied science and initial commercialization should be fostered (but not maintained) by governments.
- *Beware of alarmists* that advocate approaches that are costly—there are cost-effective approaches to address some of the most alarming threats related to the environment. If there are problems with an issue (e.g., nuclear power, tight oil fracking), fix the problems. Do not kill the industry!

It is important to implement smart strategies that improve the economy and are sustainable without subsidies. There are many good solutions and every reason to be optimistic. It is possible for the prosperous countries to continue to prosper. It is also possible this success will show others on the planet paths for them to prosperity.

REFERENCES

[1] See <http://www.eia.gov/forecasts/aeo/er/executive_summary.cfm>.
[2] See <http://hyperphysics.phy-astr.gsu.edu/hbase/astro/procyc.html>; May 15, 2002.
[3] Synthesis of isotopes of elements 118 and 116 in the 249Cf and 245 Cm + 45Ca fusion reactions. Phys Rev October 9, 2006;C 74:044602.
[4] See <http://www.aip.org/physnews/graphics/html/element118.html>; May 15, 2002.
[5] Images provided by the Uranium Information Center, Ltd., GPO Box 1649N, Melbourne 3001, Australia provided the information.
[6] See <http://www.eia.gov/electricity/annual/html/epa_01_01.html>.
[7] Electric Power Monthly (PDF). Report. U.S. Department of Energy, Energy Information Administration, 4 March 2015, see <http://www.eia.gov/electricity/monthly/>.
[8] Moné C, Smith A, Maples B, Hand M. 2013 Cost of Wind Energy Review; Technical Report, NREL/TP-5000-63267, February 2015. <http://www.nrel.gov/docs/fy15osti/63267.pdf>.
[9] Tyson K, Shaine P, Bergeron, Putsche V. Modeling the penetrationof the biomass-ethanol industry and its future benefits, September 18, 1996, bioenergy'96, Opryland Hotel, Nashville, TN.
[10] Tshiteya RM. Conversion technologies: biomass to ethanol. Golden CO: National Renewable Energy Laboratory; September, 1992. p. 3–5.

[11] See <http://www.usda.gov/nass/pubs/trackrec/track02a.htm#corn> corn production U.S.
[12] See <http://www.eia.gov/cfapps/ipdbproject/iedindex3.cfm?tid=5&pid=57&aid=6&cid=regions&syid=2005&eyid=2015&unit=BB>.
[13] Olsen T. Working with tight oil. Chem Eng Prog April, 2015, pgs. 35–39.
[14] From U.S. DOE Public Domain, see <http://www.eia.doe.gov/pub/oil_gas/petroleum/analysis_publications/oil_market_basics/Supply_Petroleum_Trap.htm>; 2003.
[15] Mississippi State University 2009 BioFuels Conference Thursday, August 6, 2009.
[16] Darzins A. Algae as a Source of Feedstocks for Biofuels, representing the *National Renewable Energy Laboratory (NREL).*

CHAPTER 3

Energy and Sustainability

Contents

POLITICS OF CHANGE IN THE ENERGY INDUSTRY

Energy conversion and utilization is a multitrillion dollar business and vital to today's society. The magnitude of these industries presents both mega opportunities and extraordinary challenges. While these industries were once predominantly driven by technology, today, politics dominates essentially every aspect. One of today's greatest challenges is to advance an industry in which technology has been displaced from the role of driving

Sustainable Power Technologies and Infrastructure.
DOI: http://dx.doi.org/10.1016/B978-0-12-803909-0.00003-8

the industry to the passive role of yielding to the political policies that drive the industry.

The politics of energy includes contributions from corporations and environmental groups. Mega corporations seek to maintain the status quo, because these corporations control billions of dollars of revenue per year. Why ruin a good thing? Environmental groups including the US Environmental Protection Agency tend to be single issue and focus on their perception of environmental impacts. Typically, true balances of benefit and risk are much more complicated than the portrayals of environmental groups. There is a quadrangle of overlapping interests and conflicts formed by industry, environmentalists, politics, and public welfare.

Technology Emerging to What End?

Several fundamental issues related to defining and discussing emerging energy technologies include:

- When the goal of proposed improvements is defined, the technical challenges may be substantial.
- When the goal of proposed improvements is not defined, the challenge borders on futile.
- Which comes first: the fuel or the engine? Change in energy technology is uncertain and filled with obstacles, but they must match.
- While new technologies may be on the sidelines ready to meet the goals, the momentum of the industry is formidable and will dominate the discussion.

In view of these obstacles in the energy industry, the task of identifying the potential of an emerging technology is difficult. Proposed objectives to be met by new technology should have a reasonable chance of surviving over time and be substantially free of speculation. To this end, one development goal stands out: *Emerging energy technology must be based on fuel price. That fuel price must be for the delivered product and, for planning purposes, need to include anticipated environmental costs.*

For the electrical power industry, the energy fuel consumer is the large utility company that generates electrical power. For the vehicular fuel industry the millions of individual owners of cars and trucks are the final consumers.

Recent trends have included an emphasis on the sustainability of an industry. A reasonable interpretation of sustainable is sustainability in the

30-year time frame. There is too much uncertainty for longer time frames in view of potential breakthroughs in science and technology.

This 30-year time frame is greater than the 3–10 year time frame that large corporations plan to get full return on investments (ROIs). It is less than the total sustainability pursued by advocates of wind and solar energy.

In the following discussion the potentials of different energy feedstocks will be reviewed. The feedstock costs provide an indicator of the potential of a fuel for meeting goals on fuel price.

COST OF FEEDSTOCK RESOURCES

Without detailed process knowledge, as is the case when evaluating the potential of emerging technology, the *potential* of a process can be estimated by calculating the *gross profit* [1]. In this approach, the cost of the feedstock fuel consumed by a process is subtracted from the value of the final product (a liquid fuel or electricity). The higher the "gross profit," the greater the economic *potential* of that process. Technology is developed with the goal of realizing actual profits that approach the gross profits.

Table 3.1 summarizes and ranks feedstock costs by their representative prices at the end of the twentieth century. The "liquid fuel" column reports the feedstock costs for the energy equivalent to that in a gallon of gasoline. The last column estimates the feedstock cost for producing one kWh of electricity.

The lowest cost feedstocks provides the best potential for low-cost energy. Municipal solid waste, spent nuclear fuel, uranium, and coal have high gross profits and are sufficiently abundant to evaluate for commercialization. In each case, technology can bridge any gaps between actual and gross profits.

In 2015, the use of natural gas is growing rapidly due to the low cost of natural gas power plants, abundant source for shale fracking technology (creating stable prices), and the ability of natural gas to start up and shutdown quickly to complement wind power. While many fuels have greater potential for lower prices than natural gas, the technology for many of these is still rather expensive.

Wastes and By-Products as Feedstocks

The greatest potentials exist with municipal solid waste and spent nuclear fuel. Historically, both vehicular fuel and electrical power

Table 3.1 Summary of feedstock costs for commonly used and considered fuels

Fuel	Price ($/MMBTU)[a]	AVG ($/kWh)	Liquid fuel[b] ($/gasoline gallon equ.)	El. conv. efficiency[c] (%)	Electricity cost ($/kWh)
Municipal solid waste (MSW)	$(−2.00)− (−4.00)	−0.0102	−0.36	20−35%	−0.037
Spent nuclear fuel	$(0.08)	−0.0003	−0.01	25−45%	−0.001
Full uranium	$0.08	0.0003	0.01	25−45%	0.001
Remote natural gas	$0.50−1.00	0.0026	0.09	N/A	
US uranium	$0.62	0.0021	0.072	30−33%	0.0068
French reprocessed uranium					0.0090
Coal	$1.20−1.40	0.0044	0.145	35−45%	0.0105
Oil sands ($10−15/barrel)	$2.00−3.00	0.0086	0.30	28−53%	0.021
Natural gas	$6.00 ($3.00)	0.0205	0.68	50−56%	0.039 (0.020)
Biomass	$2.10−4.20−6.80	0.0149	0.52	20−45%	0.044
Petroleum ($45−75/barrel)	$9.00−15.00	0.0411	1.44	28−53%	0.135

[a]Price does not include $0.42/gal motor vehicle fuel tax.
[b]Assuming 0.119 MMBTU per gallon of gasoline.
[c]Electrical conversion efficiency is the thermal efficiency of the cycle. The price of the feedstock fuel is divided by the thermal efficiency to estimate the cost of fuel consumed to generate 1 kWh of electricity.

technologies were alive and well before there was either a municipal solid waste or nuclear waste problem of any magnitude. Both municipal solid waste and nuclear waste (spent nuclear fuel) provide an opportunity to generate useful energy from a "waste" that imposes a burden on society.

Today's US nuclear power plants extract about 3.4% of the energy available in the fuel rods before they are set aside as spent fuel. The spent fuel elements are stored at each of the nuclear power plants, and transporting them to a common repository in Yucca Mountain is the subject of bitter debate that seems to have been resolved not being a good option.

Reprocessing the fuel reduces the mass of the waste requiring long-term storage and recovers about 96% of unspent fuel. This reprocessing is practiced in Europe and Japan, but not in the United States. Excess weapon grade uranium/plutonium can also be blended with natural uranium to make power plant fuel and eliminate this inventory as a weapon threat.

Similar to nuclear reprocessing, the conversion of municipal solid waste into fuel and electricity can eliminate a waste problem and provide the needed energy. Landfill corporations receive $15–60 per ton of waste to dump this waste in a properly prepared hole in the ground, and unfortunately, often see waste-to-energy projects as competition for their revenues. The municipal waste-to-energy process must overcome at least four opposing groups: (i) direct political pressure from electrical power providers and landfill corporations; (ii) air quality regulations that make it very difficult to build the new conversion facilities; (iii) the cost and availability of conversion technology; and (iv) Mafia-type control of the collection of solid waste. Full use of municipal solid waste for energy is as much a political issue as it is a technology opportunity.

In the absence of American leadership, answers are likely to come from Europe. In Japan and Europe, landfill land is becoming less available. In Germany and other countries waste-to-energy industries are prospering. Across Europe in 2010, there were about 400 waste-to-energy plants, with Denmark, Germany, and the Netherlands leading the pack in expanding them and building new ones [2].

Energy facilities designed to run on waste products typically benefit from the economies of scale made possible by adding other feedstocks to their mix. Municipal solid waste plants will be able to take in biomass produced near the plant locations where hauling distances are short. Coal could also be used in the municipal solid waste plants. Co-firing these facilities with biomass and coal would provide improved economies of scale and feedstock reliability to keep the plant in operation. In the future, more advanced facilities would be true "solid fuel refineries" with the option to produce chemicals, liquid fuels, electrical power, and recycled metals.

Liquid Fuels Market

Both nuclear power and municipal solid waste facilities are better suited to produce electrical power rather than liquid fuels. For liquid fuels, the next least expensive feedstock options are remote natural gas, coal, and oil sands (and heavy oil). This liquid fuels market is not synonymous with automobiles—a case will be presented in Chapter 7 for conversion to electrical power as the primary distributed energy for automobiles. Electrification of trains is possible, and short-haul trucks may run off batteries. Interesting options like the Hyperloop and Terreplane System are emerging as an alternative for some of the aircraft market. There is no end in sight for the end of liquid fuel dominance for farm tractors.

As Table 3.1 indicates, biomass will have a difficult time competing with oil on a cost basis. While it is true that ethanol and biodiesel may actually have periodic cost advantages over gasoline and diesel as these industries expand (2001–2008), cost advantages disappear when the volumes of ethanol and biodiesel expand by using "excess" agricultural commodities competing with increasing population food demands. A side effect of this expansion might be higher farm commodity prices—this might bring an end to government-subsidized agriculture.

Oil corporations in North America are moving toward tight oil and oil sands. Heavy oils are an option after the most accessible tight oil reserves are consumed—oil prices greater than $100 per barrel increase the recovery options. These fuels will work with incremental modification of existing refining infrastructure. Today, the use of tar sands tends to be a better long-term plan than conventional petroleum. The driver for commercialization will be corporate profitability. This type of industrialization should have little problem providing liquid fuels well past the end of the twenty-first century—for a price. Lessons of history show that the environment, military conflict, major trade deficits, and huge fluctuations in prices can be the results of industrial expansion driven by corporate profitability.

A problem with the liquid fuel industry is the lack of diversity in feedstocks—petroleum feedstocks and oil corporations dominate and control the industry. One of the paths forward in this industry is electrical power (including nuclear) and increased automobile efficiency to displace 33–50% of this 187 million gallon a year. If biodiesel and ethanol replace another 10–20% of this industry market, the dominance of petroleum will be undermined. This diversity will provide stable prices and nations can eliminate trade deficits to the extent desired. Diversification is key.

Fischer–Tropsch fuels are a fourth player in diversification of the liquid fuel industry. The growth of Fischer–Tropsch is slow, but persistent.

Fischer–Tropsch Technology

Fischer–Tropsch technology was developed in the mid-1920s in Germany to produce liquid fuels from coal. It is being used today to produce liquid fuels from both coal and natural gas. The Fischer–Tropsch technology can provide liquid fuels at prices competitive with petroleum today. Early estimates indicated that Fischer–Tropsch technology is borderline competitive with crude oil at $20 per barrel (eliminating nontechnical cost barriers); however, recent history has shown that sustained prices greater than $55 per barrel have led to a much more rapid increase in tight oil production (drilling) as an alternative to Fischer–Tropsch process.

There is reason to believe that Fischer–Tropsch technology and shale oil fracking technology are sustainable to compete on a large scale with conventional petroleum oil at prices of about $55 per barrel and higher. The primary difference between the technologies is that it only takes a few weeks to drill a well using fracking technology with an investment near $1 million and a production well that can pay for itself in less than 12 months; while a Fischer–Tropsch facility can cost hundreds of millions of dollars, take 2 or 3 years to bring into production, and then take several years to pay for itself. The difference is the risk involved and the time it takes to realize a return of the investment.

Because of the high costs for Fischer–Tropsch production facilities, growth will be slower, but once the investments are made they will impact the industry for decades in the form of steadily increasing market share. This slow but steady growth is depicted by US Energy Information Agency (EIA) projects for "natural gas plant liquids" of Figure 2.1.

The primary applications will be low-cost natural gas available at remote sites or at oil production sites where it is currently flared.

The key event that accelerated the maturing of tight oil recovery (fracking) technology was high oil prices. To some extent these prices were artificial.

Impact of Commodity Crude Bidding

Dan Dicker wrote a book titled *Oil's Endless Bid, Taming the unreliable price of oil to secure our economy* [19]. Addressing our current problems with the price of crude oil, on page X of the Introduction the following:

"WE have caused this; there is no one else to blame. We have inspired this disaster with lax regulation, blind faith in free markets, and unfettered greed. The oil market has followed a similar pattern to other modern asset markets, becoming enmeshed in more and more complex derivative products that benefit mostly the people that sell them, we encourage and reward best the people who create and squeeze profit out of these new product markets, and we invite—no warn—every investor to participate as well, lest the miss the latest and greatest money making opportunity. The result of this avalanche of activity is clear, causing prices to boom, only to bust violently before beginning the cycle over again."

"So why should I care about the swinging prices of oil? . . . The difference is that people choose to invest in stocks, therefore, they bear responsibility for their own risks and possible losses. But whether it is the heat in our homes or the fuel for our cars, even the food we eat and the clothes we wear—just about everything in our lives is tied to the costs of energy. We are all invested in oil, whether we like it or not. Business is hardly exempt. More than 50% of the companies on the New York Stock Exchange rely on energy as their single largest input cost, and that doesn't even include the energy companies themselves (some of which were being put out of business by the high price of oil!)."

These are the facts of the international crude oil market. It is very difficult to predict how the crude oil market will move. It is clear that about 15% of the energy in a barrel of crude oil will be spent refining the crude into the array of products sold in the energy and chemicals market. It is worth considering what might be done to make us less dependent on this "sole source" of energy. Estimating the economic "break even" or the profit margin in alternative energy sources is worth considering.

To its merit, commodity crude bidding the high, albeit temporary (see Figure 3.1), prices it created has provided the incentive for the advancement of technology in tight oil recovery. Like a frog sitting is a

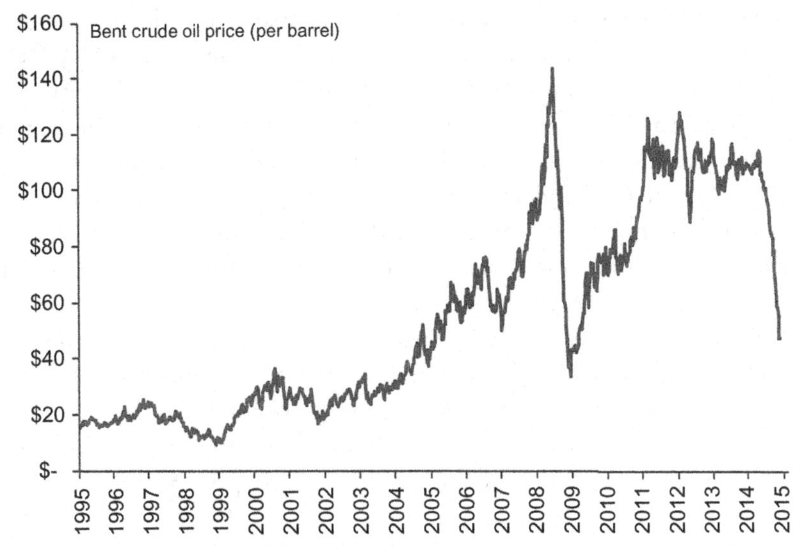

Figure 3.1 Historic crude oil prices.

pot of water over a flame, if the temperature increases gradually the frog dies a slow death; however, if the temperature increases rapidly the frog is aware and jumps from the pot. Commodity crude bidding has brought disruption and that disruption has brought evolution.

CASE STUDY ON INVESTMENT DECISIONS AND POLICY IMPACTS

Corporate investment decisions can be understood by performing the same profitability analyses the corporations use. These profitability analyses can also be evaluated under assumptions of different government (or corporate) policies to show how policies impact investment decisions. A case study was performed on the investment for a Fischer–Tropsch process to convert Wyoming coal to make synthetic oil to replace imported petroleum. The impact of technology cost was compared to the following four nontechnical barriers:

- **Petroleum reserves** are the number of years of petroleum crude oil in proven reserves held by the corporation considering an investment into an alternative fuel facility. A base case of 0 year of reserves was assumed. The typical reserves for an oil corporation are from 7–14 years, so, conservative figures of 5 and 11 years were used in this sensitivity analyses.

- **Intangible costs** are the costs of the risks associated with investing in a new technology. These costs include the risk of OPEC (Organization *of Petroleum-Exporting Countries*) lowering the price of crude oil to drive the synthetic fuel facility out of business. Intangible costs were incorporated into the sensitivity analysis by either assuming that a higher ROI and shorter payback period (20%, 6 year) would be required to attract investment capital or by assuming that the price of the synthetic fuel would decrease to a very low value ($10 per barrel) shortly after startup (3 year or 5 years).
- **US Tax structure** is the taxes paid to the US and local governments including social security and unemployment taxes that must be paid by US employees. A base case of 34% corporate income tax was assumed. The sensitivity analysis included an assumption of 0% corporate income tax and that half of the "threshold" price was due to taxes (corporate income, personal income, property, FICA, etc.) and that the threshold price would be reduced by 50% if these taxes were not selectively placed on domestic production.
- **Return on investment** is the % ROI and the time in years (year one is defined as the first year of production) for payback of the investment capital. A base case of 12.5% ROI with a 15-year payback was assumed. For comparison, investment rate of return (IRR) calculations were performed for a 5% ROI and 30-year payback since this is reflective of today's municipal bonds for civil infrastructure investment. Also considered was 10% ROI and 20-year payback.

IRR was chosen as the preferred profitability analysis for this case study because the IRR calculation provides the "threshold" price of the synthetic oil product—petroleum prices above this threshold price would justify investment [3]. The lower this threshold price the more likely the technology will be commercialized and compete with imported petroleum. In the IRR calculation, the "threshold" price for the synthetic oil is adjusted until the net present value of the process is $0 at the end of the process life (e.g., 15 years for the base case). After preparing a base case IRR, the sensitivity of the threshold synthetic fuel price to the four nontechnical barriers was determined by repeating the IRR calculation for the upper or lower limits of each of the nontechnical barriers.

Table 3.2 summarizes the parameters used in preparing the base case from which the sensitivity analysis was performed. Table 3.3 summarizes the sensitivity analysis results showing the impact of the nontechnical barriers.

Table 3.2 Conditions for base case Fischer–Tropsch facility used to perform sensitivity analysis

Property	Value	Justification
Capacity	20 billion gallons per year	This is one-sixth the amount of gasoline consumed in the United States
Capital cost	\$3.90/gal/yearly capacity (\$78 billion capital investment)	This is the lower cost estimate published by the DOE (US Department of Energy) and some companies commercializing this technology
Operating cost	\$7.50 per barrel	This is typically published for a coal facility. The cost of coal (essentially the only feedstock for this process) is about \$3.75 per barrel
Construction time	30% in year one, 70% in year two, startup in year three	Standard for facility of this type
ROI	12.5%	Standard corporation ROI for low risk ventures
Investment payback	15 years	Standard for facility of this type
Corporate Income tax	34%	Year 2002 corporate tax rates line out at 35% for taxable income over \$20 million per year
Startup/working capital	2 months operating expense	Standard practice

The base case yields a threshold price of \$41 per barrel of synthetic oil; however, this base case assumes no intangible costs and that the investing corporation held zero years of oil reserves.

By relying on oil corporations and including reasonable intangible costs, a price in excess of \$150 per barrel would be needed before investment into the needed infrastructure would meet profitability expectations. This threshold price has been known to be elusive, and this calculation confirms that crude oil imports in excess of \$600 billion per year are likely to be realized before corporate investments are justified based on current corporate investment criteria. Any plan of relying on oil corporations to take the lead developing alternatives to petroleum is flawed.

Table 3.3 Summary of changes in the sensitivity factors used in sensitivity analysis. COS is Cost of Sales

	Reserves	ROI, P	Taxes	Intangible	COS	Capital (/gal/yr)	Prices (/barrel)
Base case	0	12.5% 15 yr	34%	N/A	$7.50	$3.90	$41.00
Petroleum reserves							
Low res., P-$10	5 yr						**$115.00**
Typical res., P-$10	11 yr						**Infinity**
Return on investment							
Municipal bond		5%, 30 yr					**$20.21**
Low ROI		10%, 20 yr					**$32.72**
US tax structure							
No corporate tax			34– 0%				**$33.31**
All Taxes Gone			none				**$20.50**
Intangible costs							
High ROI		20%, 6 yr					**$73.17**
$10 Crude at 3 years				$10 @ 3			**$96.27**
$10 Crude at 5 years				$10 @ 5			**$67.50**
Technology costs							
50% Reduction					$7.50	$1.95	**$24.24**
25% Reduction					$5.62	$2.93	**$32.61**

The case study revealed that existing petroleum reserves of a corporation would provide the greatest investment deterrent to that corporation. The past decade has revealed that it is not only the conventional petroleum reserves, but tight oil and recoverable heavy oil reserves that are a deterrent.

Remaining deterrents from greatest to least impact were: intangible costs/risks, demands for high returns on investment, unfavorable tax structures, and the cost of the technology.

Hindsight suggests that oil corporations will invest in a US alternative fuel industry only when their **petroleum reserves (ΔIRR of >$100/barrel)** are depleted to about the time it takes to build the alternative fuel infrastructure, or about 2 years of reserves. Reduction of reserves to this level would be needed for large-scale investment into Fischer–Tropsch facilities and is not likely to occur in the near future.

Intangible costs (ΔIRR of $38/barrel) are the second greatest barrier to investment into a domestic alternative fuel industry. Antitrust laws fail to cross international boundaries. Needed investments into a US domestic fuel industry are not made (in part) because OPEC can flood the world oil market with low-price petroleum and drive domestic synthetic production out of business. Possible solutions are international treaties or price-dependent tariffs that effectively extend antitrust laws across international boundaries. These treaties/policies need to be in place at the time investment decisions are made, which means they need to be established now.

Corporate demands for high ROI (ΔIRR of $14.5/barrel) had the third greatest impact on the threshold price. States and communities routinely make infrastructure investments at lower ROI's such as 5% ROI and 30-year payback. The IRR spreadsheets on which corporations base investments include short-term corporate monetary gains in wealth. Long-term, noncorporate, and nonmonetary wealth generation should be included in these calculations.

Domestic taxes (Δ IRR of $14.1/barrel) have about the same impact as high corporate demands on ROI and have a greater impact on investment decisions than the cost of an otherwise good technology. Presently, about half the price of a barrel of synthetic petroleum produced from Wyoming coal would be taxes (corporate income, personal income, property, FICA, etc.) while essentially nothing (about 10 cents per barrel) is charged to imported petroleum; these import fees represent docking/harbor fees.

It is illogical to burden domestic industry with taxes while foreign producers are allowed to enter the US market without paying a similar tax. While it is true that imported petroleum "can" have similar taxes paid to a foreign country, the key qualifier is "can." If imported oil is produced in a country where taxes are primarily "value added taxes" (VAT) paid on domestic sales (otherwise known as a *consumption tax*), a foreign producer could export to the United States with little or no tax burden.

The current US tax structure is a problem that should be addressed.

All four of these nontechnical barriers to commercialization can be readily and sustainably corrected.

True Barriers to Commercialization

Each of the four nontechnical barriers to commercialization of the synthetic oil case study has a greater impact than reasonable advances in technology for Fischer–Tropsch facilities. If solutions to the nontechnical barriers were implemented, threshold petroleum prices to attract investment into domestic synthetic oil production would be less than $20 per barrel and possibly as low as $13 per barrel. However, this production would be by a nonpetroleum corporation without refineries and a fuel distribution network—this producer would have to sell to existing oil corporations or face tens of billions of dollars of additional investment for separate refineries. In view of possible agreements with foreign producers (e.g., 11 years of reserves) it is not certain that the US major petroleum corporations would displace contracted petroleum with a synthetic oil alternative.

Intervention or a nonfuel alternative is needed. The plug-in hybrid electric vehicle (PHEV) is an example of a nonfuel alternative. The Canadian oil sand industry is an example of how government intervention can be effective. These represent two possible solutions—other good solutions may exist.

The Canadian oil sand (formerly referred to as tar sand) industry is a success story that demonstrates how commercialization barriers can be overcome. At $9–15 per barrel production costs, oil sand production costs are more than Fischer–Tropsch production costs. But even at these higher costs, large-scale mining of oil sands in Canada began in 1967 [4] when oil prices were less than $10 per barrel. It took about a decade to make a profit from the oil sands (when including regional opportunity costs of not allowing imported petroleum to compete with the oil sand fuels).

The Canadian National Oil Policy introduced in 1961 made possible the oil sand commercialization. It established a protected market for Canadian oil west of the Ottawa Valley and freed the industry from foreign competition. This policy protected companies from the greater intangible costs and provided an environment for smaller companies (other than the major petroleum companies) to develop the technology. In addition, in 1974 the Canadian and provincial governments invested in Syncrude's oil sand project and provided assurances about financial terms [5]. New refineries were built (Shell Canada Limited Complex at Fort Saskatchewan, Alberta). Today, with oil prices in excess of $50 per barrel, the Canadian oil sand industry is profitable beyond most investors' expectations; it provides energy, security, and quality jobs.

The United States is at a disadvantage today because it did not heed the warnings of the oil crises in the 1970s and 1980s. However, it is possible to turn this disadvantage into an advantage. New and better options are available today. Tesla Motors was founded in 2003, but it was 2013 before profitability was realized; this is similar to the Canadian oil sand industry in that it took about a decade for profitability to be realized.

In addition to Tesla Motors becoming a sustainable company, its production and demand for electric vehicle batteries has a wide impact. Tesla Motors brings with it electric vehicles from other corporations who wish to "join the band wagon" and lower cost and better batteries for HEV and PHEV approaches. All assist with the ultimate of desired stabilities in vehicular fuel markets which is attained when grid electricity is established as a major competitor with gasoline both as a direct replacement (electric vehicles) and to increase the efficiency with which gasoline is used (HEVs).

PHEV vehicles are similar to HEVs on the market today. They use extended battery packs (e.g., 20 miles of range) that charge using grid electricity during the night providing the first 20 (or so) miles out of the garage each day without engine operation. Gasoline is fully replaced with grid electricity for most of each day's transit. Per-mile operating costs for grid electricity are about one-third the cost of gasoline. Rather than going to petroleum producers, the majority of the fuel operating revenues would go to local communities.

If a consumption tax were applied to imported fuels representative of taxes on domestic synthetic fuel production, the higher vehicle cost of a PHEV would be recovered in about 3 years. Advancing technology would rapidly reduce this time to about 2 years. Less oil would be imported, domestic jobs would be created, and the new demand for off-peak electricity would allow restructuring of the electrical power grid to include base load generation with increased efficiency for electrical power production and reduced greenhouse gas emissions. Up to 20% of gasoline might be replaced without expansion of the electrical power grid.

Petroleum Reserves and Protecting the Status Quo

The starting point for solving problems related to the deterioration of the US manufacturing base and the inability of the United States to displace its reliance on imported petroleum is to recognize that these are both artifacts of corporations attempting to maintain the "status quo."

In the middle twentieth century, the US steel industry decided not to invest in new US steel mills because such mills would replace operational and profitable steel mills they already owned—such investments were not good short-term business decisions. On the other hand, Japanese investors had a greater incentive to invest in a steel manufacturing industry because their infrastructure had been destroyed during World War II—investments were made in more efficient mills that often produced a superior quality product. Eventually, most of the US steel industry lost out to foreign competition.

From a corporation's perspective, the "upside" potential is clearly greater to invest profits in infrastructure that does not compete with existing infrastructure. Whether it be the steel industry, textile industry, or petroleum industry, the path of short-term corporate profitability was and is different than the path of long-term prosperity for countries, states, and communities.

From this observation it follows that countries should not allow energy corporate giants to decide the long-term national energy strategies.

A solution is to select energy options that do not rely on the refineries or distribution networks of major oil corporations. Increasing the fuel economy of vehicles is such an approach, but it has limited potential. A new technology referred to as "PHEV" has the potential to substantially replace oil with domestic electricity and may be a technology that displaces petroleum. Hydrogen gas would be used on fuel cell versions of PHEVs, but no hydrogen distribution infrastructure will be required. Hydrogen would be generated "onboard" the vehicle.

Intangible Risks (Costs) and International Antitrust Policies

Investment in new infrastructure brings with it the risk of losing the monetary investment. Intangible risks are those that are difficult to predict and often outside the control of the investors.

When pursuing alternatives to petroleum-based fuels, the intangible costs include: (i) the risk that oil-producing countries will flood the market with oil to maintain market share and drive competition out of business; and (ii) the risk that a better alternative will be available in a few years. The latter is an accepted risk for all investments because competition is recognized as good for the country. The former is recognized as going against the best interest of a country as documented by antitrust cases against corporations like Standard Oil and Microsoft.

Reform in the antitrust laws is needed, because John D. Rockefeller demonstrated that even when you are guilty your competition will be gone, your wealth will be great, and you will merely have to stop certain practices. Reform is needed to provide quicker response to unscrupulous business practices or to stop such practices from ever eliminating the competition.

To the credit of current antitrust laws, they are used to proactively evaluate mergers before they occur to make sure the mergers do not create a monopoly. However, these laws do not prevent corporations from lowering prices to stifle the competition.

In the case of petroleum imports, one solution is to add an "antitrust tax" to imported petroleum if the price is lowered more than, say, 25% from the recent 5-year average price (25% and 5 years are one of many possible combinations). For example, if the price of imported oil into the United States was $40 per barrel for five consecutive years and OPEC decided to reduce the price to $20 per barrel, this "antitrust tax" would place a $10 per barrel tax on imported oil. New alternative fuel industries in the United States would then have to compete with an effective price of $30 rather than $20 per barrel. A year later, the new 5-year averaged price would be $36 per barrel, and the new minimum effective price would be $27 per barrel. Sustained price decreases could occur, but the reductions would be gradual, and the intangible risk to domestic industry would be considerably less.

An "antitrust tax" on imports would only be applied if significant decreases in prices occurred. It would be proactive. It would not limit how low prices could eventually go. And it would reduce/eliminate the intangible risks that create one of the barriers to the commercialization of alternatives to imported petroleum.

US taxes take their toll on the profitability of domestic industries. Figure 3.2 summarizes US taxes on "hypothetical" domestic synthetic oil (like the case study) for comparison to the taxes on imported petroleum. For these calculations, every dollar of sales going to a domestic synthetic

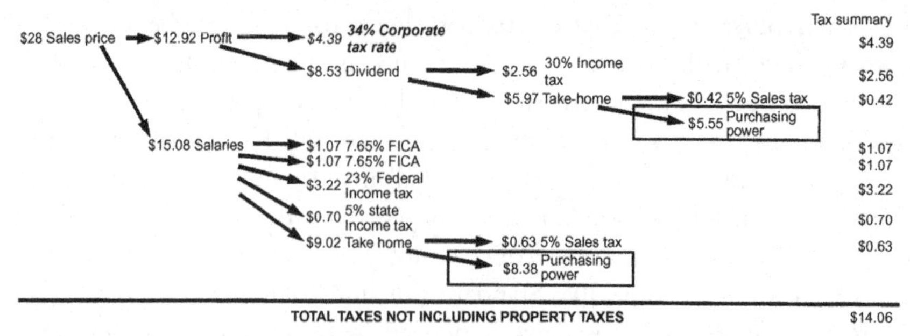

Figure 3.2 Summary of tax breakdown on $28 barrel of synthetic crude.

oil producer is interpreted as either going to the government or becoming "true" purchasing power in the hands of either a worker or investor.

As indicated in Figure 3.2, for a domestic production facility selling a $28 barrel of oil, about $14 in taxes is paid for every $14 in purchasing power that goes to either investors or workers. These numbers will vary depending upon specific examples. The inclusion of state portions of employer taxes (unemployment, etc.) as well as property taxes further increases theses taxes. Sales taxes should debatably be excluded from this analysis since it is placed on all products independent of where they are produced. This shows that taxes are about 50% of the domestic sales price of most products produced and sold in the United States—an effective domestic tax rate of 100%!

If $50 is spent on a barrel of crude oil purchased from a foreign government, essentially no US taxes, tariffs, or social costs are paid on that oil. In some cases, little or no taxes would be paid to any government. Foreign taxes may also be excluded from those items exported from foreign countries to improve international competitiveness.

Even though the domestic and imported oils are superficially equivalent purchases to a refinery in the United States (based on current US law), the fundamental question remains whether these options produce the same US societal benefit. This can be stated in a different way. From the perspective of a US citizen is $1 in US tax money going toward the purchase of a jet aircraft for the US Navy the same as $1 in Iranian tax going toward the purchase of a fighter jet for the Iranian air force? Certainly not!

Why then, does the US government continue policies that give foreign production competitive advantages and direct cash flow away from the US government and to foreign governments?

Who pays the US taxes to make up for the lack of taxes on imported crude oil?

US citizens will ultimately pay costs to keep the US government going. Mega energy corporations and several foreign governments benefit greatly by having this portion of the tax bill not show up on their balance sheet. It is clear these are shortcomings of current US tax laws.

Where, exactly, have our classical analyses of this trade gone wrong?

The philosophical father of free trade, Adam Smith, identified two legitimate exceptions to tariff-free trade—when industry was necessary to the defense of the country and when tax was imposed on domestic production [6]. Corporate income taxes are an obvious example of a tax imposed on domestic production. Personal income taxes may be a less obvious example, but, from the perspective of foreign competition, the personal income tax has the same impact as the corporate income tax—so do property taxes, unemployment taxes, dividend taxes, and FICA. Sales/consumption taxes are fundamentally different since they are applied without bias to the origin of the item sold.

Clearly, any strategy on sustainable energy technology and policies should include a critical analysis of tax policies to make sure that domestic industries are not hurt by imposing higher taxes on domestic energy supplies as compared to imported energy.

Corporate Profitability and High Investment Thresholds

Corporations tend to pursue only the most lucrative investments available when reinvesting their profits. The spreadsheet analyses of these investments often indicate the potential to recover all capital investment plus a yearly 20% ROI in the first 6 years (6-year payback) of production. For new endeavors, anticipations of this high return and quick payback are standard. The following represent typical ROI and payback periods for corporate investments:

- 20% ROI, 6-year payback, for first time processes with high intangible costs/risks—including risk from foreign competition not covered by US antitrust laws.
- 12.5% ROI, 15-year payback, for proven technology with moderate to low risk.
- 10% ROI, 20-year payback, for a proven technology with very low risk that fits in well with a corporation's current assets.
- 5% ROI, 30-year payback is not acceptable for corporate investment but is used for municipal infrastructure funded through bonds.

The case study results of Table 3.3 demonstrate that if an oil corporation owns years of petroleum reserves in excess of the time it takes to build an alternative fuel facility, the ROI simply cannot be high enough to compensate for the reduction in value of those assets. Only under the assumption of less than 2-year reserves can a corporation meet reasonable ROI investment goals.

To place things into perspective, recent petroleum oil prices (3/05) exceeded $55 per barrel. If imported petroleum is taxed at the same rate as the case study's domestic production of synthetic fuel from coal, the refineries would be paying about $110 per barrel. As summarized in Table 3.3, if a corporation that was not vested in petroleum reserves were to commercialize this technology using municipal bonds, the cost of domestic production would be $20.21 per barrel (including taxes). In this hypothetical environment of "equitable taxes on both foreign and domestic products" and "investment using municipal bonds"—the facilities would be built and actual ROIs would be much higher than the minimum expectations of the investors.

The Canadian oil sand industry is an example of a community and government making what was perceived as a low ROI investment in the 1970s. This has resulted in exports of fuel (rather than imports), regional prosperity, and increased national security. With crude oil prices at $55 per barrel and production costs from the oil sand fields at about $12 per barrel, the returns on investments are huge and the industry has self-sustained growth/expansion.

The irony of the Canadian oil sand industry example is that a providence made an investment based on anticipation of minimal monetary gains and received great monetary and nonmonetary returns. At the same time, corporation after corporation holding out for high ROI investments have gone out of existence.

Today, states in the United States are attempting to reduce this nontechnical barrier by providing tax credits or land for plant sites. For example, Alabama provided Hyundai $253 million in "economic incentives" to build an automotive plant there—these incentives included sewer lines, highway paving, and tax breaks [7,8].

Land and tax credits are effective and low-risk approaches for states and municipalities to attract corporate investment. A more direct approach would be for local and state governments to provide capital with bond-based funding. If this direct investment approach was used, communities would be able to match their capabilities and needs with the industry the community is trying to attract. The underlying message here is that communities realize value from local industry beyond the ROI realized by the corporation and that it is often good policy to provide incentives to attract industry. In this approach, there must be assistance to communities to help them make smart incentive decisions.

A particularly intriguing opportunity exists for state and local communities to provide bond funding for local investment with a "balloon" interest rate that pays off well after the 10 or 20 years. Low-bond rates for the first decade allow the corporation to receive the quick payback while high interest rates in the long term make the communities/states winners.

It is important to recognize that profitability (10%, 20%, or higher) can continue for decades after the payback period used to determine the threshold prices that warrant investment. A community can position itself for this long-term upside of a good manufacturing infrastructure investment.

TAXES AND SOCIAL COST

The Cost of Driving a Vehicle

Essentially no tax is applied to imported crude oil; however, substantial taxes are applied to gasoline at the pump. The taxes at the pump for maintaining the nation's highway system should not be confused with the taxes that increase the cost of producing domestic synthetic oil presented in the case study.

With crude oil selling at about $55 per barrel, about 20% of the price of automotive gasoline is taxes averaging about $0.42 per gallon. The crude oil feedstock is $1.30 ($55 per 42-gallon barrel, March of 2005) of a $2.01 gallon of regular unleaded gasoline. Premium gasolines would have higher refining and sales component costs amounting up to an additional $0.20 per gallon. Table 3.4 summarizes federal taxes corresponding to the $0.184 federal tax on automotive gasoline for commonly used fuels.

For the gasoline price summarized in Figure 3.3, typical price contributions are: federal tax ($0.184), average state tax ($0.236), crude oil ($1.30), refining ($0.133) and distribution/sales ($0.150).

Federal and state taxes on gasoline go toward the building and maintenance of our nation's highways totaling about $72 billion per year, and most agree that these moneys are well spent. An average of $0.42 per gallon is applied at the pump to gasoline sold for use on highways. For

Table 3.4 Highway taxes on fuels [9,10]

Types of fuels	Cent/ gallon	Types of fuels	Cent/ gallon
Gasoline	18.4	Special motor fuel (general)	18.4
Gasohol		Liquid petroleum gas (LPG)	13.6
10% gasohol	13.1	Liquid natural gas (LNG)	11.9
7.7% gasohol	14.3	Aviation fuel (other than gasoline) noncommercial	21.9
5.7% gasohol	15.4	Aviation fuel (other than gasoline) commercial	4.4
Gasoline removed or entered for production of:		Gasoline used in noncommercial aviation	19.4
10% gasohol	14.6	Inland waterways fuel use tax	24.4
7.7% gasohol	15.5	Diesel fuel	24.4
5.7% gasohol	16.3	Diesel fuel for use in trains	4.4
Kerosene–highway	24.4	Diesel fuel for use in buses	7.4
Kerosene–aviation fuel	21.9	Including 0.1 LUST tax	
100% Methanol (natural gas)	4.3	Compressed natural gas (egg)	5.4
100% Methanol (biomass)	12.3	Liquefied natural gas	18.3
100% Ethanol (biomass)	12.9	Propane	18.3

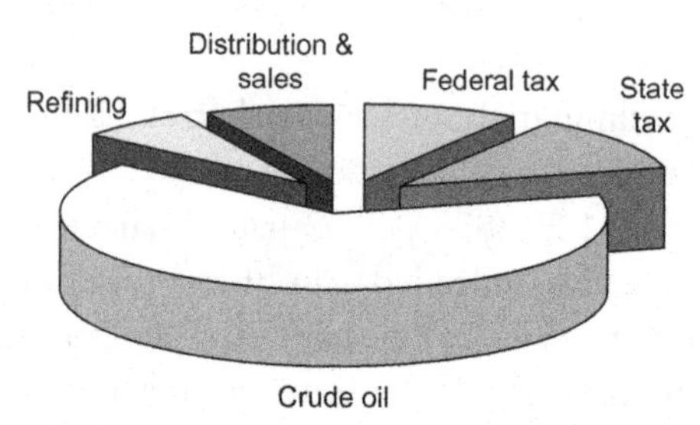

Figure 3.3 Summary of price contributions on a gallon of gasoline on a 201 cents per gallon of unleaded regular gasoline.

diesel, a red dye is placed in the off-highway diesel (no taxes) to distinguish it from diesel for which highway taxes have been paid—law enforcement can sample the fuel tank to verify that diesel intended for farm tractors is not used to haul freight on highways.

A Fischer–Tropsch fuel (fuel of case study) would be completely compatible with petroleum-based diesel and could be distributed in the same pipelines. The $0.42 per gallon highway tax is applied to all fuels (imported and domestic, alike) at the pump.

For a hypothetical gallon of Fischer–Tropsch fuel and based on the estimate of Figure 3.2, half of the $1.30 per gallon ($0.65 per gallon) would actually be a compilation of additional taxes collected prior to the fuel reaching the refinery. As the American Petroleum Institute [11] points out, imported crude oil is taxed multiple times prior to the refinery as summarized by the following per barrel taxes:

- Import duty: $0.05–0.11
- Merchandise processing fee: up to $0.06
- Harbor maintenance fee: up to $0.025
- TOTAL: up to $0.19, but typically about $0.10 per barrel.

These taxes on the imported petroleum total to about 0.25 cents per gallon as compared to the projected 65.00 cents per gallon for the case study fuel.

In addition to these taxes, approximately half of the refining, sales, and distribution costs (assuming refining occurred in the United States) are also a compilation of taxes. Here, about 14 cents is applied to fuels based on both imported and domestic (e.g., Fischer–Tropsch fuel from Wyoming Coal) oils.

If a $0.79 ($0.14 + $0.65) per gallon consumption tax were applied to all gasoline as an alternative to Figure 3.2 compilation of taxes, the Fischer–Tropsch gasoline would remain at a price of $2.01 per gallon while the import-based gasoline would cost $2.66 per gallon. It is possible that under this tax structure the production of Wyoming-based Fischer–Tropsch fuel would increase to totally displace imported oil. It is possible that the price of imported oil would precipitously fall to about $28 per barrel to compete with domestic fuels. It is possible that something between these two extremes might also occur.

How Does the Cost of Fuel Stack Up

Few people actually sit down and calculate the cost of owning and operating a car. These costs include fuel cost, parking (in some instances including the cost of buying and maintaining a garage parking space), car depreciation, insurance, and maintenance (oil changes, new tires, new brake pads, etc.). Two scenarios are summarized below with different miles traveled each year and different miles per gallon for the vehicle. A gasoline price of $1.79 per gallon is assumed.

Fuel: 12,000 miles/30 miles/gal × $0.42/gallon highway tax =	$168
Fuel: 12,000 miles/30 miles/gal × ($1.179–0.42)/gallon =	$304
Fuel: 12,000 miles/30 miles/gal × $0.20/gallon for premium =	$80
Parking at work and parking in home garage	$500
Car cost: $30,000 × 12/120 =	$3000
State/local property insurance ($10,000 vehicle/3*6.13%)	$204
Insurance: $700 per year	$700
Maintenance	$150
Total	$5106

Fuel: 24,000 miles/13.1 miles/gal × $0.42/gallon highway tax =	$384
Fuel: 24,000 miles/13.1 miles/gal × ($1.179–0.42)/gallon =	$694
Fuel: 24,000 miles/13.1 miles/gal × $0.20/gallon for premium =	$183
Parking at work and parking in home garage	$500
Car cost: $30,000 × 24/150 =	$4800
State/local property insurance ($12,000 vehicle/3*6.13%)	$245
Insurance: $700 per year	$700
Maintenance	$250
Total	$7756

This analysis shows that the fuel component of the fuel cost is typically less than 10% of the total cost of owning and operating an automobile. If the entire country were using ethanol fuel costing $1.50 per equivalent gasoline gallon to produce and distribute, the additional burden would be equivalent to a 5–10% increase in the per-year operating expense of an automobile.

Alternative fuels like Fischer–Tropsch fuels would not increase the cost of owning and operating an automobile because Fischer–Tropsch fuels can be produced for the same price as petroleum at $20 per barrel. Domestic Fischer–Tropsch fuels would actually save consumers about $375 if the increased taxed revenues from domestic production were returned to consumers in the form of reduced income taxes.

Corporate Lobbying Retrospect

The cumulative impact of past corporate lobbying on energy policy is far greater than the impact of Washington's oil lobbyists on today's pending legislation. Years of legislation that has been created under their influence is a problem.

Legislation providing essentially tax-free crude oil production, especially in foreign countries, gives foreign governments and oil companies competitive advantages of more than $25 per barrel of crude oil (about 50% of the price of imported petroleum) over alternative fuels. Legislation inhibiting the development of reprocessing technology has created nuclear waste disposal problems and increased the price of electricity.

As a first step to developing longer term and sustainable energy strategies, it is import to acknowledge that government policies on both taxes and inhibiting reprocessing need to be changed.

Certainly, it is better to have the $28 per barrel stay in the United States as taxes and buying power rather than to go to foreign countries with unknown agendas. At the very least, if $28 of a domestic barrel including $14 in US federal/state/local taxes, then of the $28 spent on imported petroleum, $14 should also go to the US taxes. The most likely scenario of a policy promoting commercialization of **competitive** alternative fuels in the United States is lower pre-tax gasoline prices. This translates to lower costs to consumers under the assumption that total taxes collected by the government remain constant.

The gross profit analysis of Table 3.1 indicates several options with feedstock prices less than $0.40 per equivalent gasoline gallon. For coal and oil sands, proven technology is ready to implement. For hydrogen from nuclear power, smart approaches could be implemented cost-effectively (see section on Energy Wildcards later in this chapter).

The mega corporations, by definition, are on top of their industry. For them, change is interpreted as reduction status. They do and will resist alternatives to petroleum and also resist nuclear power that will decrease the value of natural gas and coal reserves and facilities.

In response to corporations that serve their shareholders, government leaders must recognize that these corporations will not be objective in their advice. Government leaders must understand energy options and represent the people in their actions. Commercialization of alternatives to petroleum would be much easier with one or more of the corporative giants on board. The one thing that should bring them on board is recognition that a substantial part of more than $160 billion (2005) per year of oil moneys going to foreign producers could become corporate revenue.

The electrical power industry takes a more modular approach to systems than the liquid fuel industry, so greater versatility exists leading to healthy competition. For example, the paper industry produces a byproduct called black liquor. Factories are able to custom-design boilers to burn the black liquor. The steam produced by these boilers can be used to drive steam turbines and produce electrical power. Similar sustainable and economical liquid vehicular fuel applications are rare.

Because of the versatility, viability of localized operations, and domestic nature of the electrical power industry, problems similar to the inequitable tax structure of crude oil are less common. Foreign competition is not a big issue. Federal regulations that inhibit development and commercialization of nuclear waste reprocessing are another exception.

Reprocessing technologies that tap into 20%, potentially 100%, of the energy contained in uranium are potentially the low-cost option for sustainable electrical power production. Uranium reprocessing technology could take markets away from the coal and natural gas industries. The role of reprocessing in the US electrical power infrastructure needs to be reconsidered in the absence of the influence of special interest groups.

Nuclear power appears to be gaining favor with increasing global warming concerns. The major objection to nuclear power is on what to do with spent fuel. Reprocessing technology holds the potential to solve problems related to waste disposal by recovering fuel values from the spent fuel while reducing the total mass of radioactive waste. Valuable uses are already known for some of the fission products (ruthenium, e.g., rare in nature) found in nuclear wastes. Valuable uses may be found for sources of the fission products that are at present only recognized as waste. Alternatively, these wastes can be converted to benign materials through nuclear processes (currently considered too costly). When considering the hundreds of years of energy available in the stored spent fuel, uses for available fission product metals may be found long before current stockpiles of spent fuel are reprocessed.

One hundred years ago, nuclear energy and most of the chemical products we use every day were unknown. Advances of science and technology will continue to bring valuable by-products. Reprocessing spent nuclear fuel is now practiced in Europe. Reprocessing produces about 750 kg of nuclear waste from a 1000 MW power plant as compared to 37,500 kg of waste produced by a current US nuclear power plant without reprocessing. Current US policy is to bury the 37,500 kg per GW year of spent fuel even though 36,200 kg of this is unused nuclear fuel.

Pressure is mounting (by those who do not understand and those vested in other energy sources) to bury the spent nuclear fuel and excess weapon grade material. This would be an expensive disposal process and make this energy resource impossible to retrieve. If this plan is successful, it will be a strategic victory for those wishing to eliminate nuclear energy as an option for electrical power production. It would make access to one of our nation's greatest sources of energy even more inaccessible.

DIVERSITY AS A MEANS TO PRODUCE MARKET STABILITY

There is no substitute for competition to create and maintain low prices and increase the consumer buying power. Even in the era of strict government regulation, US consumers have benefited from healthy competition in the electrical power industry. The most important and impacting competition has been among energy sources. Figure 3.4 summarizes the distribution of electrical power by energy source [12] in the United States. For example, in the 1970s, when petroleum became more expensive, coal replaced petroleum (with a slight delay to build new power plants capable of using coal) and electrical power prices were relatively stable. In more recent years, wind power and natural gas have replaced coal in response to concerns on emissions from coal combustion—the result has been an even more balanced diversity in electrical power utilization.

It's evident from the source graph of Figure 3.4 that two factors have contributed to relatively constant electricity prices in the United States for the past two decades. The primary factor is diversity—if coal prices go up natural gas gains a larger market share and vice versa. No single source dominates even half the market. The second factor is the strong foundation provided by an abundant supply of coal.

Significant price fluctuations have not occurred in consumer prices for electricity in recent history. Furthermore, prices have been cheap by

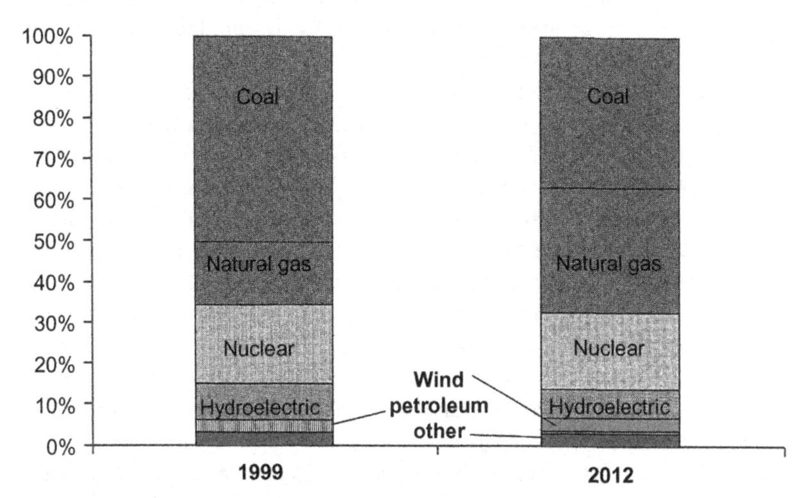

Figure 3.4 Comparison of electrical power generating capacity by fuel sources for electrical power generation in the United States between 1999 and 2012.

most standards. An exception is the failed attempt to convert California to a "free-market" region (see insert The 2001 California Electrical Power Debacle).

The 2001 problems in California were caused by strict government regulation on prices and emissions. The problems became shortages in electrical power. California purchased electrical power because some environmental groups promoted policies that eliminated building of new nuclear or fossil fuel power plants. The mistake was the free-market bidding for power by distributors required to sell electrical power at a fixed price. The distributors were forced to sell electricity at a loss. Even though this situation was only temporary, the financial situation has burdened the distributors with long-term debt and increased the cost of electricity to the customers.

The 2001 California Electrical Power Debacle

In the past, the Federal Trade Commission controlled the price of electricity to customers. Private power producers got approval to use a "free-market" model in California with a promise that the price to consumers would be reasonable. In this model, a power distribution utility would request bids for power from the producer. This would assure low-cost power to the customers.

Historically, the large population centers of Southern California placed strict environmental restrictions on fossil fuel power plants. A few fossil fuel power plants were built. They also refused to allow nuclear power plants to operate near the geological faults in the interest of safety. The region soon became dependent on imported power, electricity was produced in surrounding states and sold to California.

The new California free-market model contained the seeds of disaster. The distribution utilities were required to set low, fixed rates that they could charge the thousands of customers. They purchased power from public and privately owned electric utilities on a "low-bid" basis. The bid price was adjusted daily (later, hourly). When a power production shortage developed, the distribution utility had to bid more to meet customer demand, but could not recover the cost because the sale price was fixed low. The bid price for electricity suddenly spiked to nearly $1 per kWh, a factor of 10 above the selling price to the customer. In a matter of days, a multibillion dollar distribution utility went bankrupt and a second was salvaged by the California legislature. The huge energy conglomerate, ENRON Corporation, reportedly made huge profits in this electricity bidding war.

The Governor of California recommended to the legislature that they underwrite long-term contracts to purchase electricity. These contract prices were well above the cost of electricity before the failed free-market experiment. It will take several years to complete the analysis of this free-market failure. Reliable, affordable sources of electricity are as important to a local economy as water is to life. Construction permits for natural gas–fired power plants have been approved in Southern California and with this plan, public confidence should return with local ownership.

To the extent that diversity is a strength of the electrical power industry in the United States, the lack of diversity in the vehicular fuel industry is a weakness. Horizontal drilling, fracking technology, and accessible tight oil reserves are changing this situation for vehicular fuels. The problem is that a few corporations control much of the vehicular fuel supply.

The development of at least one large volume and sustainable vehicular fuel alternative is needed to counter this threat to security. Such an alternative could stabilize fuel prices and the national economy, and provide an incentive for otherwise hostile states to be good neighbors. Neither ethanol nor biodiesel have the low cost or capacity to fit this task. Oil fracking technology and Fischer–Tropsch conversion of coal-to-liquid fuels have the capacities to stabilize liquid fuel prices for decades. Maintaining good relations with Canada substantially increases this stability both because of oil sand reserves and shale oil and shale gas reserves.

Fischer–Tropsch fuels have the additional built-in none-sole-source capability since they can be produced from coal, natural gas, biomass, and municipal solid waste. This can provide further stability. A still better solution is to power a significant fraction of vehicles from the electrical grid through batteries and direct electricity options.

ENVIRONMENTAL RETROSPECT

Sustainability

It is certain that science and technology will continue to produce yield discoveries as in the past. The inability to predict future breakthroughs is the reason a 30-year sustainability goal is a better planning criterion than perpetual sustainability. While the wind will blow and sun will shine for thousands of years, any particular wind turbine or solar receiver may only be functional for 20 years. Which is more sustainable, a wind turbine that must be replaced in 20 years or a nuclear power plant that must be replaced in 40 years? Which is more sustainable, a solar receiver that will take 5 years of operation just to produce as much energy as was consumed to manufacture it or a nuclear power plant that produces as much energy in 4 months as it took to manufacture the facility?

No technology should be developed simply because it appears to offer perpetual sustainability. Few scientists are presumptuous enough to say they know what energy options will be used in 100 years, so why should we assign a premium value to an energy source having perpetual sustainability over an energy source with a 100 plus year reserve? On the other hand, an energy source having 100 years of reserve should be considered premium relative to an energy source having a 20-year reserve.

Basing energy decisions on single-issue agendas like sustainability is not productive. Rather, hidden costs associated with nonsustainable technologies should be evaluated and included in the economic analysis of a technology. An economic analysis that includes hidden costs and analysis based on cost alone has broader implications. For example, what medical breakthrough did not happen because the resources were spent on very costly solar receivers? If there is a hidden cost associated with potential greenhouse warming due to carbon dioxide emissions, should an appropriate "CO_2 tax" be placed on a worldwide basis? At the same time, no reasonable technology (e.g., nuclear reprocessing) should be barred by federal law. Such restrictions are subject to abuse by special interests such as corporations vested in alternative energy technologies.

At any point in US history, total sustainability could have been put in place. For example, if in 1900 environmentalists were successful in persuading the US government only to use sustainable energy sources, today's world would be substantially different. There is no reason to believe that single issue environmentalism is any more appropriate today than it would have been in 1900. The total energy program moves forward on many fronts. The environment must be protected, but other factors must be part of the future which is consistent with the slogan, "be aware of alarmists."

It is possible to solve the problems identified by alarmists using approaches that are different than those proposed by the alarmists. Ultimately, it is not who proposes a solution but (i) does the solution address a core problem such as greenhouse gas emissions; and (ii) is the technology cost competitive (or can it be cost competitive either by fostering the technology with temporary incentives or by stopping incentives going to competitive technologies).

Environmentalism History

Environmentalism has a rich tradition of keeping industry under control. History bears witness to the devastation caused by deforestation. As early as 6000 BC, the collapse of communities in southern Israel was attributed to deforestation. In southern Iraq, deforestation, soil erosion, and salt buildup devastated agriculture in 2700 BC. The same people repeated their deforestation and unforgiving habits in 2100 BC, a factor in the fall of Babylonia. Some of the first laws protecting timbering were written in 2700 BC.

Advances in citywide sanitation go back to at least 2500 BC and can be attributed to people uniting in an effort to improve their environment against the by-products of civilization. In 200 BC, the Greek physician Galen observed the deleterious acid mists caused by copper smelting. Lead and mercury poisoning was observed among the miners of the AD 100 Roman Empire. High levels of lead may have been a factor in the fall of the Roman Empire. The bones of aristocratic Romans reveal high levels of lead likely from their lead plates, utensils, and in some food [13].

Poor sanitation, including raw sewage and animal slaughter wastes dispersed throughout cities, during the dark ages contributed to both the bubonic plague and cholera—the insects and stench must have been horrific. In these ancient examples, the factors that allowed civilization to attain its magnificence, also presented new or reoccurring hazards. In this early history, the lives saved by the benefits of agriculture and metal tools far outweighed those lives lost or inconvenienced due to adverse environmental impacts.

The dark smoke of coal burning became evident as a significant problem in the thirteenth century. In 1306, King Edward I forbade coal burning in London. Throughout history the tally of deaths attributed to air pollution from heating and other energy-related technology accumulated. The better documented of these cases are reported prior to government regulations that finally brought the problems under control.

On October 26 and 31, 1948, the deaths of 20 people along with 600 hospitalizations were attributed to the Donora Smog incident, Pennsylvania. A few of the other smog incidents in the next few years include 600 deaths in London from "killer fog" (1948), 22 dead and hundreds hospitalized in Poza Rica (Mexico) due to killer smog caused by gas fumes from an oil refinery (1950), 4000 dead in London's worst killer fogs (December 4–8, 1952), 1000 dead in a related incident in London in 1956, 170–260 dead in New York's smog (November 1953), and in

October 1954 most of the industry and schools in Los Angeles were shut down due to heavy smog conditions (a smart, proactive measure made possible by the formation of the Los Angeles Air Pollution Control District in the 1940s, the first such bureau in the United States). In the 1952 London incident, the smoke was so thick that buses required a guide to walk ahead of the bus with all London's transportation except subway traffic coming to a halt on December 8, 1952.

In 1955, the US congress passed the Air Pollution Research Act. California was the first state to impose automotive emission standards in 1959 including the use of piston blow-by recycle from the engine crankcase. The automakers united to fight the mandatory use of this modification that cost $7 per automobile. Subsequent federal legislation has been the dominant force on changes in US energy infrastructure during the last 25 years.

The late 1960s has been characterized as an environmental awakening in the United States. Prior to 1968, newspapers rarely published stories related to environmental problems while in 1970 these stories appeared almost daily [14]. Sweeping federal legislation was passed in 1970 with the Clean Air Act (CAA) establishing pollution prevention regulations, the Environmental Policy Act (EPACT) initiating requirements for federal agencies to report the environmental ramifications of their planned projects, and the establishment of the Environmental Protection Agency. The CAA was amended in 1990 specifically strengthening rules on SO_x and NO_x (sulfur and nitrogen oxides) emissions from electrical power plants to reduce acid rain—this legislation ultimately led to closing some high-sulfur coal mines.

Starting with 1968-model automobiles, the CAA required the EPA to set exhaust emission limits. Ever since, the EPA has faced the task of coordinating federal regulation with the capabilities of technology and industry to produce cleaner running vehicles. Since the pre-control era, before 1968, automotive emissions (gasoline engines) of carbon monoxide and hydrocarbons have been reduced to 96% while nitrous oxide (NO_x) emissions have reduced to 76% (through 1995, the result of 1970 CAA) [15]. The phasing out of lead additives from gasoline was a key requirement that made these reductions possible.

On February 22, 1972, the EPA announced all gasoline stations were required to carry unleaded gasoline with standards following in 1973. Subsequent lawsuits—especially by Ethyl Corp., the manufacturer of lead additives for gasoline—ended with the federal court confirming that the EPA had authority to regulate leaded gasoline. Leaded automotive gasoline was banned in the United States in 1996. In 2000, the European Union banned leaded gasoline as a public health hazard.

The removal of lead from gasoline was initially motivated by the desire to equip automobiles with effective catalytic converters to reduce the carbon monoxide and unburned fuel in the exhaust. The lead in gasoline caused these converters to cease to function (one tank of leaded gas would wreck these converters). The influence of energy corporations was obvious when the US Chamber of Commerce director warned of the potential collapse of entire industries from pollution regulation on May 18, 1971. This has been viewed as a classic example of *industrial exaggeration*.

Corporate influence was again seen in 1981 when Vice President George Bush's Task Force on Regulatory Relief proposed to relax or eliminate US leaded gas phase out, despite mounting evidence of serious health problems [16].

Since the banning of lead in gasoline, scientific communities are essentially in unanimous agreement that the phasing out of lead in gasoline was the right decision. In addition to paving the way for cleaner automobiles, these regulations have ended a potentially greater environmental disaster. All the lead that went into automobiles did not simply disappear—it settled in the soils next to our highways. Toxic levels of lead in the ground along highways continue to poison children and result in mental retardation even today (see insert Lead (Pb) and its Impact, as Summarized by the EPA).

In perspective, the air quality in our cities is good and generally improving. While the federal government monitors emissions and works to reform emission standards, the public and media tend to follow other issues more closely. Issues such as oil spills and global warming make the news.

Lead (Pb) and Its Impact, as Summarized by the EPA

http://www.epa.gov/oar/aqtrnd97/brochure/pb.html

Health and environmental effects: Exposure to Pb occurs mainly through inhalation of air and ingestion of Pb in food, water, soil, or dust. It accumulates in the blood, bones, and soft tissues. Lead can adversely affect the kidneys, liver, nervous system, and other organs. Excessive exposure to Pb may cause neurological impairments, such as seizures, mental retardation, and behavioral disorders. Even at low doses, Pb exposure is associated with damage to the nervous systems of fetuses and young children, resulting in learning deficits and lowered IQ. Recent studies also show that Pb may be a factor in high blood pressure and subsequent heart disease. Lead can also be deposited on the leaves of plants, presenting a hazard to grazing animals.

Trends in Pb levels: Between 1988 and 1997, ambient Pb concentrations decreased 67%, and total Pb emissions decreased 44%. Since 1988, Pb emissions from highway vehicles have decreased 99% due to the phaseout of leaded gasoline. The large reduction in Pb emissions from transportation sources has changed the nature of the pollution problem in the United States. While there are still violations of the Pb air quality standard, they tend to occur near large industrial sources such as lead smelters. Between 1996 and 1997, Pb concentrations and emissions remained unchanged.

The Exxon Valdez oil tanker spill (March, 1989) is one of the infamous oil spill incidents. This oil tanker ran aground in Price William Sound, Alaska, spilling 11 million gallons of petroleum. It is the more infamous because of the costly remediation/penalties (over $1 billion in fine with Exxon claiming $3.5 billion in total expenditures) that Exxon was required to perform as a result of this incident. Five billion in punitive damages was also awarded against Exxon, but this remained to be collected after almost a decade. In 1992, the supertanker Braer spilled 26 million gallons of crude oil in the Hebrides islands. Both of these incidents are dwarfed by the Amoco Cadiz wreck off the coast of France in 1978 with a spill of 68 million gallons.

In view of the Amoco Cadiz incident and the cumulative tens of thousands who died in London's killer fogs, it is easy to understand the increased environmental consciousness in Europe compared to the United States. For example, the European governments are aggressively addressing potential global warming issues while the US government tends to withdraw from international cooperation on the issue. Neither the US nor the European governments dispute the fact that carbon dioxide levels are increasing in the atmosphere. They do have varying opinions on the implications of these increasing carbon dioxide emissions.

On June 23, 1988, NASA scientists warned Congress about possible consequences from global warming with potential effects of drought, expansion of deserts, rising sea levels, and increasing storm severity. On December 11, 1997, the Kyoto Protocol was adopted by the US President Clinton (a democrat) and 121 leaders of other nations. The republican-dominated US Congress refused to ratify the protocol. More recent comments by President Bush concerning the Kyoto Protocol sound like the comments and actions of the US Chamber of Commerce director and former Vice President George H. W. Bush on the phaseout of lead from motor gasoline.

CLIMATE CHANGE

The scientific community is in agreement that carbon dioxide concentrations have increased in the atmosphere as a result of combustion of fossil fuels. Increases have been measured and the reasons for the increases are generally understood. There is less agreement about the consequences of these increased carbon dioxide levels in the atmosphere, with cries ranging from "catastrophe," to "forget it."

Since the impact of higher carbon dioxide levels in the atmosphere is unknown, the benefits of slowing or reversing trends in carbon dioxide are also unknown. The corresponding risk-benefit analysis of using technology to reduce greenhouse gas emissions has even greater uncertainty since the cost/risk associated with reducing emissions on individual national economies is also unknown.

The consistent arguments put forward by American politicians opposed to reducing greenhouse gas emissions are the high costs and adverse national economic consequences of any new technologies to address the problem. The facts suggest that these arguments have no basis; since there is a range of options that include electrical energy storage, hybrid cars, nuclear energy, wind power, and use of rechargeable car fuel cells. Unfortunately, good solutions to greenhouse gas emissions could compromise competitive advantages of mega corporations that have major influence in national politics.

Figure 3.5 summarizes carbon dioxide emissions by sector sources [17]. At 34%, the electricity generation produces the most carbon dioxide. Upon recognition that the predominant source of greenhouse gas emissions from the residential and commercial sectors are Heating, ventilation, and air conditioning (HVAC) and hot water heating, this 34% increases to about 45% of the greenhouse gases due to electricity and heating. Another 27% comes from transportation.

It is important to note that Figure 3.5 indicates both the industrial and agricultural sectors have had essentially constant or even decreasing carbon dioxide emissions since 1990. Concerns about the impact of reducing greenhouse gas emission on industry are put to rest by simply recognizing that neither industry nor agriculture are the source of

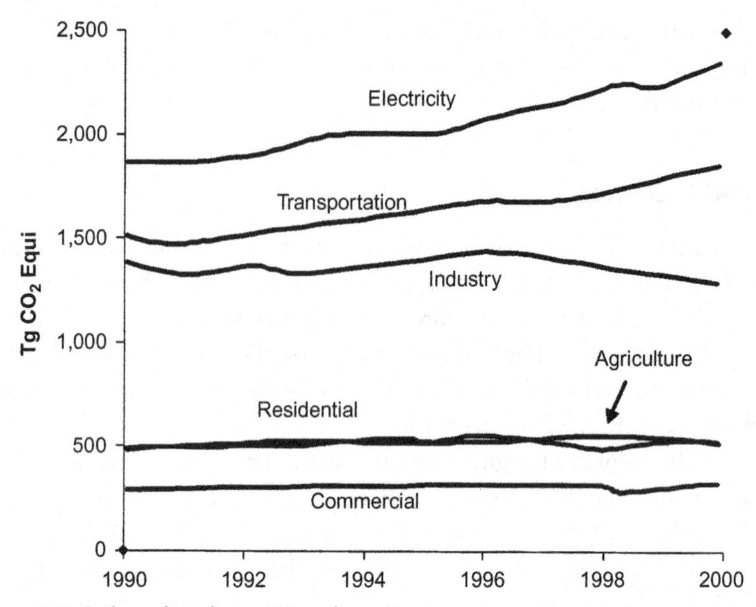

Figure 3.5 Carbon dioxide emissions by sector.

increasing carbon dioxide emissions. If the goal is to maintain 1990 carbon dioxide emission levels, industry and agriculture have achieved the goal and should not be burdened with further reducing emissions.

Reprocessing spent nuclear fuel could cost-effectively increase the share of wind and nuclear power generation from about 17–42%. This would reduce carbon dioxide emissions by about 10% (34% reduced to 24% from electrical power generation). Increases in efficiency of coal-fired facilities could provide another 5% reduction. Replacing fuel-fired heating of residences and commercial buildings with heat pumps could provide another 5% reduction. Diesel engines could provide another 5% reduction in the transportation sector. Use of rechargeable fuel cells in hybrid cars could provide another 3% reduction through use of electrical grid recharging and another 5% from increased miles per gallon. All of these technologies promise to be low-cost alternatives. They represent a 33% reduction in total carbon dioxide emissions with minor improvement in the technology supporting these alternatives.

The "propaganda" states how costly it would be to reduce greenhouse gas emissions in the United States. These greenhouse gas reductions could be made while saving consumers' money, reducing oil imports, and putting the country on an energy track that is sustainable for the next few centuries.

Table 3.5 Estimates of emissions per electrical power produced

Pollutant	Hard coal	Brown coal	Petroleum fuel oil	Natural gas
CO_2 (g/GJ)	94,600	101,000	77,400	56,100
SO_2 (g/GJ)	765	1361	1350	0.68
NO_x (g/GJ)	292	183	195	93.3
Particulate matter (g/GJ)	1203	3254	16	0.1

New Coal Regulations

The European Environment Agency (EEA) documented fuel-dependent emission factors based on actual emissions from power plants in the European Union. [8] summarized in Table 3.5. For every gigajoule (GJ) of electrical power produced, coal may produce up to twice as much carbon dioxide. Coal produces more carbon dioxide because more energy is stored in carbon–carbon chemical bonds in coal rather than carbon–hydrogen chemical bonds in fuel oil and natural gas. Natural gas combustion energy can be converted to electrical power at efficiencies above 50%. Natural gas is a "clean fuel"—very little sulfur dioxide or particulates are produced.

The conclusion that is broadly accepted is that if the goal is to reduce carbon dioxide emissions associated with global warming, the use of coal for electrical power generation is a bad choice. Pollution from coal combustion tends to be high, and China in particular is paying a high price related to health and environment for using coal as a primary energy source.

These trends are well known, and in the United States there is no need for alarmist overreaction. New coal-fired power plants are more efficient than older plants and they must have toxic gas and particulate emission control units that are not required on older power plants should be allowed to yield the return on the investment made into those facilities.

The regulatory approach that has been used for targeted emissions such as nitrous oxides is known as emissions trading. In this approach, the government imposes penalties for exceeding certain levels of emissions; however, "good-performing" facilities can trade or sell credit to "bad-performing" facilities. A merit of this approach is that it can encourage the building of new production facilities that have emissions below regulated limits, and these new facilities can be used to balance older facilities in a company's (multiple new and old plant) portfolio.

Similar regulatory strategies have failed for carbon dioxide emissions in the United States, where:

> *The American Clean Energy and Security Act (H.R. 2454), a greenhouse gas cap-and-trade bill, was passed on June 26, 2009, in the House of Representatives by a vote of 219–212. The bill originated in the House Energy and Commerce Committee and was introduced by Representatives Henry A. Waxman and Edward J. Markey. Although cap and trade also gained a significant foothold in the Senate via the efforts of Republican Lindsey Graham, Independent Democrat Joe Lieberman, and Democrat John Kerry, the Legislation died in the Senate. (Wikipedia, emissions trading)*

While specific legislation on carbon dioxide emissions have failed, the threat of such possible emissions has caused many corporations to include a "carbon dioxide penalty" in their profitability estimates when considering investment options in new facilities. Hence the perceived "threat" and recent availability of relatively abundant and inexpensive natural gas has provided a significant slowdown in new coal facilities in the United States without actual regulatory mandates.

One could speculate that as long as the corporations do not get too ambitious building new coal-fired facilities, that trends will continue to reduced-CO_2 options even in the absence of specific regulations.

Carbon Dioxide Sequestration

Carbon dioxide sequestration refers to an approach of removing carbon dioxide from the air or flue gas from power plants as a means to reduce carbon dioxide emissions in the atmosphere. Figure 3.6 shows the US IRS (US Internal Revenue Service) form used to receive credit by a qualified facility for disposal of carbon dioxide in a "secure geological storage." An example of a secure geological storage is an underground geological formation that previously contained natural gas. The natural gas has been removed making space for carbon dioxide to be pumped into that formation.

Ironically, one of the best ways to securely store carbon dioxide is to separate the oxygen from it, create a carbon-rich solid, and to bury that carbon-rich solid. The final resting place for that carbon dioxide is essentially coal. The approach of removing carbon (coal) from the ground, burning the carbon for energy, removing the carbon dioxide from the exhaust gas, compressing the carbon (carbon dioxide) so that it can be

Form **8933**

Department of the Treasury
Internal Revenue Service

Carbon Dioxide Sequestration Credit

▶ Attach to your tax return.

▶ To claim this credit, the qualified facility must capture at least 500,000 metric tons of carbon dioxide during the tax year.

▶ Information about Form 8933 and its instructions is at *www.irs.gov/form8933*.

OMB No. 1545-2132

2014

Attachment Sequence No. **165**

Name(s) shown on return | Identifying number

Qualified carbon dioxide captured at a qualified facility, disposed of in secure geological storage, and not used as a tertiary injectant in a qualified enhanced oil or natural gas recovery project.

1a	Metric tons captured and disposed of (see instructions)			
b	Inflation–adjusted credit rate	$21.51		
c	Multiply line 1a by line 1b		1c	

Qualified carbon dioxide captured at a qualified facility, disposed of in secure geological storage, and used as a tertiary injectant in a qualified enhanced oil or natural gas recovery project.

2a	Metric tons captured and used (see instructions)			
b	Inflation–adjusted credit rate	$10.75		
c	Multiply line 2a by line 2b		2c	
3	Carbon dioxide sequestration credit from partnerships and S corporations		3	
4	Add lines 1c, 2c, and 3. Partnerships and S corporations, report this amount on Schedule K. All others, report this amount on Form 3800, Part III, line 1x		4	

Figure 3.6 US IRS carbon dioxide sequestration credit form.

injected into a secure formation, and then forcing it into the ground is as stupid as it appears in this description.

The removal, purification, and compression of carbon dioxide take energy. Estimates indicate that to remove the carbon dioxide emissions from a 1 GW power plant, half of the electricity would be for sale or a second GW power plant must be built for the removal and disposal process. At least the technology would tend to double the cost of coal-based power production.

While technically possible, the superficial analysis indicates that it does not make sense to take carbon out of the ground to burn it and then to put it back into the ground.

There are other solutions to the carbon dioxide emissions problem. It is the opinion of the authors that carbon dioxide sequestration is an opinion that only "muddies the waters" for good approaches (like nuclear power and electrification of transportation) that can make meaningful reductions in carbon dioxide emissions. Federal support for carbon dioxide sequestration technology is a result of corporate lobbying at its worst!

As far as IRS credit for sequestration, it is only a good approach if such credits are fully paid for by those emitting the carbon dioxide emissions. This translates to an IRS carbon dioxide tax being used to pay for IR carbon dioxide sequestration credits.

EFFICIENCY AND BREAKTHROUGH TECHNOLOGY

An important goal of a strategic energy technology is to have competitive commercial technologies that allow electricity to directly compete with liquid fuels. This includes but is not limited to battery technologies and vehicles that run directly from grid electricity like electric trains. These technologies are examples of win–win approaches when they enhance the stability of price and demand for both liquid fuels and electricity.

A problem, and weakness, of grid electricity is that consumption tends to be dominated by on-again/off-again machines. These include devices such as air conditioners, electric stoves, and clothes dryers that tend to run between 9:00 AM and 9:00 PM. The "peak demands" caused by these uses is a problem and opportunity.

Meeting peak demand for electricity by the most economical means is the responsibility of electricity providers. The typical approach involves the use of special power plants as "peak demand facilities." The peak demand facilities are powered by natural gas or petroleum and are less efficient, operating at 25–30% efficiency rather than the 35–53% of baseline power systems. They are cheap to build but are expensive to operate for a few days per year. They also produce more greenhouse gases per kWh of electrical energy.

Nuclear and solid fuel facilities have problems meeting peak power demand because it can take hours for them to go from 0% load to full load service. Solar and wind power options supply electricity only when the sun shines and the wind blows and do not operate at our convenience. Production facilities for meeting peak demand are necessary, and these facilities tend to be expensive because they may only be used the equivalent of a few weeks during the year.

An alternative to converting fuel to electricity upon demand is to store electrical energy during low demand and return it to the grid during high demand. Battery storage is currently too costly and lower cost batteries to make it affordable are a national research priority. Other options are used to a limited extent. Pumped water storage is an alternative that has shown commercial acceptance. Not included in Figure 3.4 on fuel sources is *pumped water storage*.

Hydroelectric power contributes about 6.8% of the electrical power generated in the United States. An additional 2–3% of hydroelectric power is provided by water that is pumped to higher elevations by excess baseline power during off-peak demand periods. One of the disadvantages of water storage is that energy is used during the pump for storage and recovered by a turbine during energy recovery. For example, energy that is initially generated at 50% thermal efficiency will contribute at a thermal efficiency of about 30% if it is used to drive a pump to store water and later converted back to electricity by a turbine. The overall efficiency is better than a peak power demand facility, but the pump storage system is capital intensive because it requires real estate with a favorable elevation change.

Thermal Energy Storage

During summer, a refrigeration system can be used to freeze ice at night using off-peak power rates. During the day, ventilation air passed over the ice (high cost, peak power) provides cooling as an alternative to running the air conditioner during the day. It has been decades since such a primitive system has been used; however, modern versions are in use. At Curtin University in Western Australia, air conditioners are run at night to chill water with the cool water stored in a large tank. This chilled water is circulated to buildings throughout campus to provide cooling during the day. Such systems use less-expensive nighttime electricity (typically available to commercial customers but not to residential customers) and require smaller air conditioners since the air conditioners can be run at a constant load during the 24-hour cycle rather than just during the day.

Thermal energy storage can be used to make both heating and air conditioning systems operate in better harmony with electrical power supply. For solar heating, energy storage is particularly important. The need for supplemental heating is at a minimum during the day when solar devices are operating at their greatest efficiency. Heating is needed when the sun is down and the solar device is providing no heat. Thermal energy storage allows the heat to be stored during the day for use during the cool nights.

A successful and common method for reducing peak demand is to provide lower electrical rates to industrial customers for off-peak hours. Especially in industrial settings, some energy-intensive operations can be scheduled for off-peak operation. In these cases, the equipment is turned off during peak demand hours to avoid the higher penalty rates of peak-demand periods. In one form or another, price incentives promote a number of solutions to peak demand energy needs. Price incentives are successful but, alone, cannot bring electrical demand to constant levels throughout the day.

Better options are needed for electrical energy storage. A disadvantage of converting electrical energy into either chemical or potential energy and then back to electrical energy is the lost efficiency at each conversion. Usually, each conversion is less than 90% (often less than 85%) efficient, and the conversions to storage and from storage has a combined efficiency of less than 75% (often less than 70%). This reduction in efficiency means 25% more fuel consumption, 25% more air pollution, and 25% more greenhouse gas emissions for that peak power that would be avoided if the electricity demand was level.

A promising technology for shaving peak demand is thermal energy storage. Thermal energy (stored heat or stored refrigeration) is a lower form of energy than electricity or stored hydraulic (potential) energy. In principle, conversion to thermal energy is irreversible; however, for heating and air conditioning applications, thermal energy is the desired form of energy.

Thermal energy storage systems can approach 100% efficiencies and can be used by all customers using electricity for heating or air conditioning. Since heating and air conditioning represent a major component of peak demand loads, this technology can have a major impact.

Modern thermal energy storage options include ice storage, chilled water storage, and use of phase-change materials. Phase-change materials are chemicals that freeze near room temperature. For example, a material that releases energy at 74°F as it freezes and takes in energy at 76°F as it thaws might be used to keep a house at 78°F in the summer and at 72°F in the winter. A material with a large heat of fusion is ideal for this application.

The utility of thermal energy storage goes beyond converting inefficient peak demand electricity to more efficient baseline load. Solar energy storage is an example of the need for auxiliary heating can be partially eliminated. During parts of the air conditioning season, nighttime temperatures are often sufficiently cool to cool a medium (water, solid grid, etc.) to offset air conditioning demand during the heat of the day. This is especially true for commercial buildings that often require air conditioning even during spring and fall days.

Another promising technology is the PHEV [18] that charges batteries and produces fuel cell hydrogen during the night for use during the day (the engine on the vehicle provides backup power). Night hours are the best time to recharge the vehicle batteries since it is typically not in use. Also, electrical power is used to replace imported petroleum. The diversity of the electrical power infrastructure can be used to stabilize and decrease the prices of petroleum. Major investments are not necessary in this approach since hybrid vehicles are on the market and advances can be made incrementally.

It is projected that up to 20% of gasoline could be replaced without building additional power plants; however, when base load electrical power demand increases, history shows there will be investors willing to make investments. These investments will be for new generation, more efficient electrical power plants. If coal were used as a fuel source (see Table 3.1, $0.011 per kWh), the coal costs would be less than $0.02 per mile compared to $0.05–0.10 per mile for gasoline ($2.00 per gallon).

Anticipated Breakthrough Technologies

Breakthroughs in grid-based electrical power production are not likely to occur in the next couple of decades. Areas of breakthroughs do exist for community electric power networks, solar devices, and electric-powered vehicles. Considering that (i) about 10% of electricity is lost during distribution; (ii) much of the cost of electricity is associated with grid maintenance and accounting; and (iii) the many "remote" locations are in need of electricity; there are opportunities for off-grid power.

- Can a photoelectric roofing material be made that costs slightly more than normal roofing material?
- Can cost-effective thermal storage units be manufactured for home use?
- Could cogeneration (using the waste heat of electric power generation) be more widely used?
- Are there cost-effective ways to use electric power for vehicles?
- Can PHEVs enhance the viability of remote and community electricity networks?
- Are there approaches to small nuclear reactors that overcome the high risk associated with larger nuclear reactor facilities?

The answers to all of these questions are yes. Will these transformations occur in the next decade? Some are likely to happen in the next decade; others may take longer. The next chapters cover these options in greater detail.

REFERENCES

[1] Seider WD, Seader JD, Lewin DR. Process design principles. New York: Wiley; 1998. p. 47.
[2] See <http://www.nytimes.com/2010/04/13/science/earth/13trash.html?pagewanted=all&_r=0>.
[3] Seider WD, Seader JD, Lewin DR. see Chapter 10 Process design principals. Wiley; 1998.
[4] Deming. D. Oil: Are we running out? Second Wallace E. Pratt Memorial Conference Petroleum Provinces of the 21st Century, January 12–15, 2000; see <http://geology.ou.edu/library/aapg_oil.pdf>.
[5] See Canada's Oil Sands and Heavy Oil. Petroleum Communication Foundation, Available at http://www.centreforenergy.com/EE-OS.asp "browse option", 2001.
[6] Smith A. p 361 Wealth of Nations facsimile of 1776 original. Prometheus Books; 1991.
[7] See <http://www.conway.com/ssinsider/bbdeal/bd020401.htm>.
[8] See <http://www.sb-d.com/issues/winter2002/currents/index.asp>.
[9] Lazzari S. The Tax Treatment of Alternative Transportation Fuels, CRS Report for Congress, Library of Congress, March 19, 1997.
[10] See <http://api-ec.api.org/filelibrary/ACF763.doc>; May 21, 2002.
[11] See <http://api-ec.api.org/filelibrary/ACF763.doc>.
[12] See <http://www.eia.doe.gov/cneaf/electricity/ipp/html1/t17p01.html>.
[13] See <http://www.radford.edu/~wkovarik/envhist/>; May 19, 2002.
[14] De Nevers N. Air pollution control engineering. New York: McGraw Hill; 1995. p. 2.

[15] Changes in Gasoline III, published by Technician's Manual, Downstream Alternatives, Inc. P.O. Box 190, Bremen, IN 46506-0190 1996, Page 8.
[16] See <http://www.runet.edu/~wkovarik/envhist/10eighties.html>; May 20, 2002.
[17] See Emissions by Economic Sector: Excerpt form the Inventory of U.S. Greenhouse Emissions and Sinks: 1990–2000. U.S. Greenhouse Gas Inventory Program, Office of Atmospheric Programs, U.S. Environmental Protection Agency, April, 2002. <http://www.epa.gov/globalwarming/publications/emissions/index.html>.
[18] See <http://www.iags.org/pih.htm>.
[19] Dicker D. Oil's Endless Bid: Taming the Unreliable Price of Oil to Secure Our Economy. New York: Wiley; 2011.

CHAPTER 4

Energy Conversion and Storage

Contents

The story of energy conversion is the story of life. It is the story of people coming out of the caves to harness the environment. It is the story of conquest. And it is the story of the rise and fall of empires.

As illustrated in Figure 4.1, the human body and the steam engine are both parts of the remarkable conversion process that propagates life and drives machines. Starting with the nuclear fusion of hydrogen atoms in the sun, energy is transferred to Earth and converted to useful applications, and ultimately ending in a useless form of waste heat. The industrial revolution came of age when we began to understand how all these forms of energy are related.

The use of fire to cook food, to win metal from ore, and to work metal into useful tools is described in ancient historical documents. In the late eighteenth century, methods began to appear that converted the heat of a fire (thermal energy) to work that could replace humans or animals to perform tiresome daily tasks. The industrial revolution of the nineteenth century was fueled by fossil fuels feeding steam engines. The twentieth

Sustainable Power Technologies and Infrastructure.
DOI: http://dx.doi.org/10.1016/B978-0-12-803909-0.00004-X

Sun converts mass to energy through fusion.
Energy in transit, by radiation.
Photosynthesis converts radiation (energy in transit) to chemical energy in the form of wood or fruit.
Energy is stored for various lengths of time, including transformation to coal or petroleum over hundreds of thousands of years.
PATH 1: The human body (all animals) converts chemical energy to muscle movement (force acting through a distance).
PATH 2: A locomotive burns wood to power a steam engine—the engine produces movement (force acting through a distance).
Radiation and molecular motion transfer waste heat the surroundings. In the end, and all through the process, essentially all of the energy is converted to waste heat that is of no economic value. The energy used by the muscles is ultimately expended overcoming frictional resistances as is the energy driving the locomotive.

Figure 4.1 On Earth, most energy comes from the sun and ultimately becomes heat. This is a fascinating story of trial and error with the successful inventions providing the many devices we use every day.

century brought the nuclear age where the burning of coal could be replaced by the fission of uranium and the disaster of nuclear bombs elevated nations to superpowers. The opportunities of the twenty-first century are limited by what we choose to do rather than what can be done.

This is a fascinating story of trial and error with the successful inventions providing the many devices we use every day.

USE OF THERMAL ENERGY

It was the middle of the nineteenth century when experiments were performed convincing some scientists that the energy associated with heat was exactly equivalent to the energy that produces work. It took a new generation of scientists to finally end the arguments of those who could not accept this principle of energy conservation. This idea has now elevated to the postulate *energy is conserved*. This is a cornerstone of classical physics that has been elevated to the *first principle of thermodynamics*.

Classical thermodynamics is an exact mathematical structure that describes the behavior of gases, liquids, and solids as their temperature, pressure, and composition change. The simple form of these mathematical relationships was established using experiments performed to improve the design of engines that convert thermal energy to work. The theory applies equally to energy conversions involved in heating and cooling a house, conversion of energy from coal combustion to electricity, combustion of gasoline to power an automobile, and the conversion of food to energy used to walk and talk.

Our narrative will proceed with some examples of tasks we understand as work. These examples will give us a quantitative definition of work and we can then assign numerical units to energy. We will show how any gas can be used to do work. We then show how we can use water as a liquid or a vapor (steam) as the fluid to accept thermal energy, do work, and then reject the remaining thermal energy to complete the work-producing cycle. Today's machines that convert energy to work use this basic principle.

THE CONCEPT OF WORK

Everyone understands that work is the use of energy to perform a task. It is work to lift a box from the floor to a table. Shoveling sand or snow is work. When we carry a bag of groceries from the store to the parking lot, it is work. All tasks we do that have us moving an object from one place to another we understand to be work. When we repeat the task several times, we need rest, and we need food to restore the energy we used for performing those tasks. This qualitative description forms the basis for the scientific definition of work.

The scientific definition of work used in physics must be mathematically exact. This is necessary because the numerical value of the work we assign to a task must be the same for every scientist or technologist who does the analysis. This allows scientists to communicate efficiently. The definition we use in physics follows: *work is the numerical value of a force multiplied by the distance over which that force is applied.* We can use this definition to compute the work required to lift a box from the floor to a table.

Earth exerts a force on all objects near it. This force is gravity. We measure this force in pounds. When a 10-pound box is lifted 2 feet from the floor to the top of a table, 20 foot-pound work is done against the force of gravity.

The analysis of the work of shoveling snow or sand involves replications. The weight of each shovel of snow times the distance the shovel travels upward against the force of gravity is the work per full shovel.

Multiply the work per shovel by the number of times you must load the shovel and this is the total amount of work required to clear the sidewalk. It is more work to shovel snow into the back of a truck than to shovel snow aside in the driveway because the truck is higher.

If we threw 100 pounds of sand instead of 100 pounds of snow into the truck, the same amount of work would be required to load snow or sand. This physics definition of work allows us to calculate the numerical value of the required work and the work is independent of the material we move. We can replace the sand with corn, coal, or cucumbers, but it is the weight in pounds times the height of the truck bed that determines the work required to load the truck.

When we move groceries from the supermarket counter to our car in the parking lot, the definition of work appears to fail. In particular, if the faster/longer counter, the shopping cart, and car are at the same level, the groceries have not moved upward against the force of gravity, yet we know it takes work to push the grocery cart from the counter to the car. To make this definition of work more general, early scientists identified that gravity was only one of many forces that are routinely overcome to perform work.

Where Did the Work Go?

You have certainly noticed that pushing the grocery cart up an inclined parking lot to your car is more difficult than on a flat parking lot. In this case, you must provide the work it took to push the cart on a flat parking lot plus you are "lifting" the groceries against gravity from the store counter to the height of your car. Where does the work we do pushing the grocery cart go?

Let's consider a box sliding across a table. If you push on the box to slide it across the table, you apply force to move the box times the sliding distance. The box did not move up or down so there was no work raising or lowering the box. You must push with enough force to overcome this friction between the box and the table as the box slides. Where did that work energy go?

Rubbing your hands together to keep them warm on a cold winter day is an example of frictional force. When you press them together harder, or rub fasterlonger, your hands get even warmer. Your muscles do work to overcome the friction between your hands, and this work is converted to heat through the frictional contact. This shows that you can convert work to thermal energy (heat), since rubbing your hands back and forth requires a force 3-quarts times the total distance you rub them, and this is our definition of work.

For the grocery cart, the energy expended in performing the work ended up in two forms. The energy overcoming gravity was stored by its location—specifically by its height. The potential to recover this energy remains and this is called potential energy. This potential energy can be converted to kinetic energy by letting the cart speed up as it rolls down the hill.

The energy expended to overcome friction of the cart is like the friction of the box, it becomes thermal energy. The box actually becomes warmer. The work done sliding the box is converted to heat. Our grocery cart is designed to reduce the friction, wheels with bearings turn easily, and this reduces the work required to move groceries to the car. You could have put the groceries in a box, attached a rope, and dragged the box to the car. This would take much more work than using the cart, even if one of the cart's wheels doesn't turn very easily.

What would happen if the cart were flung off a cliff or loading dock? Literally, the cart would go faster and faster as it falls. When it hits the ground, the speed of the cart would be converted to thermal energy, as it stopped at the end of the fall.

Notice, the work required to push the cart to the car can be computed if we measure the force, in pounds, required to push the cart times the 200 feet to the car. The resisting force of the shopping cart is an example of a frictional force. In this case, the work required depends on how easily the wheels on the cart turn. This is the way one can compute the work required to move a load on a level road using a cart, truck, or a train.

If work can be converted to heat, is it possible to convert heat into work? The first steam engine demonstrated this could be done. It is this puzzle that scientists solved during the middle of the nineteenth century. They were able to show that work could all be converted to heat, but there is no machine we can design that will convert all of the thermal energy (heat) into work. A fraction of the thermal energy (heat) must be transferred from the "work producing" machine or engine.

Converting thermal energy to work—we are all familiar with fire. When wood, coal, oil, or natural gas burns it produces thermal energy. The energy is released when the molecules that make up the fuel break apart and chemically combine with oxygen in the air (combustion) to form new molecules. This oxidation reaction destroys the fuel chemicals to produce mostly carbon dioxide and water vapor releasing heat (thermal energy) in the process. The ash remains of a campfire are the oxides of the minerals that were in the wood. The thermal energy that is released is diffused, it moves in all directions with equal ease. The development of heat engines to convert this diffuse thermal energy to useful work sets us on the path of the industrial revolution.

Early experiments were done with air trapped in containers to develop the science that describes how to build heat engines. Consider the following modification of their experiments to show how to produce work from thermal energy.

An open can is equipped with a free-sliding piston that seals at the walls so that heating the can from 30 to 60°C causes the piston to move upward. Any weights on the piston would move up or down with the piston. Movement is caused by the pressure force of the trapped air on the bottom side of the piston. This force in pounds times the displacement of the piston in feet is the work done lifting the weights in units of foot pounds. For all practical purposes, the work performed by this primitive heat engine is the same as the work performed by a person lifting the same weights over the same distance.

In addition to the ability to perform work, fundamental observations were noted for this primitive machine comprised a free-sliding piston in a can. First, the volume inside the can at 60°C is always 1.1 times the volume at 30°C. Also, the pressure of the contained gas remains at a constant value during heating. Placing more weight on the piston increases the pressure.

If superglue is put on the sides of the piston and the piston is locked into place, our constant pressure experiment is converted to a constant volume experiment. When heated from 30°C to 60°C at constant volume, the pressure increases. In fact, the pressure at 60°C is 1.1 times the pressure in the can at 30°C—the same 1.1 multiple that described the volume change at constant pressure for the same temperature change.

You can use a bicycle tire pump as "modern" laboratory equipment to show that the temperature of air increases when it is compressed. Increase the bicycle tire pressure from about 30 psi (pounds per square inch) to 60 psi. When you have completed a few quick strokes of the tire pump, touch the pump cylinder near the bottom. It will be warm. Had you continued to operate this pump to fill a large tank from 55 to 60 psi, completing many quick strokes over several minutes, the cylinder of the pump would be hot to grasp and hold.

The tire pump experiment is a qualitative demonstration that the work used to compress a gas raises its temperature as the pressure increases. The pump requires work. Is it possible to reverse the process starting with a hot gas at some high pressure and allowing the gas to expand moving a piston against a force to do work? The answer is, "Yes."

Start with the piston near the bottom of the cylinder. Fill the space with a hot gas at an elevated pressure. Reducing the force on the piston a little allows the hot gas to expand causing the temperature and pressure of the gas to decrease. Continue reducing the force and the expansion will continue until the piston gets to the top of the cylinder. The work done will be the force on the piston times the piston displacement. This work-producing stroke that expands hot gas to produce work is exactly the reverse of the compression stroke that required work. Both the temperature and pressure of the gas will decrease as the gas expands to produce work.

Early investigators cleverly designed experiments to demonstrate that the compression/expansion (work in/work out) cycle is reversible under ideal conditions. Is the total work required to perform the compression stroke recovered in the expansion stroke? In the most ideal of circumstances, the answer is, "Yes." However, total work recovery cannot be attained in a real machine. There will always be friction between the piston and the cylinder in a real engine. Work wasted overcoming friction can never be recovered.

EARLY ENGINE DESIGNS

Like the free-sliding piston in a can, the "ideal engine" is important in visualizing how heat is converted to work in a practical engine. The ideal engine operates under the following rules:

- The piston moves in the cylinder without friction.
- There can be no heat transfer to (or from) the gas from (or to) the piston/cylinder during the expansion (compression) stroke.

While both of these rules are not realistic, real engine performances can approach these "ideal engine" specifications. For example, lubricating oil and Teflon O-rings can reduce friction, or piston compression can be made to occur so fast that the heat simply does not have time to escape through the cylinder walls.

The free-sliding piston in a can (cylinder) suggests that a gas such as air may be used in an engine to produce work. The gas is a fluid for doing work and is referred to as a *working fluid*. This work-producing machine should operate in cycles with each cycle producing a small

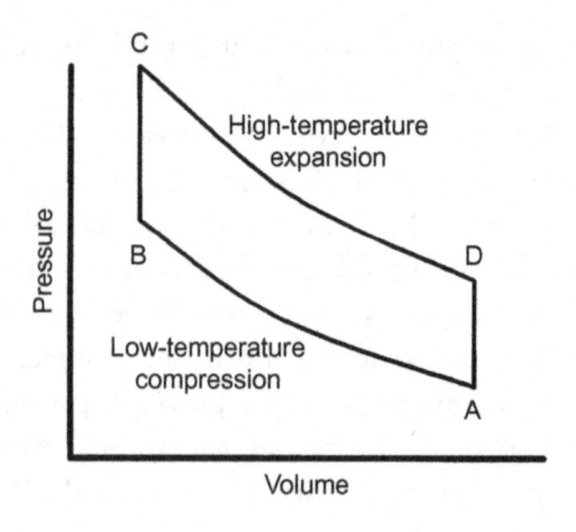

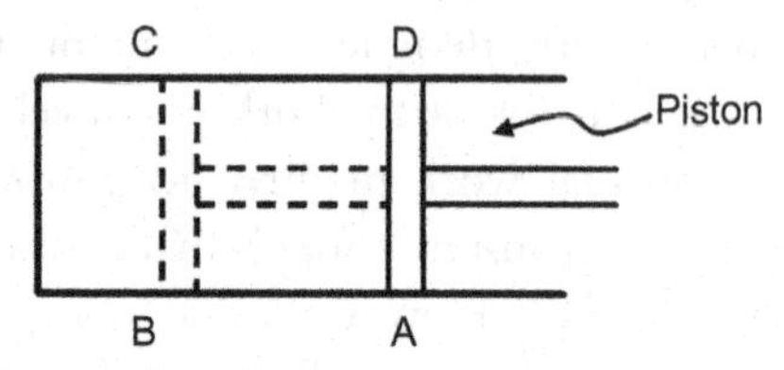

Figure 4.2 Illustration of how pistons perform work.

increment of work. An important theoretical result was obtained from early analysis of this "ideal engine cycle." Such an ideal cycle is traced schematically in Figure 4.2. The line drawing represents the position of a piston in a cylinder as the engine performs the work producing cycle. Focus your attention on the gas. Everything else is mechanical equipment required to contain the gas and transfer the work from the gas to a task we assign. It is the gas that does all of the work.

Suppose we start with a piston fully extended so the volume is at maximum—assume this corresponds to a volume of 3 quarts (point A, see Figure 4.2). Fill this volume with a gas at room temperature and pressure. This will fix the quantity of gas in the airtight cylinder. Here are the four steps for our engine cycle:

Step 1. Do work on the gas by pushing the piston from 3 quarts to 1-quart volume (from point A to point B). Like the tire pump, work is put into the process and the working fluid (gas) heats in response to taking in this work.

Step 2. Heat the gas at constant volume—the 1-quart volume position. Similar to when the free-sliding piston was super glued to the can wall, the pressure increases and no work is done heating the gas.

Step 3. Expand the gas to its initial volume (position D). Work is done by the gas as it moves the piston against a force.

Step 4. Cool the gas at constant volume to the initial temperature and pressure—the same condition as at the start of the cycle.

The gas is exactly at the same temperature and pressure at the start of step 1, so the cycle can be repeated again and again. More work is done in step 3 than is required in step 1 since the expansion step 3 has greater force (greater pressure) than the compression step 1; so net work is produced in each cycle. Modern engines convert the back and forth motion of the piston to rotating shaft motion using a crankshaft.

There is a problem operating this simple hot gas expansion engine. A heat engine must take in thermal energy to produce the gas at high temperature and pressure. This hot gas then expands to produce work as the temperature and pressure decrease. At the end of this power stroke, the gas must be cooled before it is compressed to start the next cycle. If we do the gas heating and cooling through the walls of the cylinder, it takes a long time to complete one cycle. This piston/cylinder combination is an obvious choice for an engine, but we must design the engine so we reduce the time required to complete each four-step work-producing cycle.

Two approaches are available that are superior to heating the working fluid through the walls of the cylinder. For steam engines, valves can be used to control the flow of high temperature and high-pressure steam into an engine as the piston moves to produce work. Alternatively, heat can be generated in the engine by actually burning a mixture of air and fuel in the engine cylinder. The latter, "internal combustion engine," requires fresh air and fuel in the engine at the start of each cycle.

The Science of Heat Engines

The details of the engine cycle in Figure 4.2 will provide an improved understanding of this cycle.

During step 1, the temperature and pressure of the gas will increase and the volume will decrease to the minimum volume defined by the piston stoke length. The work required for this step will be the force on the piston times the distance the piston travels from the initial to final position. No heat is transferred to or from the gas during this step according to the ideal engine assumption.

During step 2, the temperature and pressure of the gas increase during this constant volume step to the pressure shown at position C. Since there is no change in volume, there is no work associated with this step. During step 3, the temperature and pressure of the gas will decrease. The work produced will be the force on the piston times the distance the piston travels from C to D. No heat is transferred to or from the gas. During step 4, no work is performed since the volume is constant.

What can we learn from this ideal engine cycle? If we did not transfer heat to the gas in step 2, the work required to compress the gas from A to B is exactly equal to the work produced in step 4, expanding the gas from C to D. With no heat transfer, the points B and C would be identical to the points A and D. This engine produces no network because the work produced is equal to the work required to compress the gas.

Both the input of heat to the engine and release of heat by the engine are necessary for the engine to produce work. The heat input could come from the burning wood. Outside air could be used to cool the engine during step 4—this heat is rejected or lost to the environment.

When we transfer thermal energy (as heat) to the gas in step 2, the path from A to B is separated from path C to D. The work required to compress the gas in step 1 is now less than the work produced in step 3. This is true because at each position of the piston (volume of the gas), the force on the piston along the expansion path from C to D (producing work) is always greater than it is along the compression path from A to B (requiring work). Since energy is conserved (i.e., thermal energy and work are equivalent), the network done by this ideal engine will be exactly equal to the difference between the heat added in step 2 and the heat extracted in step 4. We have assumed that there is no thermal energy transfer to the gas during the compression stroke, step 1 and the expansion stroke, step 3. The gas starts at the temperature, pressure, and volume of point A and when step 4 is completed, the temperature, pressure, and volume of the gas are exactly the same as the starting values.

The Steam Engine Operating Cycle

The steam engine was developed over several decades. Working engines using steam were in operation before the theory of the work producing cycle described above and was developed in the nineteenth century. Water was the working fluid of choice for heat (steam) engine design starting in the mid-seventeenth century.

One of the first devices that used steam was designed to lift water from mines. In 1698, Thomas Savery was issued a British patent for a mine water pump based on a simple device. It was known that when steam condenses in a closed container, a vacuum forms. The mine pump consisted of a metal tank and three valves. The only moving parts were the valves.

Ideal Engine Work and A Newly Defined Temperature Scale

The "ideal engine" cycle can be repeated any number of times, each cycle producing work from the thermal energy we add to the gas. The fraction of the thermal energy we put in at step 2 that is converted to work by this ideal engine is represented by the simple ratio:

$$\text{Frac converted} = \frac{\text{Heat in} - \text{heat out}}{\text{Heat in}} = \frac{\text{Work}}{\text{Heat in}}$$

The thermal energy removed in step 4 is discarded and this ratio (work/heat in) is the fraction of the thermal energy converted to work.

The remarkable conclusion of these experiments is that the maximum work produced by any heat engine is independent of the working fluid. One can calculate the maximum fraction of thermal energy in a fluid that can be converted to work and it depends only on the difference between the high temperature at which thermal energy passes into the engine and the low temperature at which thermal energy is taken out.

$$\text{Frac converted} = \frac{\text{Thot} - \text{Tcold}}{\text{Thot}}$$

The temperatures in this expression must be modified to correspond to the absolute temperature scale of thermal physics by the addition of 273.15 + T°C (or 459.7 + T°F) to complete the calculation. This simple expression is the upper limit of the work that an engine can produce per unit of thermal energy transferred to that engine at high temperature. The materials used to build the engine cylinder and piston will set this maximum working temperature. The cold temperature will be close to the temperature of the atmosphere since the low temperature must be above the ambient temperature (temperature of a nearby river or the air) for thermal energy to pass from the cylinder.

With the vent valve open, a second valve was opened to admit steam to the tank from a steam generator (boiler). The steam and vent valves were closed when the tank filled and water sprayed on the outside of the tank to condense the steam producing the vacuum. A valve in the pipe that extended from the tank down to the water in the mine was opened and the vacuum drew water from the mine into the evacuated tank. When the steam was condensed, the water flow stopped and this valve was closed. The vent valve was opened to drain the mine water and the condensed steam. The cycle was then repeated. This pump is not very efficient, but feeding wood and water to the steam generator and operating the valves requires much less physical effort than manning the mine water pumps to lift the water from the mine.

The Savery mine water pump could only raise water about eight or nine feet. Pumps had been used for many years to raise water from deep wells by placing the piston/cylinder of the water pump close to the water surface in a well. When the piston is pulled up from the bottom of the cylinder, water flows into the cylinder through a flapper valve in the bottom of the cylinder. At the end of the stroke, the valve at the bottom of the cylinder closes and a valve in the piston opens so water can flow through the piston as it moves down to complete one pump stroke. A rod connected to the piston is placed inside the pipe that brings the water to the mine or well surface. A pump handle moves the connecting rod and piston up and down to lift the water from the well. This pump can lift water many feet. The diameter of the piston, the length of the stroke, and the number of strokes per minute determine the volume of water removed from the mine. The first pumps were hand operated and it was tedious work.

The design of the water pump suggested the next design that used steam to operate the mine water pump. In 1712, Thomas Newcomen developed a pump that replaced the steam chamber of the Savery pump with a piston/cylinder. This engine is illustrated in Figure 4.3.

When the piston was at the top of the cylinder, the space was filled with steam and the steam valve closed. Cold water was sprayed into the cylinder and the vacuum pulled the piston down. This pulled down the rocker beam that raised the water pump piston lifting water out of the mine. Water from the condensed steam was drained from the steam cylinder as fresh steam was blown in. A weight on the rocker arm pulled the piston to the top of the steam cylinder completing one pump cycle. This engine cycle was slow but by making the diameter of the steam piston and the pump piston large, the engine could move lots of water from the mine in a day. Operating the valves and putting wood in the boiler was easier than using a pump handle and operating the pump by hand.

Figure 4.4 illustrates the next significant improvement. By this time, designers had learned to produce steam from water in a closed vessel.

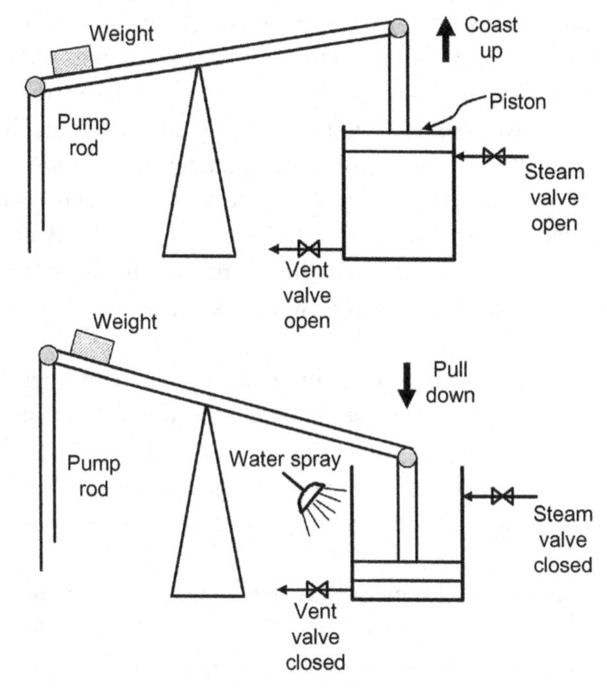

Figure 4.3 Condensing steam used to move a piston.

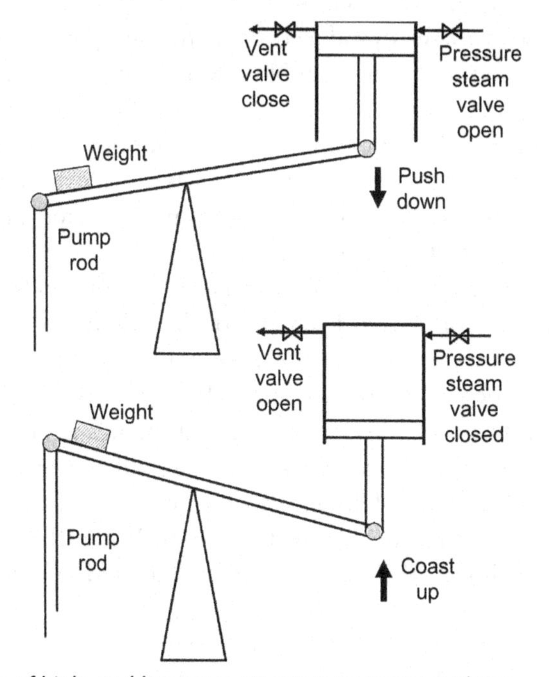

Figure 4.4 Use of high- and low-pressure steam to power a piston.

This boiler produced steam at pressure, a hot gas that can "push" a piston on the power stroke of a heat engine rather than using a vacuum produced by condensing steam to "pull" the piston. Using steam pressure to push the piston, the connecting rod on the engine is attached to the rocker arm from the top and pushes the pump handle down rather than pulling it down in the vacuum pump design. At the end of the power stroke, the steam valve is closed and a vent valve opened to allow the steam to escape from the cylinder. The counterweight on the rocker arm returns the steam piston and the water pump piston to the starting position.

Closing the vent valve and opening the steam valve begins the next cycle. This operating cycle repeats as quickly as the valves can be opened and closed. It was no longer necessary to do the slow step of condensing the steam in the cylinder. This new design increased the number of strokes per hour and greatly increased the volume of water pumped per day.

The next major development was the steam engine design described in a British patent issued to James Watt in 1769 [1]. Watt's design included three features that led to the modern steam engine: (i) he used a boiler to produce steam at an elevated temperature and pressure to drive the steam engine; (ii) steam was alternately introduced on opposite sides of the piston so work was produced on both the "push" and the "pull" stroke of piston motion. This "double action" piston was the standard in most steam engines; (iii) his engine converted the oscillating motion of the piston cylinder into continuous rotational motion using a connecting rod that turned a crank attached to a flywheel. This engine produced continuous rotation which could be used to pump water; but even more important, to turn shafts on machines.

Watt's engine design used a cylinder, closed at both ends, fitted with a piston attached to a drive rod. Each end of the cylinder had a steam valve and an exhaust valve. To operate the engine, the steam valve on one side of the piston is opened and the exhaust valve is closed. On the other side of the piston, the exhaust valve is opened and the steam valve closed. The piston moves toward the low pressure (toward the open exhaust valve) with a force equal to the difference in the pressure on the two sides of the piston multiplied by the area of the piston. When the piston reaches the end of the cylinder, the position of the steam and exhaust valves have reversed and the piston moves in the opposite direction. The work produced is the product of the net force on the piston times the stroke distance the connecting rod moves.

The drive rod on the piston is coupled to a connecting rod, which turns a crank that turns a shaft connected to the load. The piston applies a pushing force on the crank as it moves to the right and a pulling force as it moves to the left. A flywheel on this shaft keeps the crank turning at nearly constant speed at the end of each piston stroke when no work is produced as the piston changes direction. The steam and exhaust valves are connected to the engine crankshaft by a cam so they are "timed" to open and close in proper sequence with the position of the piston. A pressure control valve in the steam supply line adjusts the steam pressure so the rotational speed of the engine can be held nearly constant as the load on the engine changes.

The basic design of the Watt steam engine was the workhorse of the industrial revolution. This was the design used in machine shops, on board ships, for train locomotives, farm tractors, to generate electricity, and any application that required a rotating shaft. The mechanical parts of the engine were customized to satisfy the requirements of the many applications. Improved materials of construction and the design of the steam boilers improved the efficiency and safety of the conversion of thermal energy to steam using the locally available fuels. The art and technology of the steam engine design could never overcome the problem of vibration from the rapid oscillation of the piston and the enormous weight required for large engines with high power output. The steam turbine gradually replaced the steam engine when it was an advantage to increase the rotational speed.

For these early engines, boiler design improved with engine design. The very early engines used steam at low pressure. The boilers that generated low-pressure steam were relatively safe. When the steam engines began to use pressurized steam well above the normal boiling point of water, the boiler became an explosion hazard. Boiler failures were common with injury and death often the result. The shape of the boilers, new metal alloys used for construction and replacing riveted joints with welded seams improved boiler safety. Boiler construction experience written into steam generator design codes and pressure testing procedures for new and "in-service" boilers has nearly eliminated steam boiler explosions in modern steam power plants.

TURBINE-BASED ENGINES

The Steam Turbine

In its simplest form, a turbine is a windmill enclosed in a tube that directs the air flow across the "windmill" blades.

Wind has been used as an energy source to produce work for centuries. Ancient art show sails were used to assist or replace men rowing ships. Much later, "sails" were set on wooden spokes that turned windmills centuries before the invention of the steam engine. Those who developed the steam engine knew that steam vented from a boiler through a nozzle produced a "wind." When the pressure in the boiler increased, the exit steam velocity also increased. A logical question, why not use this "steam wind" to drive a windmill? Certainly, one real advantage is this wind direction and velocity will be constant and controlled by the temperature, pressure, and flow rate of the steam to this "steam windmill."

The first practical steam turbines were built about 100 years after Watt's steam engine. An early turbine design by Charles Parsons was patented in 1884. Parsons' turbine used a jet of steam to turn several turbine wheels mounted on a single shaft. The turbine blades were placed on the outside edge of a wheel and stationary vanes and were set between each of the rotating wheels that redirected the steam flow onto the blades of the next turbine wheel. Steam flow is parallel to the turbine shaft (perpendicular to the turbine wheels) just like a classic windmill. This makes it possible to gradually extract energy from the steam as the temperature and pressure of the steam drop as it passes through each turbine wheel.

Parsons showed that the most efficient conversion of the thermal energy in steam to work occurred when the pressure decrease across each turbine wheel was the same. Furthermore, at each turbine wheel, the pressure drop should be equally divided between the stationary vanes and the turbine blades.

Sufficient progress had been made with the design and operation of the Parson turbine by 1894 that a syndicate was formed to test it on a small ship. This ship, named the Turbinia, ultimately attained the then spectacular speed of 34.5 knots. This test established the steam turbine as the power plant of choice for marine applications.

The Parsons steam turbine used the flow of steam as a high velocity wind to turn the turbine blades much like a windmill. A second practical design changed the shape of the turbine blades and had the steam pass over them much like water passing over a water wheel. This turbine was designed by

Carl G. deLaval in 1887 and also was very successful. The turbine wheel has "bucket shaped" blades that are "pushed" by the steam flowing from one or several nozzles directed onto the turbine buckets mounted on the turbine wheel. The steam nozzles were designed to increase the velocity of the steam as it approaches the turbine wheel and add to the force the steam passes to the turbine wheel increasing the power generated. This turbine was developed to drive a cream separator, which requires very high rotational speed. One of these turbines that developed about 15 horsepower was actually used for marine propulsion before Parson's turbine.

In 1895, George Westinghouse obtained the American rights to the Parsons turbine. He built a turbine to generate electricity for his Westinghouse factory in New York City. Electric motors were used on each machine in the shop. This was the first fully electric powered manufacturing plant in the United States. The steam engine which turned all of the drive shafts, with their belts and pulleys, were all replaced by electric wires and switches to electric motors mounted on each machine. This is the model that dominates the design and layout of all modern machine shops, assembly lines, offices, and homes today. A steam turbine drives a generator to produce electricity that is easily distributed to the point of use.

The first turbines did not efficiently convert the thermal energy in steam to power on a rotating shaft. Much of the steam "blew" past the turbine blades without producing work. Modern turbine design uses very close spacing between the multiple rows of turbine blades and the stationary blades that redirect the flow of steam as it passes through the turbine. The shape of each blade is as carefully designed as an airplane wing to reduce the energy loss to friction and maximize the power generated as the steam passes through the turbine.

The power output of a turbine increases as the temperature and pressure of the inlet steam increases. Higher temperatures require special metal alloys for the turbine blades and for the turbine casing that must contain the steam as it flows through the turbine. The high steam pressure and rotational speed of the turbine rotor place force the turbine blades that must be designed to withstand the stress without bending or breaking. The rotating blades must stay perfectly aligned so there is no contact with the stationary blades. The blades must not only be flexible enough to avoid "brittle fracture," but they must also be designed so they do not flutter (like a Venetian blind in an open window on a windy day), even a small vibration could destroy the turbine.

Material science and technology research funded by the US military developed and demonstrated that new high performance metals could meet the demands on Navy ships. The military turbine technology was soon available to private industry because better performance reduces the cost of operating the turbine. The firms that produced the military turbines also produced turbines for industry, and it was natural to transfer the technology from the military to the private sector.

Process Steam and Its Use

Steam turbines are very versatile. They have been designed to operate with very hot, dry, high-pressure steam or with lower temperature, low-pressure wet steam. Large central heating plants produce steam that is used in petroleum refineries and chemical plants to heat process streams. Controlling the pressure of condensing steam controls the temperature that the thermal energy passes to the process. We get high processing temperature when the condensing steam pressure is high and low temperature at low pressure. It is simplest to build a steam boiler that operates at constant temperature and pressure so the steam plants are usually designed to produce all of the steam required at a pressure greater than required for any process in the plant.

Low-pressure steam is obtained by passing all of the steam at high pressure through a turbine and drawing off part of the steam at lower pressure a few turbine wheels into the turbine. Additional side streams can be withdrawn as the steam pressure decreases through the turbine. The turbine drives a dynamo to produce electricity as it provides steam at the desired pressure for the process. Central heating plants for large buildings or building complexes—a university is an example—always produce electricity with the steam before it is distributed to heat rooms or buildings.

Steam turbines that deliver 50–10,000 horsepower have been designed to drive pumps and blowers and are distributed throughout chemical plants and petroleum refineries. These turbines are designed to operate with "available steam." The steam discharged from these turbines is then used in the plant as process steam (a heat source for processing chemicals). This combination of producing work for rotating machines and providing process steam is an efficient way to use more of the thermal energy from the fuel that produced the steam.

The steam turbine is the primary source of power to drive dynamos in electric power stations. There are many power stations with electricity as the only product. In these power plants, the steam from the turbine is condensed and the low-temperature thermal energy is discarded. The

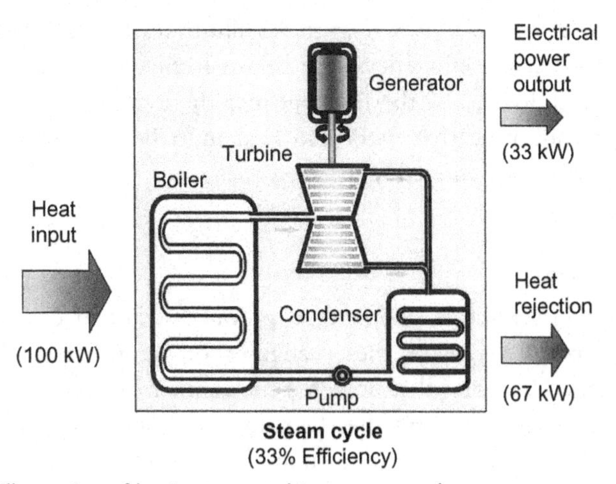

Figure 4.5 Illustration of basic steam turbine power cycle.

power plants that use fossil fuel and the nuclear power plants are different. Fossil fuel plants use higher pressure steam and higher temperatures, and nuclear plants are limited to lower temperature "wet" steam.

Figure 4.5 illustrates the steam cycle and the key components of that cycle. The plants that use fossil fuel such as coal, oil, or natural gas generate high temperature (typically 1000°F or 538°C), high-pressure (3550 psi or higher) steam. The steam actually passes through three or four turbine stages all mounted on one shaft. As the temperature and pressure of the steam decrease, the turbine wheels get larger and the metal alloy in the turbine blades changes to match conditions of the steam and the rotational stress on the blades. The initial hot, dry steam expands through the turbine until the pressure is very low (actually a vacuum) and the temperature is about 120°F. After expanding to about 120°F, the steam passes to a condenser where it condenses to liquid water. The liquid water is pumped back to the steam boiler at the pressure of the boiler.

After a century of design improvements, most of the modern fossil fuel power plant turbines turn at 3600 revolutions per minute, and the largest units produce more than two million horsepower to generate 1000 MW (one billion watts) of electricity. Steam flows through the turbine at the rate of about 90,000 pounds per minute and it takes about 12,000 tons of coal per day to generate the steam to run the turbine. These are huge power plants!

The thermal efficiency of a modern coal-fired power plant is about 40%—about 40% of the fuel's chemical energy (thermal energy) is converted to electrical power. Figure 4.5 illustrates the energy flow for a cycle at 33% thermal efficiency. The thermal efficiency is defined as the work produced divided by the heat put into the cycle—it is a dimensionless number indicating that work and heat must be in the same units to compute the thermal efficiency.

Nuclear Power

The first nuclear reactors designed to produce steam to drive a turbine were developed to replace diesel engines on submarines. The diesel engines charged batteries that powered the submarine underwater, but air was required to power these diesel engines. Surfacing to run the diesel engines, or even using "breathing tubes" to recharge the batteries while submerged, revealed the location of the submarine. Modern nuclear powered subs remain submerged for 90 days or more, thereby, reducing the risk of detection as they cruise deep underwater. These nuclear power plants are now standard on many military surface ships. This increases their range and eliminates the difficult (dangerous) task of refueling at sea. The nuclear reactor technology to produce steam to drive a turbine soon passed to domestic power plants. The US government with the "Atoms for Peace" program of the 1960s provided additional financial incentives to speed this domestic use of nuclear energy.

Water-moderated nuclear reactors are the most common for commercial electric power production. Liquid water is required in the nuclear reactor to "slow down" the neutrons produced by nuclear fuel fission so the reactor continuously produces the thermal energy to make steam. The water also serves to extract the thermal energy produced by the nuclear fission process.

There are two types of water-moderated reactor designs used commercially. The boiling water reactor (BWR) has the reactor core and water contained in a pressure vessel. The water is allowed to boil and the steam goes directly to the turbine. Figure 4.6 illustrates a BWR.

The pressurized water reactor (PWR) has the reactor core and water contained in a pressure vessel, but the pressure is high enough that the water never boils. The hot, high-pressure water is pumped through a heat exchanger that produces the steam that goes to the turbine. The PWR design keeps the water that comes in contact with the reactor core isolated from the steam that goes to the turbine. Figure 4.7 illustrates a PWR.

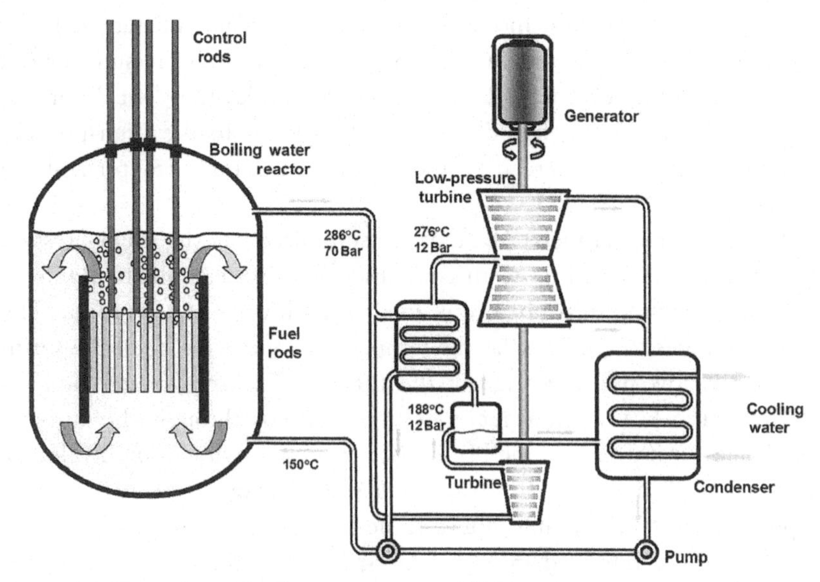

Figure 4.6 Illustration of boiling water reactor (BWR) and steam power cycle.

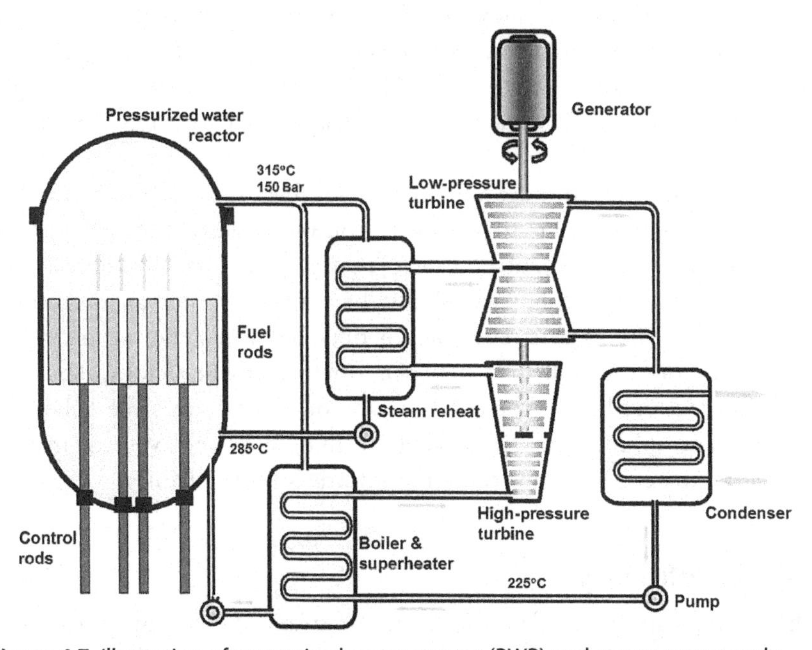

Figure 4.7 Illustration of pressurized water reactor (PWR) and steam power cycle.

The steam turbines in modern nuclear power plants operate at lower steam temperature (about 560°F, about 290°C) and pressure (about 1000 psi) than the fossil fuel plants. The steam produced is "wet," formed in contact with boiling liquid water, and it passes to the high-pressure stage of the turbine where the pressure drops to about 150 psi and the temperature to 350°F (177°C).

As the steam temperature and pressure drop in the high-pressure turbine, about 14% of the steam condenses. This water must be removed from the remaining steam before it goes to the low-pressure turbine. The steam continues to condense as the temperature and pressure of the steam drop in the low-pressure stage of the turbine. The small water droplets that form move at high velocity with the steam through the turbine. They collide with the turbine blades that are also moving at high velocity. The turbine blades must be made of a special alloy to keep the water droplets from "sand blasting" the turbine blades away. Liquid water must be continuously removed from the low-pressure turbine casing to avoid damage to the turbine blades.

The turbines in nuclear power plants turn at 1800 revolutions per minute and develop over two million horsepower to produce 1000 MW of electric power. They run slower because the diameter of the turbine wheels must be larger to allow much more steam to flow through them. The lower energy content of low-pressure steam requires about 160,000 pounds per minute of steam flow. This is a huge machine that weighs about 5000 tons with a rotating shaft that is approximately 73 meters long.

These huge turbines are remarkably reliable. It is necessary to shut down the power plant to perform routine maintenance on all of the mechanical equipment. Nuclear power plants usually run on an 18-month refueling schedule and this becomes the interval for turbine maintenance. The greatest efficiency is obtained when the turbine operates continuously at full load between reactor refueling stops. The turbines in coal-fired power plants operate on about the same schedule. These steam turbines are a remarkable technological achievement matching special materials of construction with precision mechanical design.

Thermal Efficiency

The primary measure of performance for steam engines and turbines is the thermal efficiency. The thermal efficiency is defined as the amount of shaft work produced divided by the amount of heat taken from the boiler. The best of our modern natural gas power plants are able to convert about 55% of the heat provided by the natural gas into work. The remaining 45% is lost to the surroundings in the form of hot exhaust gases leaving the power plant stacks or steam leaving the cooling tower. The coal-fired power plants have thermal efficiencies up to 45%. The thermal efficiency of the nuclear plant is 30–33% because the steam temperature and pressure are lower. The first steam engines were doing well to convert 10% of the fuel energy into shaft work.

Gas Turbines

The gas turbine has a long history. The theory of gas turbines and how they should work was known long before the first one was built. It appears the first gas turbine patent was issued to John Barber in 1791. There is no record that this gas turbine was ever built, but it did establish a basis for future development.

The first US patent that described a complete gas turbine was issued to Charles G. Curtis in 1895. The first turbine that ran was built in France in 1900, but the efficiency of this unit was about 3%, not very encouraging. The early gas turbine builders tried to use the steam turbine wheel because it was available and it worked well with steam. It is a simple matter to mix fuel and air and burn it in a chamber, but how do you increase the pressure of the hot gas to make it flow through a turbine to produce work the same as steam turbine?

On paper, the gas turbine promised to deliver considerably more power than a gasoline or steam engine of similar size. Early workers observed that higher initial steam temperatures and pressures provided more work per gallon of fuel oil—efficiency increased with increasing temperatures and pressures. The large quantity of thick metal alloy pipes necessary to take steam to higher temperatures and pressures was costly, but a small, inexpensive combustion chamber could provide vast amounts of hot combustion gas for a gas turbine. The design of the gas turbine created new challenges. First, the pressure of the air and fuel had to be increased before combustion. Next, the steam turbine metals would soften and deform at the high temperatures in a gas turbine. Metallurgist began the search for new, high-temperature alloys and ceramics to make the turbine blades durable at higher temperature, a search that continues today.

In principal, the same pistons and cylinders used to compress air in the gasoline engine could be used to compress air for the gas turbine. However, in practice much of the pressure gained was lost in the tortuous flow path through the many valves, turns, and manifolds needed to direct the air to a combustor chamber. The gas turbine engine needed a simple compression process that did not rely on valves, pipes, and manifolds. While piston-in-cylinder (reciprocating) compressors work well when compressing to pressures greater than 150 psig, reciprocating compressors did not work well for lower pressures of about 60 psig that were needed in gas turbines.

The axial flow gas compressor (the flow of air is parallel to the compressor shaft) provided this simpler compressor that was effective at lower compression ratios. (Pressure ratio is defined as the pressure after compression divided by the pressure before compression. A pressure ratio of 5 corresponds to increasing pressure from 15 psia to about 75 psia.) It was the mid-1930s when the aerodynamic theory developed to design aircraft wings was applied to the shape of the axial gas compressor blades assuring the commercial future for gas turbines.

An axial flow compressor operates on the same principle as the steam turbine. The design engineers changed the shape of the blades in the turbine so that the gas pressure increases as the wheel turns. You supply the power to spin the compressor shaft and the pressure increases at each set of rotating and stationary blades, and you have an axial flow compressor. This axial compressor avoids the use of valves and gas flow through tortuous manifolds by placing the compressor on the same shaft as the gas turbine. With fuel delivery and a well-designed combustion chamber located between the compressor and turbine the major engine components are in place. Some of the compressed air bypasses the combustion chamber to reduce the temperature of the gas passing into the turbine. Good design adjusts the size of the compressor and the turbine to deliver maximum power at the turbine driveshaft.

The turbine blade is the critical mechanical part in a gas turbine. The aerodynamic shape of the outside surface of each row of blades must be matched to the conditions of the hot gas flowing past them. The first row is exposed to the highest temperature and pressure because the gas cools and the pressure drops as it expands through the turbine; each turbine wheel after the first is exposed to cooler exhaust gases. In some modern designs, air passageways are machined inside each turbine blade to allow cool, compressed air to blow through the blade to cool it as the hot gas flows by on the outside. Ceramic coatings can be applied to the outside surfaces of the blade to give additional heat protection to the turbine blade. These are technical fixes to "beat the heat" and keep the turbine running at higher gas inlet temperatures. Manufacturing these complex blades increases the cost of the turbine, but running at slightly higher temperatures increases the fuel efficiency. For example, increasing the gas inlet temperature from 900°C to 1000°C increases the thermal efficiency of the turbine about 10%.

The greatest boost to gas turbine development came from the development of jet engines for aircraft propulsion. The British and the Germans both tested jet aircraft engines from the mid-1930s. The Germans had jet powered fighter planes in service during the final stages of World War II. The development of the US military jet engines was classified "secret" and advanced technology remained classified following World War II.

The gas turbine as an aircraft engine was successfully demonstrated during World War II. Immediately following the war there was a period when redesign and testing produced the first commercial turboprop airliner. It was Vickers Viscount turboprop that was first flown in 1949 with conventional twin propellers powered by gas turbines. It entered commercial service in 1953. This was the beginning of the end for the reciprocating engine on all but small aircraft. Commercial aviation was powered by gas turbines spinning the propellers (turboprops) and gas turbines providing vast amounts of hot exhaust gases that led to the advanced jet engines in service today.

Gas turbines exceed in two performance criteria:

1. A high-power output per pound of engine weight makes the gas turbine ideal for aircraft service.
2. A gas turbine can be started and run at full power in a few minutes.

Gas turbines also work well with a variety of fuels; natural gas, fuel oil, waste gas in an oil refinery, etc. As long as the fuel combustion does not produce solid particles that "sand blast" the turbine blades, it can be fuel for a gas turbine. Given a fuel supply, the gas turbine combustion chamber can be designed to optimize performance for that fuel.

The period from 1945 to 1970 represents consolidation of the gas turbine as a power source for many applications. The first commercial gas turbine train locomotive was put in service in 1950. The low thermal efficiency of the gas turbine was a disadvantage relative to diesel units. The simple gas turbine is less efficient than either the gasoline engine or diesel engine with maximum efficiencies of about 30%, 37%, and 50%, respectively. Interest in developing turbine technology for locomotives disappeared as freight traffic moved from railroads to the highways.

The first trials placing a gas turbine in a personal boat came in 1950. Again, the gas turbine offered no real advantage over the conventional diesel powered ocean fleet. It's only recently that gas turbines have replaced steam turbines on navy ships.

Pumping natural gas through pipelines from the gas wells to customers all over the country requires lots of power. A gas turbine fueled with natural gas driving a centrifugal compressor powered by a 1800 horsepower gas turbine was installed on a large, 22-inch diameter pipeline in 1949. Gas turbines have found a place on oil platforms located at sea to provide electricity and to turn pumps and compressors. They use the oil or gas produced on the platform as fuel.

It was in 1949 that the first gas turbine electric generating unit was placed in service in the US rated at 10,000 kW represents an important achievement in the development of electric power generation pointing toward modern electric power plants.

The first gas turbine installed in an automobile traveled across the United States in 1956. Chrysler Corporation built 50 gas turbine–powered cars in 1963–65 to be used by typical drivers on daily trips. The advanced technology of the internal combustion engines at that time offered better fuel efficiency and reliability. The market for the gas turbine car did not develop and the project died. There was no reason to develop the gas turbine for this application.

Gas turbines are now widely used to supply electrical power because they are the least expensive engines for a given power output in the large engine market. However, the simple gas turbines' disadvantage of low fuel efficiency (i.e., high fuel costs) can more than offset the money saved in purchasing the engine when placed in continuous service. Obviously, the gas turbine is not best for all electrical power applications.

Electric power demand changes in predictable ways based on season and time of day. The long-term use cycle corresponds roughly to the seasons, and a short-term 24-h cycle corresponds to variations in human activity during the day as compared to night. A large quantity of electricity cannot be stored, so it must be produced at the rate it is used. There are three levels of electrical power production designed to meet this cyclic demand.

Base load power must be continuously supplied every day of the year. The base load power generators must show efficient conversion of fuel to electricity, must be very reliable, and they should provide the lowest cost per kilowatt-hour of electricity produced. These are usually efficient coal-fired power plants or nuclear plants—the low fuel costs for these plants compensates more for their high initial capital cost for continuous power production. They run at constant power output for months without shutting down or throttling back.

Intermediate load generation units operate about 75% of the year and they take care of the swings in seasonal load. They also assume the load when a base load unit is shut down for routine maintenance or for an emergency shut down of any generating unit on the power grid. These plants should be capable of operation at partial load. The energy conversion efficiency of any power plant is reduced when it operates at partial load because the steam boiler and turbine are most efficient at full load.

Peak load generation units are used in the summer for a few hours afternoon when all the air conditions are turned on. Peak power production will be required about 5% of the year (certainly less than 10%) and usually only part of the 24-h day. The thermal efficiency of these units is not as important as the ability to start them quickly, run them for a few hours, and shut them down. The gas turbine is best suited for this peak load assignment. There is a wide range of horsepower ratings available, and there is often an economic advantage to use several small gas turbine units for peak demand service. This allows the use of just enough units operating at full power to cover the peak demand for that day. Gas turbines can be started and brought to full power in minutes. This is in contrast to coal-fired steam plants that take many hours to start or shut down.

The recent strategy for intermediate electric power production is to use a *combined cycle* power plant. The exhaust gas from a gas turbine is still very hot. This hot gas can be passed through a boiler to produce steam to turn a turbine and generate electricity. Combining the power output of a gas turbine with that of the steam turbine increases the total thermal efficiency of the combined cycle plant to over 50%. This compares to efficiencies of about 30% for "simple" gas turbines. Improved thermal efficiency becomes really important when fuel costs are high, but the flexibility of "quick" start and shutdown is lost because boilers are slow to heat up and produce steam.

Natural gas is often the fuel of choice for gas turbine plants. There is the added advantage that the carbon dioxide released to the atmosphere per kilowatt is less than for a coal-fired power plant. Combined cycle plants are about half the price of coal-fired plants due to elimination of costly coal handling facilities, use of smaller boilers (only part of the electrical power is provided by steam), elimination of precipitators to remove the particulates in coal exhaust gases, and the elimination of scrubbers to remove sulfur oxides from coal exhaust gases. Natural gas is considered a "premium fuel" because it does not require the costly solids-handling and exhaust-treatment facilities. Control of the supply of natural gas by a few pipeline distributors has created price fluctuations that make it difficult to estimate the future production cost of electricity from natural gas. This makes investment planning for the natural gas plants difficult. Government loan protection is given to utility investors in some regions of the United States where electricity is in short supply and demand is increasing.

There will always be demand for specialized uses for gas turbines. Today, electric power generation and aircraft propulsion are featured. Any device designed to power a commercial aircraft must satisfy many performance criteria. It must meet the performance demands of the aircraft, be fuel efficient, low weight, reliable ("never" fail in flight), and easy to maintain with maximum number of operating hours before the engines must be replaced. These are the problems gas turbine design engineers continue to work on.

Government regulations regarding safety must be satisfied by the design. The exhaust emissions are federally controlled. The amount of carbon monoxide, unburned hydrocarbons, and nitrogen oxides in the exhaust gases must meet EPA standards. Noise standards have been established that require special tests to find the source of the noise and the mechanical design changes to reduce the noise. Each of these requirements places demands on the design of the combustion chambers and mechanical components of the gas turbines on an aircraft.

FUEL CELLS

Practical fuel cells were designed during the 1960s and were first used during the NASA Apollo program. The fuel cells were fueled with stored hydrogen and oxygen. This was a particularly valuable technology for space travel since the process that produced electricity also produced water for the astronauts. Transporting drinking water into space could almost be eliminated, by matching the minimum fuel cell hydrogen consumption to the drinking water needs.

Similar to steam turbines, rockets, and jets, the military (specifically NASA) paid for fuel cells during the early and costly development stage. It is this technology adapted for use in automobiles that is being developed for domestic use with financial support from the US Department of Energy.

Fuel cells are like batteries, but instead of storing energy using the chemistry of lead/acid they are powered by a fuel and oxygen. Both fuel cells and batteries convert the energy of chemical bonds into electrical power. When connected to electric motors, fuel cells can be used to power vehicles much like gasoline or diesel engines.

Fuel cells weigh less than batteries and they produce electricity as long as the fuel (hydrogen and oxygen) is supplied. The chemical reaction that produces electricity in a battery is reversible. The quantity of chemical is limited by the size of the battery. When the electrical current slows or stops, an electrical current is passed through the battery to recharge it (reverse the chemical reaction). This "remakes" the chemicals that produce electricity. The discharge/charge cycle can be repeated many times, but there must be a source of electricity to recharge the battery. This is a clear advantage for the fuel cell when a reliable source of low cost fuel is available.

Fuel cells are also considerably lighter than batteries. For a car with a 100-mile range it takes about 1100 pounds of lead/acid storage batteries to power the car [2]. Depending upon design, fuel cell systems complete with on-board fuel to achieve the same range would weigh less than 100 pounds. That's why fuel cells worked so well for space travel.

At the beginning of the twenty-first century, much attention is focused on fuel cells. The excitement comes from the potential of fuel cells to achieve high fuel efficiency and low emissions. Table 4.1 lists several performance categories to compare the advantages of fuel cells to gasoline and diesel engines, gas turbines, and batteries.

Table 4.1 Performance strengths of different mobile power sources

Property	Fuel cell	Gasoline engine	Diesel engine	Battery	Gas turbine
Fuel efficiency potential	E	G	G	E[a]	F
Present and near-term fuel efficiency	F	F	G	E[a]	F
Ability to deliver good fuel efficiency with varying loads	E	F	F	E[a]	P
Will work with variety of fuels	F	E	E	P	E
Power output per system weight	G	G	G	P	E
Engine cost per power output	P	E	G	F	G
Durability and expected performance life	I	F	G	F	E
Emissions at vehicle	E	G	F	E	G
Emissions not at vehicle	F	E	E	G	E
Feasibility of practical rechargeable systems	E	P	P	E	P
Performance synergy with hybrid vehicles	E	G	E	E	G
Lack of hidden performance problems	P	E	E	E	E

E, excellent; G, good; F, fair; P, poor; I, insufficient data.
[a]While batteries are very efficient for converting their stored chemical energy to electrical power, the efficiency of converting an available energy source (coal or nuclear) to the chemical energy in the battery is poor. The reported efficiency is based on electricity to charge the battery and not the fuel to produce electricity.

How Does a Fuel Cell Work?

An understanding of fuel cell operation is much like understanding combustion.

Hydrogen (H_2) and oxygen (O_2) will burn to form water (H_2O). This is a chemical reaction depicted by the following chemical equation.

$$H_2 + O_2 \rightarrow H_2O + \text{energy release}$$

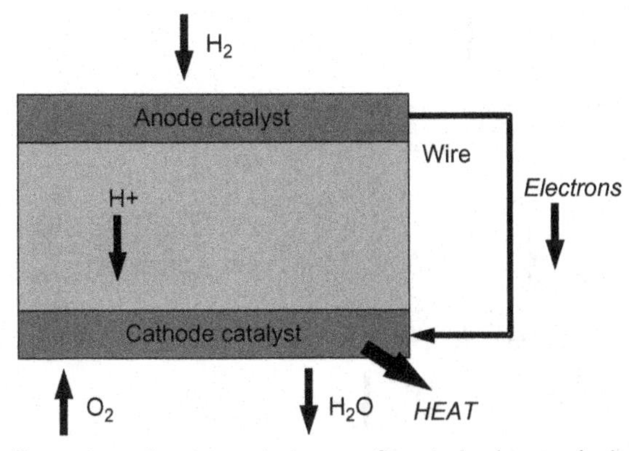

Figure 4.8 Illustration of atom and electron flow in hydrogen fuel cell.

Combustion, as with all energy releasing reactions, occurs because the final product is more stable than the reactants. The progression of compounds going from less stable higher energy states to more stable lower energy states is the natural route for all energy technology.

We want the reaction to occur when and where the energy can be used. For hydrogen combustion, heat or a spark applied to a mixture of hydrogen and oxygen will cause the reaction to proceed quickly and release lots of heat. Are there other ways to control the rate of this reaction to produce electricity rather than heat?

Consider the following series of steps which occur when platinum dust is coated on a membrane surface (see Figure 4.8).

The hydrogen molecule, H_2 is pulled apart to form two H+ (hydrogen cations) and two electrons. It is the platinum powder that makes this happen.

The H+ ions are able to travel to an oxygen molecule by passing through a membrane that does not pass electrons.

The electrons travel in a wire contacted to the platinum anode through an external circuit where it can turn a motor and end up at the cathode platinum dust on the other side of the membrane where there is oxygen.

On the oxygen side of the membrane, two of the electrons and two hydrogen cations combine with an oxygen atom to form a very stable water molecule. The driving force for forming water is so strong that this reaction literally "pulls" the electrons through the wire to the oxygen/cathode side.

As improbable as this process sounds, it works for proton exchange membrane fuel cells. This process does occur at room temperature because platinum is such a good catalyst (see Figure 4.9).

If this occurred for the system shown by the diagram, the same amount of heat would be released as when you burn hydrogen. If the wire is connected to the armature wires of an electric motor, the oxygen side of the

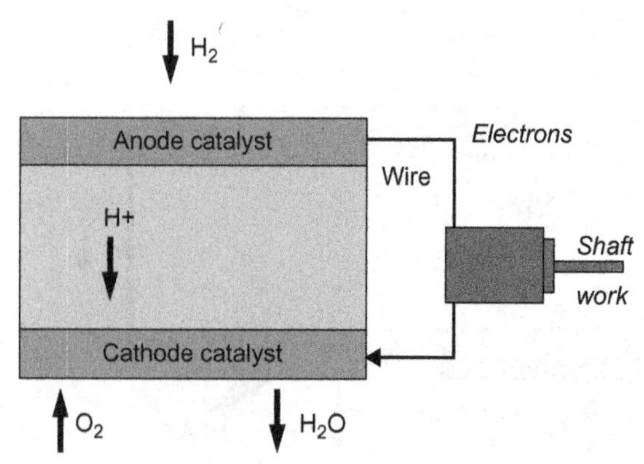

Figure 4.9 Illustration of fuel cell circuit powering an electric motor.

membrane acts as though it is "pulling" the electrons through the wire/motor due to the electromotive force produced by the reaction of hydrogen and oxygen to form water. The current passing through the electric motor produces work.

Work and Efficiency in Fuel Cells

In an ideal fuel cell, 100% of the energy that would otherwise go to heat during combustion goes to electrical power (e.g., to a motor). The ideal fuel cell converts 100% of the chemical energy into electricity.

In the worst case, essentially all the chemical energy would go to heat—this could burn up the fuel cell. The heat must be transferred from the fuel cell to keep it from being damaged.

The best practical fuel cells operate at about 55% efficiency. This means that 55 BTUS of work are generated for every 100 BTUS of combustion energy released by "burning" hydrogen in the fuel cell. The other 45 BTUS of heat are released in the fuel cell and must be removed to prevent the fuel cell from overheating.

Issues that have pushed fuel cells into the spotlight are fuel efficiency, low emissions, and compatibility with the hybrid motor vehicle. Fuel cells avoid the thermodynamic limitations of combustion by directly converting chemical energy into electricity at about ambient temperatures. Practical fuel cells do operate at about 80°C or higher to allow the waste heat to be easily removed.

Fuel cells appear to be the "natural choice" as a power source for hybrid vehicles since they store energy as fuel and use electricity to turn the wheels.

There is little doubt that the potential of fuel cells will eventually outperform conventional engines in essentially every category. Keep in mind that the people who describe fuel cell technology in the technical literature have vested interests in fuel cells. It is easy (almost natural) for them to be optimistic about the progress that can be made in the next one or two decades.

There is no corporate incentive to change from internal combustion engines to fuel cells as long as petroleum is plentiful and cheap. It took legislation that set pollution standards (eliminate lead and sulfur from motor fuels) and increase the miles per gallon (fuel efficiency standards) to move the auto and fuel industries to reach performance levels available today.

For a state or nation, the economic incentive to change from internal combustion engines to fuel cells rests on the simultaneous replacement of an imported fuel with a locally produced fuel. By replacing imported gasoline with electrical power produced from any low-cost fuel (uranium?), a state could keep billions of dollars spent on imported fuel. This could create thousands of jobs. The last chapter of this text will discuss this economic picture.

Fuel Efficiency

An ideal fuel cell running on pure oxygen and hydrogen produces 1.229 volts of electromotive force to push electrons through the electrical devices when there is no load on the circuit (the switch is open). Since the chemical reactions of a fuel cell provide the same flow of electrons independent of the fuel cell efficiency, this voltage is used to compute the efficiency. For example, when the fuel cell is connected to an external load, the "open circuit" voltage of the fuel cell is split between the resistance to anion flow in the cell and the resistance to electron flow in the external load. If the fuel cell produces 0.615 volts when connected to a light bulb, the efficiency of the fuel cell with that load is 50% $\left(\frac{0.615}{1.229} \times 100\right)$. The conversion efficiency improves when the external load is decreased. It takes careful design of the fuel cell to handle the variable power requirements of commuter car travel and maintain high efficiency.

At the start of the twenty-first century, the efficiency of a good hydrogen fuel cell running with atmospheric air rather than pure oxygen varied from about 45% at maximum design load to about 65% when operating at low power output corresponding to the "most efficient" load [3]. If electrical power is used to make the hydrogen, and this electric power is generated with the efficiency varying from 30% to 55%, this gives an overall fuel cell efficiency of 13.5–36%. If the fuel cell operates on hydrogen produced by converting gasoline to hydrogen on the vehicle, the conversion of energy from gasoline to hydrogen is about 70–80% which gives an overall efficiency of 31.5–52%. The technology to use gasoline to power a fuel cell car requires the technology development to reform the gasoline to hydrogen. The gasoline reformer and fuel cell cars must be road tested before they come to the dealer's showroom.

The reality for fuel cells in vehicles today is that they are less efficient than the best diesel engine. The current goal of an overall fuel cell efficiency of 70% is not realistic neglecting the energy loss during hydrogen production. Operation at maximum efficiency requires a large fuel cell to keep the electrical current produced per unit area of the cell low. A fuel cell does not respond to the quick starts and acceleration required for urban driving. The fuel cell power output can be "leveled out" by using a battery pack that provides surge power and is recharged during cruise and stop phases of the trip. The gasoline powered hybrid cars on the market today provide an ideal test for an electric powered car. Design a fuel cell to replace the gasoline engine, the electricity generator, and battery in the hybrid and you have a fuel cell car.

Battery and Fuel Cell Options

Whether we use a fuel cell or battery to power a car, the following components are necessary:

- A wire that conducts electrons (an electrical current) to the external load. The load can be the electric motor that turns the wheels, the lights, the CD player, etc.
- A membrane or "salt bridge," a solution that conducts cations (cations are the positively charged particles that remain when the electrons are stripped from a molecule during the chemical reaction) or any other medium that conducts cations but not electrons.
- A chemical reaction that will proceed by producing cations.
- A reaction surface/catalyst that promotes the reaction and that frees the electrons to flow to the wire and the cations to flow to the membrane (salt bridge solution or membrane) to complete the formation of products of the "burned" fuel. For the hydrogen fuel cell, this is hydrogen and oxygen combining to form water.

Different chemical reactions will produce different fuel cell voltages. The total current a fuel cell produces depends on the total catalyst surface area and the number of electrons produced when a fuel molecule goes to its reaction products.

We have the technology to make a fuel cell work with hydrogen, methanol, and natural gas. However, the technology available today requires that gasoline, diesel fuel, or coal be converted to one of the fuels used in fuel cells.

The active research associated with each of the four steps listed above for an improved fuel cell include: (i) develop better and cheaper membranes to conduct cations in the fuel cell; (ii) catalysts that are less expensive than platinum; (iii) improved design to efficiently remove the heat that is generated when the fuel cell is operating. Fuel cells can operate at theoretical efficiencies in excess of 90% but the typical practical efficiencies are 30–55%. This means that 35–70% of the energy available from the hydrogen fuel must be removed as heat. Other practical issues include preventing the fuel cell membranes from drying out, how to keep the water in the fuel cells from freezing in cold weather, and some process-specific problems associated with hydrogen generation and the use of hydrogen that contains impurities.

Methanol fuel cells rather than hydrogen fuel cells have great potential because they use a liquid fuel. It is much easier to carry liquid fuel on a vehicle. The methanol fuel cells are not as efficient as hydrogen fuel cells and they are more expensive to build.

One of the advantages of fuel cell–powered vehicles is they produce few emissions. Using methanol or hydrogen as fuel essentially eliminates emissions associated with gasoline engines since water is the primary emission with carbon dioxide added from the methanol cell. When gasoline is auto-reformed to produce hydrogen for fuel, it is a high-temperature process. Carbon dioxide, some nitrogen oxides, and particulate matter will be produced much like a gasoline engine.

It is easy to exaggerate the true emission benefits of fuel cells. The history of gasoline and diesel engines show that when high pollution levels were tolerated, cities did actually prosper. Claims of premature deaths due to auto emissions probably are true, but the case for these claims was not strong enough to lead to drastic measures such as banning cars from the city. Individual vehicle emissions in the late 1960s were less than in the 1950s, and since then emissions have been reduced by more than a factor of 10. Based on current regulations emission standards for gasoline and diesel engine emissions have been reduced to 1% of the late 1960s levels. For diesel vehicles, implementation of tier 2 standards in 2007 reduced emissions for new vehicles to 10% of the engines manufactured in 2002.

It is reasonable to ask the value of reducing vehicular emissions to 1% of previous levels. In September 2002, the EPA reported [4] that "the exposure-response data are considered too uncertain" to produce a confident quantitative estimate of cancer risk to an individual. The EPA reported the "totality of evidence from human, animal and other supporting studies" suggests that diesel exhaust is likely to be carcinogenic to humans by inhalation, and that this hazard applies to environmental exposure." While these data on cancer risk were less than definitive, the EPA reports state that long-term exposure has been shown to be a "chronic respiratory hazard to humans." A 100-fold reduction from levels that produced less than definitive cancer risks and require long-term exposure to cause chronic respiratory problems should be sufficient to protect the public. What is the value of going to an impossible to attain zero emissions from a level very near zero that is currently in the emission standards? The answer is not obvious. If emission reductions require a change to fuel cell systems that cost more to purchase and operate, expect consumer resistance.

The perceptions and priorities given to fuel cells today speak to the current status of energy politics in the United States. While there is potential in fuel cell technology, the general public and politicians responsible for setting energy policy have been subjected to distribution of selective information. The Federal Government has approved financial support to auto manufactures to develop a fuel cell car by this year. This new legislation relaxes the increased miles-per-gallon standards for new cars and continues to exempt trucks from the standards. This strategy assures that petroleum will remain the only source for transportation fuel. Great expectations for fuel cell powered cars in 15 years are used to replace any efforts to introduce new technology to address the critical issue of imported oil.

While the benefits of fuel cell technology have been exaggerated, one thing is certain. Fuel cell technology has potential to bring societal and economic benefits. If fuel cells are improved and grid electricity is used to power automobiles, the impact would be great. Fuel cell research and technology development should be continued.

FLOW BATTERY TECHNOLOGY

If a hydrogen fuel cell is set up so that hydrogen and oxygen feed the fuel cell and the water discharge is collected in a separate tank, one could say

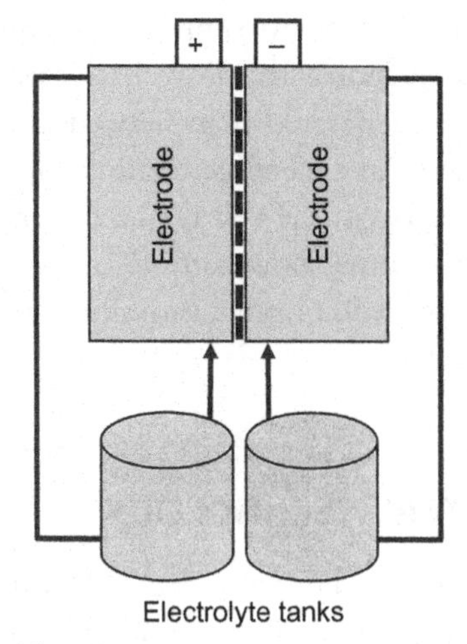

Figure 4.10 Illustration of flow battery.

the tanks are "full" when hydrogen tank is at maximum pressure before any is used and that the tanks are "empty" when the hydrogen pressure is zero. Water discharge tank would then be full.

This fuel cell system can be designed to run in reverse. Electricity is produced and goes in and gradually fills the hydrogen tank by hydrolysis of the water. The process of filling the hydrogen tank is referred to as "charging the system"; while the process of consuming hydrogen is discharging. Because the overall process is much like a rechargeable battery, this system can be referred to as a "flow battery." Figure 4.10 illustrates the critical components of a flow battery and the flow patterns during charge and discharge operations.

A flow battery is really a rechargeable fuel cell system. In traditional batteries, the tanks are not used to hold the fuel (hydrogen) and product (water); rather, both fuel and product are solids stuck to the electrodes of the battery. In a flow battery, the fuels and products are liquids or gases, and the term "flow" is used as part of the battery name to indicate that the fuels and products flow to and from the electrodes.

All it takes to store more energy in a flow battery is larger tanks of fuel and product. To provide more power output, just use a larger electrode.

A disadvantage of the majority of flow batteries relative compared to traditional batteries is that flow batteries are larger and weigh more at the same power and energy levels. For this reason, flow batteries are not considered for use on vehicles. But they are being evaluated to store grid electricity at low demand levels to supplement peak electricity demand. In these stationary applications, the flow batteries are charged at night when more electricity is generated than needed and used during the day when electricity demand can exceed generating capacity.

The regenerative hydrogen fuel cell is one of a number of flow battery technologies. Examples of other flow batteries include bromine—hydrogen, sodium sulfur, vanadium redox, zinc—bromine, and lead acid (methane sulfanate) flow batteries.

CONVECTION BATTERY TECHNOLOGY

Battery technologies have traditionally been identified as falling into one of the three categories; traditional diffusion cell batteries, flow batteries, and hybrid batteries. Diffusion cell batteries store products and reactants in solid phases where the electrolyte of the cathode and anode can mix. Flow batteries store products and reactants in liquid electrolyte phases where mixing of anode and cathode electrolytes is catastrophic. Hybrid batteries store only some of the reactants or products in solid phases where mixing of the electrolytes is catastrophic. A fourth category of batteries is known as convection batteries.

Convection battery technology targets both stationary grid and electric vehicle applications. It is compatible with the same chemistry used in traditional diffusion cell batteries. The liquid electrolyte is pumped from the anode to the cathode. Pumping lowers the energy losses associated with the limits to diffusion mass transfer in traditional batteries. For larger batteries, the added cost of pumping the electrolyte is recovered by eliminating the diffusion mass transfer losses. This motivated the research and development of the convection battery with benefits that exceeded expectations.

One advantage of pumping liquid electrolyte through the battery is ease of heat removal by routing this electrolyte through a heat exchanger. For large high-energy density batteries that are recharged in less than 20 min, the weight penalty of cooling fins and cooling water needed to cool the battery can cut the energy density of the overall battery system in half. For these systems, the heat removal needs to mimic the radiator and water circulation of a typical automotive gasoline engine. The convection battery approximately doubles the energy density for these large high-energy density batteries with fast recharge rates.

A second advantage has to do with details of lithium metal battery chemistry known as "dendrite short circuit." This occurs when lithium metal batteries are put through many charge-discharge cycles. The lithium metal on the anode forms "needles" that pierce the separator between electrodes and causes a short circuit. For this reason, rechargeable lithium batteries are referred to as "lithium-ion" batteries because lithium in the anode is not actually stored as a metal; it is stored as ions in a nanoscale grid that stabilizes the lithium ions in the solid anode. The technical description of the nanoscale grid mechanism is "intercalation chemistry."

These nanoscale grids can, also, cut the energy density of a lithium battery in half with additional cost associated with those grid materials. The electrolyte flow of the convection battery can eliminate the dendrite modes of failure.

The advantages of reducing diffusion mass transfer limitations, providing easy heat transfer, and eliminating dendrite modes of failure lead to major increases in energy density and lower the cost of batteries. Continued successful development has the potential to be breakthrough battery technology.

REFERENCES

[1] Thurston RH. Ithica, NY: Cornell University Press; 1939.
[2] Lave LB, Griffin WM, Maclean H. The ethanol answer to carbon emissions. Issues in science and technology, Winter 2001–02, pp. 73–78, 74.
[3] See <http://www.e-sources.com/fuelcell-intro.htm>.
[4] EPA: Long-term diesel exposure can cause cancer, September 4, 2002 Posted: 8:28 AM EDT (1228 GMT) at <cnn.com>.

CHAPTER 5

The New Electric Vehicle Society

Contents

Today, transportation power is dominated by petroleum-derived fuels: gasoline engines, diesel engines, jet engines, and electric-powered trains/streetcars. Electric cars are gaining on market share where, ironically, electric cars were in use before the gasoline engine was invented. In 1828, Ányos Jedlik, a Hungarian who invented an early type of electric motor, created a small model car powered by his new motor with the first practical version available in England in 1884 [1].

The primary reasons liquid fuels dominate transportation today are that liquid fuels are lighter (for a given range in miles) and that batteries are expensive. However, batteries are becoming cheaper and lighter weight. Electricity will eventually replace liquid fuels.

Sustainable Power Technologies and Infrastructure.
DOI: http://dx.doi.org/10.1016/B978-0-12-803909-0.00005-1

PETROLEUM FUELS: THEIR EVOLUTION, SPECIFICATION, AND PROCESSING

Petroleum, vegetable oils, animal fats, and alcohol-turpentine mixtures were sold for premium fuel prices in the nineteenth century. Two properties made these fuels more valuable than alternatives like wood or coal. They released considerable energy when burned (high heating values) and they were liquid. A reasonably high heating value is needed for any fuel to work in combustion applications, and this higher heating value translates to traveling farther on a tank of fuel. Liquid fuels were much easier to handle when refueling and metering the fuel to the internal combustion engines.

Drilling for petroleum (1859, Colonel Drake at Titusville, PA) [2] became common and liquid fuels so abundant in the United States that the price of petroleum dropped from \$20–0.10 per barrel in 1 year. This abundant liquid fuel was absolutely the best source of energy to power the developing internal combustion engines.

Fuel prices for liquid fuels were reliably low following World War II until the first oil crisis of 1973 crisis where the Organization of Arab Petroleum Exporting Countries (OAPEC, consisting of the Arab members of the OPEC plus Egypt, Syria, and Tunisia) proclaimed an oil embargo. During the first oil crisis, the price of oil went from \$3–12 per barrel. The 1979 oil shock provided an even greater impact on fuel prices. Since the 1970s, a few Middle East countries have been able to control the price of oil by controlling supply. In 2008, China realized a large increase in demand of oil resulting the first demand-driven spike in oil prices with oil prices going over \$100 per barrel.

Since 2008, new oil on the market from horizontal drilling and oil fracking combined with Canadian sand oil and some Fischer–Tropsch production has reduced the ability of a few countries to cause large sustained spikes in oil prices. These and other technologies become highly profitable at oil prices above about \$70 per barrel and sustainable at prices above about \$55 per barrel. These supply-side technologies have brought with them a new era in the oil economy and a nice window of opportunity for other technologies to even further reduce the impact of unreliable foreign oil supplies.

Liquid fuels are used in vehicles because they can be handled with inexpensive pumps, carburetors, and injectors. Interestingly, diesel engines were developed to run on coal dust. These modified diesel engines were expensive to build and the pistons wore out quickly from coal ash that acted like a grinding compound when it mixed with the lubricating oil. The coal-fueled diesel engine simply was not economically competitive.

Gaseous fuels can also be used to power internal combustion engines. Natural gas is commonly used to fuel gas turbines for stationary power plants. For mobile applications, the heating value of a gas fuel per gallon (per unit volume) is considerably less than petroleum fuels. This generally makes gaseous fuels inferior for transportation applications. There are special applications where natural gas and hydrogen can be used in trucks and buses.

In the early nineteenth century, engines were custom-built or modified to operate with locally available liquid fuels. As certain engine designs gained favor, refiners began to provide liquid fuels specifically prepared for the engines. During the military conflicts of World War I, the need to standardize fuel specifications became apparent. During the technology-rich World War II, fuels and engines developed together. The designs of the fuels and engines have been closely coupled since World War II.

Liquid fuels can be broadly classified by application: spark ignition (gasoline) fuels, compression ignition (diesel) fuels, and gas turbine fuels (both stationary and jet engines). Turbines use combustion chambers where the fuel burns continuously. The power output is adjusted by the rate of fuel consumption. This is much easier to achieve than the spark or compression ignition engines where the combustion must start, burn at a uniform rate, and end in a very short time interval (usually milliseconds).

The first petroleum fuels were fractions of the crude oil produced by distillation. The five fractions from the most to least volatile included: (i) liquefied petroleum gas; (ii) naphtha spark-ignition fuel; (iii) light distillate (jet fuel); (iv) middle distillate (diesel fuel); and (v) residual fuel (thick gummy oil) and road tar. There is a sixth fraction that natural gas cannot be liquefied at room temperature. This fraction is typically separated at the crude oil wellhead or in a gas plant located near the oil field.

Figure 5.1 summarizes the natural breakdown of petroleum using the slightly different European nomenclature. These fractions are liquefied petroleum gas (not shown), gasoline, kerosene and gas oil, and fuel oil. By European notation, jet fuel is kerosene and diesel fuel is called gas oil. Figure 5.1 has kerosene and gas oil lumped as one fraction.

During the early history of petroleum use, refineries sold the useful fractions and dumped everything else into any available market. This natural fraction process often did not match the demand for a specific product.

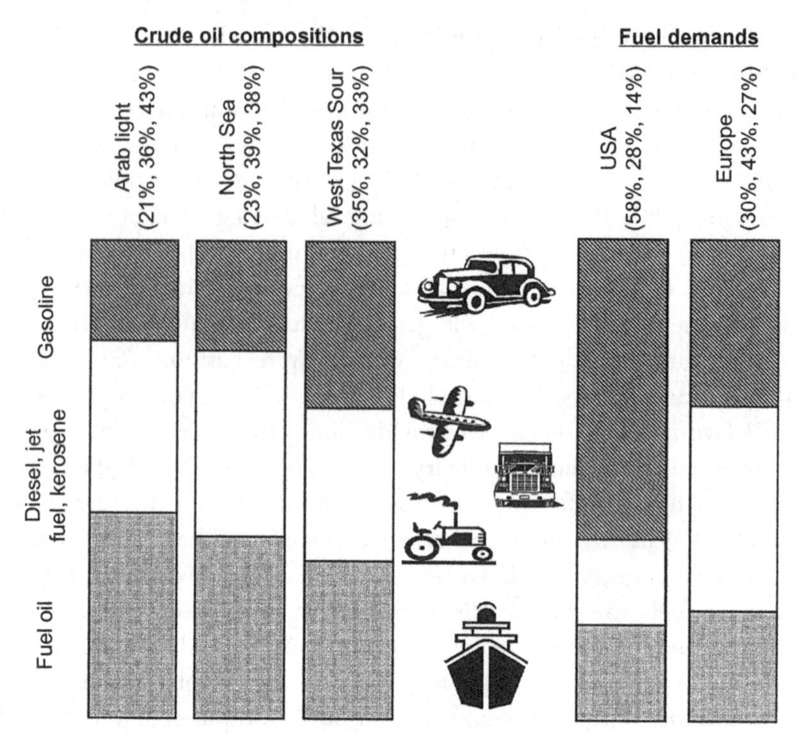

Figure 5.1 Crude oil fractions and market demands.

To paraphrase Owen and Coley: [2] In the years immediately following the exploitation of petroleum resources by drilling (initiated by Colonel Drake at Titusville, Pennsylvania in 1859), kerosene lamp oil was the most valuable petroleum fraction. Surplus gasoline was disposed off by burning, surplus heavy residue was dumped into pits and the "middle distillate" was used to enrich "town gas," which explains why it is often still referred to as gas oil. Only with the invention of the diesel engine was a specific role found for the middle distillate fraction.

Practical spark-ignition engines (and the use of spark-ignition fuels) date back to 1857 with three in service in 1860 [3]. The gasoline fraction was the middle distillate fraction. Dr. Rudolph Diesel developed the compression-ignition engine (the diesel engine and diesel fuel are named after the inventor) that was both more energy efficient than the gasoline engine and the fuel was this middle distillate fraction of petroleum.

Following World War II, there was a demand for a fuel for jet aircraft. Diesel fuels would not ignite easily and they became gummy or froze at the low temperatures of high altitude. Gasoline would boil off (vaporize) at low pressure at high altitude. A narrow-boiling fraction with properties between that of diesel fuel and gasoline was developed to meet these jet fuel requirements.

Today, petroleum is too valuable to dump unused fractions to low-value markets. The modern refinery meets the challenge by converting the available crude oil (crude oil from different oil fields have different compositions and flow properties) to the exact market demands for the wide array of products. They use a complex of chemical reactors to change the chemical composition and distillation columns to separate these into the desired products. Petroleum refining technology research has produced many of the twentieth-century developments in chemical science and engineering.

When entering the twentieth century, petroleum marketers had little concern for the composition of the fuel. The focus was to make a kerosene lamp oil that was volatile enough to ignite easily but not so volatile to burn out of control. By the end of the twentieth century, the modern refinery takes in the crude oil then tears apart and reassembles the molecules so they are in one of the classifications of a desirable product. Engineering at this molecular level makes the desired products; it uses all of the feed petroleum and improves the performance of these new fuels over the old fuels made from the "natural fraction processes."

At the beginning of the twentieth century, an engine was designed to operate on the local fuel supply. Today, the fuels are produced to perform in an engine designed for that fuel. This matching of the engine and the fuel comes from the 100 years of experience building motor vehicles. A description of each of the fuel categories will help us understand how alternative fuels can replace those petroleum fuels.

Spark-Ignition Fuels

Spark-ignition engines use gasoline but can be fueled with propane, ethanol, methanol, or natural gas. Each of these fuels has four important performance characteristics: (i) each fuel has a high-energy release during combustion; (ii) each of the fuels vaporizes (or is a vapor or gas) at about the boiling point of water; (iii) when mixed with air and compressed to about one-sixth the initial volume, the heat produced by compression will not ignite the fuel/air mixture; (iv) a spark ignites the compressed mixture and it burns without detonating (sudden complete combustion or explosion). The fuel is vaporized and mixed with air in the proper ratio before it is drawn into the cylinder of a spark ignition engine.

When the spark ignites the fuel/air mixture in the engine, the heat released increases the pressure in the cylinder and it is this force that turns the crankshaft of the engine. Best engine performance is obtained when the fuel continues to burn as the piston moves down keeping the gas hot as it expands. Combustion is complete at about half-stroke and the piston continues to move down producing work as the gas cools until the piston reaches the end of the power stroke. The burned fuel/air mixture is then expelled through an exhaust valve as the piston moves up and a new charge of fuel/air is drawn in as the piston moves down to complete the cycle. Some of the work produced is used to compress the fuel/air mixture and to turn all of the moving parts in the engine. It is the rotating crankshaft attached to the load that produces the useful work.

The important performance characteristic of this engine is to avoid preignition of the fuel during the compression stroke. The octane number, posted on the fuel pump at the gas station, is the index of resistance to preignition. A high octane number corresponds to a fuel that preignites at a high temperature (produced by a higher compression pressure). Since preignition occurs before the piston reaches the end of the compression stroke, the cylinder pressure increases too soon and it takes more work to complete the compression stroke. Part of the fuel is burned before the spark ignites it for the power stroke. Preignition produces "knock" or "ping" and it increases the mechanical stresses on the engine.

It is essential to match the octane number of the fuel (careful preparation of the fuel), a proper mix of fuel and air, and control the compression ratio of the engine (proper design of the engine). The mechanical design fixes the engine compression ratio.

The power output of the engine is controlled by the quantity of fuel/air mixture that is drawn into the engine. At low power (during idle or cruise), the pressure of the fuel/air mixture is kept low by throttling the air flow. Then, the amount of fuel and the pressure of the mixture in the cylinder are low as the compression begins so the pressure during the power stroke is also lower. This lowers the power produced by the engine. At full power (during acceleration), the engine operates at more revolutions per minute and pulls in more fuel/air. These adjustments are made on modern cars by a computer that monitors all of these functions as you drive in stop-and-go traffic or on the highway.

A typical gasoline engine in an automobile requires an octane number of about 87. This number was selected because it can be obtained by modern refining methods converting about one half of the petroleum into gasoline. Automobile engines are now designed with a compression ratio of about 6.5:1, low enough to avoid preignition using this fuel. During World War II, the gasoline-powered aircrafts were designed with a high compression ratio to increase the power with the low weight, air cooled engines. Tetraethyl lead was added to the gasoline to give an octane number of 100 (adding more lead will increase the octane number to a maximum 115). High compression, high-performance automobile engines were common following World War II until the lead additive was banned to eliminate toxic lead from the engine exhaust gas. The federal requirement to eliminate lead in fuel resulted in the industrial response, lower engine compression ratio that works just fine on 87-octane gasoline.

The American Society for Testing Materials (ASTM) is an independent group that prepares and publishes standardized tests and specifications for a wide range of industrial materials and consumer products. ASTM Standards D2699 and D2700 are used to certify that gasoline has an 87 octane number. These standards are cited and usually required in federal and state laws controlling the quality of gasoline. Customers have assurance that the product will work independent of the supplier. Different suppliers often use the same pipeline to move the fuel from the refinery to local distribution terminals. The difference between gasoline purchased from Texaco and Shell, for example, is the additives they blend into their gasoline before it goes to the filling station.

Compression-Ignition Fuels

A compression-ignition engine is usually powered by diesel fuel and recently by biodiesel. Some desirable performance characteristics of diesel fuel are: (i) a high heat release during combustion; (ii) a volatility that keeps it liquid until the temperature is well above the boiling point of water; (iii) rapid compression ignition (without a spark) when the compression ratio is about 15 to 1 or higher; and (iv) formation of a fine, uniform mist when pumping the fuel through the fuel injectors in each cylinder.

Diesel fuel specifications are almost the opposite of those for gasoline. Gasoline is designed to readily evaporate into air and not to ignite during compression in the engine cylinder. Air is compressed in the diesel engine cylinder before the fuel is injected so there can be no preignition. Diesel fuel evaporates as the fine mist particles from the fuel injectors ignite in the hot, compressed air. The fuel also lubricates the fuel injector pump. The cetane rating of a diesel fuel characterizes the tendency of the fuel to ignite. US standards for diesel fuel require a minimum cetane number of 40. The mechanical difference between a diesel engine and a gasoline engine is that the spark plugs are replaced with fuel injectors.

It is not a good idea to put gasoline into a diesel fuel tank or vice versa. Many gas stations sell both fuels. The nozzle on the gasoline pump is larger than on the diesel fuel pump. The hole below the fuel cap on the diesel fuel tank is smaller than the gasoline fuel nozzle, so you can't fill a diesel tank with gasoline. The diesel fuel nozzle will fill a gasoline tank, so buyers beware.

When designing alternative fuels, the fuel scientist/engineer first translates physical properties such as volatility and ease of ignition into molecular properties such as the size and shape of molecules. Designing a fuel becomes a manageable task since the molecules primarily contain carbon, hydrogen, and oxygen atoms with few exceptions.

Small molecules containing 10 or less carbons are more volatile and make gasoline, spark-ignition fuels. The word octane in the "octane scale" is the chemical name for an eight-carbon molecule that is found in gasoline. It is a good representative molecule for gasoline. Pure isooctane is assigned an octane number of 100 and it was used to establish the empirical octane scale about 1930.

Diesel fuels contain molecules with eight or more carbons and are less volatile than gasoline. They have cetane numbers that characterize good compression-ignition fuels. The word cetane in "cetane scale" is the name for a 16-carbon molecule that is representative of a "good" diesel fuel. Molecules with the carbons arranged in straight chains have high cetane numbers and are better fuels for compression-ignition engines. Molecules where the carbon atoms form rings (benzene or toluene) or branched chains (isooctane is an example) tend to be better spark-ignition fuels.

Refineries today use molecular rearrangement (catalytic reforming) to produce six to eight carbon atoms with branched configurations. This increases the fraction of gasoline produced per barrel of crude oil, and the gasoline has a higher octane number than can be obtained by simple distillation. Diesel fuel specifications are easier to reach by simple refinery processes, so little molecular design is necessary to produce diesel fuel. Diesel fuel is a hodgepodge of different refinery streams sent to a mixing tank and blended to give the right volatility and cetane number to be a "good" diesel fuel.

Turbine Fuels

Gas turbines have burners where combustion takes place continuously, very different than the spark-ignition or compression-ignition engine. Ignition characteristics of this fuel are much less important. The fuel does not lubricate the fuel injection system. This means a wide range of fuels can be used to power a turbine. Turbines require a fuel that burns clean without producing an ash that can erode the turbine blades. There are two properties absolutely required for aircraft jet engines: (i) jet fuel must not freeze at low temperatures that we find at high altitudes, often minus 40°F; (ii) the fuel should have a high heating value per pound of fuel to reduce the ratio of fuel weight fuel to total load on the aircraft.

Diesel fuels can and do power turbine engines, but the high freezing points make them unsuitable for jet aircraft. If the fuel freezes, the jet loses power when the fuel pumps cannot deliver the jelly-like fuel to the engine.

Gasoline fuels would work in a jet engine, but these fuels are too volatile and the vapors mixed with air lead to explosions under certain conditions in the fuel tank. The low atmospheric pressure of high altitude causes this volatile fuel to "boil off" and be lost. TWA flight 800 exploded and crashed near Long Island on July 17, 1996 as a result of fuel volatility problems. The aircraft was filled with the proper fuel, but it was heated above design temperature on this summer day due to a takeoff delay on the runway. The extended time on the runway allowed heat from air conditioners and other equipment on the aircraft to warm the fuel tank. The explosion occurred when a spark, perhaps from a short in a fuel pump, ignited the fuel-air vapor mixture above the liquid in the fuel tank. This is the reason low-volatility fuel is preferred for aircraft. For military aircraft subject to enemy arms fire, low-volatility fuel is certainly important.

Low volatility and low freezing points are conflicting specifications for a fuel. Jet fuel is designed to provide a reasonable compromise between these two important properties.

ALTERNATIVE FUELS

Modern fuels are designed at the molecular level to meet the demands of twenty-first-century engines. There are processes developed to convert available feedstocks that include coal, natural gas, and vegetable matter into modern fuels. Electricity can be used to produce hydrogen that can be used in fuel cells or spark-ignition engines. Hydrogen air burns very fast—not a very good fuel for spark-ignition engines.

Liquid fuels contain almost exclusively carbon, hydrogen, and oxygen atoms. The ratio of hydrogen to carbon atoms should be greater than one since ratios less than one tend to be solids (like coal). All potentially competitive alternative transportation fuels fall within these guidelines with the exception of hydrogen. This means we are limited to feedstocks that contain these atoms to make the alternative fuels. The useful feedstocks include natural gas, coal, municipal solid waste, and biomass. Biomass is another term for materials produced by plants and animals and includes wood, vegetable oils, animal fat, and grass.

Liquid Fuels from Coal

Coal resembles a soft rock that contains mostly carbon and hydrogen, the two components needed to form a liquid fuel. These two components provide the high-heating value to coal. To convert coal to a liquid fuel, the coal molecules with between 20 and 1000 carbon atoms must be broken down and rearranged to form molecules between 4 and 24 carbon atoms. These new molecules must have at least as many hydrogen atoms as carbon atoms. At least half of the energy in the coal should end up in the final liquid product or the process will be too expensive. Such transformations are possible with two approaches that have been used.

Liquefaction provides a low-tech conversion where the larger molecules of coal are torn apart by heating the coal. We can visualize that the atoms in the large coal molecules are held together by rubber bands (actually, these are interatomic forces that bind the atoms to form the coal molecule). As the temperature increases the atoms "jiggle" more and more rapidly and the weakest bands start to break forming smaller pieces (small molecules). As the temperature continues to increase, the small molecules become even smaller. Liquids are formed at low temperature (large molecules) and vapor is formed at high temperature. If some air (or oxygen) is present, some of the vapors burn and supplies energy to increase the temperature. If insufficient oxygen is present, the broken-off molecular segments rearrange into smaller particles that are mostly carbon and form a black smoke. You have probably seen this black smoke from a smoldering wood fire or a diesel engine that was not running properly.

How is coal liquefied? Heating coal in a vessel with little air (no air) forms vapors. The vapor is made up of smaller molecular segments of the coal. As the vapor is cooled, it will condense to form a liquid or paste. The properties of the liquefied coal will vary depending upon: (i) is the coal in large chunks or a fine powder? (ii) how quickly the coal was heated? (iii) how much air (oxygen) is present? (iv) how much water vapor was present? This process is called "pyrolysis" *chemical change caused by heat.*

Many processes have been developed for the different steps of preparing the coal, heating the coal, processing the vapors, and condensing the vapors. Processes have also been developed that improve the quality of the liquid or paste formed by pyrolysis. Pyrolysis is one of the simplest coal conversion processes but it has shortcomings. The quality of the liquid formed is substantially inferior to petroleum crude oil. A major fraction of the energy in the coal remains in the char, the part that does not form a liquid. The char is much like commercial coke and contains the coal ash and a major fraction of the carbon. This char can be used as fuel. Technologies to improve the fraction of coal converted to liquid using pyrolysis are not competitive with alternative processes.

Coal gasification is the complete destruction of the coal molecules followed by chemical recombination of the gas mixture to form liquid fuels. This is a better alternative to pyrolysis liquefaction. This process has been established as a sustainable industry. The coal is fed to the reactor as a fine powder together with steam (to supply hydrogen to the product gas) and some air (oxygen enriched air or pure oxygen) to supply oxygen. The oxygen quickly burns some of the coal to supply the energy to keep the temperature high to break up the coal molecules. At this high temperature, the carbon reacts with the water to form carbon monoxide, carbon dioxide, and hydrogen; high temperatures break apart the coal molecules similar to liquefaction. The fraction that is carbon monoxide increases as the reactor temperature increases. It is carbon monoxide and hydrogen that will be used to make the liquid fuels and this mixture is called *synthesis gas*. Coal passes through these reactors in milliseconds, so the equipment to gasify lots of coal is relatively small.

The two important advantages of gasification over pyrolysis are: (i) essentially all of the coal is consumed during gasification while there is only partial conversion during pyrolysis liquefaction; (ii) the well-defined product composition from the gasification process simplifies the process to obtain the molecules that will become the liquid fuel. In the past, liquefaction products rich in benzene, toluene, and xylene provided potential competitive advantages for coal liquefaction. Recent trends away from these compounds in fuels (they are toxic) have reduced the value of liquefaction processing. Most experts see gasification as the future way to process coal into liquid fuels.

Carbon monoxide and hydrogen are the desired products of gasification. This mixture is referred to as a synthesis gas because it can be used to synthesize (make) a wide variety of chemicals.

Two things are required to make useful chemicals from synthesis gas: (i) lower temperatures (less than 600°C) so the larger molecules are "thermodynamically" favored over the carbon monoxide and hydrogen building blocks; (ii) a catalyst to increase the rate of formation and the fraction that is converted to the desired liquid chemical products. Good chemical process engineering seeks the optimum combination of temperature, pressure, reactor design, and the ratio of hydrogen to carbon monoxide that will efficiently produce the desired liquid fuel product.

A variety of chemicals can be made from synthesis gas including alcohols (e.g.,, methanol and ethanol), ethers, and hydrocarbons. Synthesis gas produced from coal or natural gas is the primary source of commercial methanol which can be used directly as a spark-ignition engine fuel. The octane number of methanol is considerably higher than 93. When it is blended with gasoline it increases the octane of the fuel since methanol has an octane rating of 116.5 [4].

The early 1990s represented a turning point for methanol. The demand was close to the production capacity that depended on a few large natural-gas-to-methanol facilities. When one of these facilities was shut down for maintenance, the price of fuel methanol went from about $0.50–1.50 per gallon [5]. The cost to keep a bus running on $1.50 methanol was then more than twice the cost of using gasoline. The methanol fuel industry still suffers from the poor impression created by price fluctuations combined with a lack of political support for this alternative fuel.

Ethers can be used in compression-ignition engines and were commonly used to assist during "cold start" for diesel engines prior to the 1990s. Diesel engines require substantial modification to use ether fuels since the common ethers must be stored in pressurized tanks to reduce losses due to evaporation.

Hydrocarbons are made commercially using Fischer–Tropsch synthesis. German scientists Frans Fischer (1877–1947) and Hans Trospch (1889–1935) initiated their hydrocarbon synthesis work in 1923. The process starts with synthesis gas and produces hydrocarbons by adding CH_2 groups to make larger molecules. Like building blocks, these groups are added to a molecule until the Fischer–Tropsch fuel mixture is similar to the most useful fuel fractions of petroleum. During World War II, the Germans used this technology to produce quality fuels for aircraft and tanks. There are commercial plants using this process in South Africa today. A commercial plant in Malaysia starts with natural gas to make synthesis gas followed by Fischer–Tropsch synthesis to make liquid fuels.

Many in the petroleum industry consider Fischer–Tropsch synthesis to be the alternative fuel process that has been commercialized without subsidy (this view would consider Canadian oil sand fuels as petroleum of low volatility and therefore not a synthetic fuel). Based on the number of patents filed by major oil corporations related to Fischer–Tropsch synthesis, they see this process to be the best option for replacing petroleum. As long as there is crude oil, energy corporations will not invest in this process.

Natural Gas

Natural gas can be used as a spark-ignition fuel or it can be used to produce synthesis gas. Natural gas as a transportation fuel is gaining momentum for bus fleets and it is used in US postal service vehicles. The excellent natural gas pipeline infrastructure across the United States makes natural gas a candidate to replace gasoline as a transportation fuel. Disadvantages of natural gas as a vehicle fuel include: (i) refilling at room temperature will require pressurized gas storage; (ii) it costs more to store natural gas on vehicles; (iii) the energy content of natural gas per unit volume is less than a liquid fuel so it will be necessary to refill tanks frequently; (iv) there will be concern about safety with pressurized gas on a vehicle, especially when involved in an accident.

Natural gas is the primary feedstock for commercial production of synthesis gas and then converted to a variety of liquid products. Gasification and conversion to liquid fuels is a commercial method to produce "remote" natural gas fields not connected to a gas pipeline system, converting the gas to a liquid fuel.

Natural gas has a pipeline distribution infrastructure across the United States that makes it readily available as an alternative transportation fuel. For fleet owners, such as municipal buses or the US post office, the installation of refueling stations can be justified. The vehicles are more expensive since they must be modified to store and handle high-pressure natural gas, but the combination of EPACT law requirements and a desire to use alternative fuels make natural gas a candidate alternative fuel.

Solid Biomass Utilization

Biomass is a term used to identify any product produced "naturally" in the past few decades (there are annual crops and trees that take decades to mature). Fossil fuels are biomass materials that have been transformed and stored in geological formations over thousands of years. Examples of biomass fuels include wood, grass, corn, and paper. Biomass differs from coal in that it is less dense, has a higher oxygen content, and its molecular structure is recognized as food by animals and microscopic organisms.

Biomass materials can be used as feed for liquefaction and gasification processes to produce pyrolysis oils and synthesis gas. Because microscopic organisms recognize biomass as food, their enzymes can convert biomass to other chemicals. Yeast converting sugars to alcohol is one of the oldest commercial biochemical processes. Microorganisms that produce antibiotics represent another established technology. Beverage and fuel alcohol represents a large volume, low-cost product (if you subtract the beverage tax). Antibiotics are high priced; they are low-volume products commonly using microorganisms to "do the chemistry."

Ethanol is the most common vehicular fuel produced from biomass. It can be used directly in a spark-ignition engine. The ethanol octane rating is about 110. When blended with gasoline, ethanol behaves as if it had an octane rating of 115 [4]. The starchy parts of corn and sorghum are the most common feedstocks for the ethanol process that is similar for beverage alcohol. Enzymes and heat are used to convert starches to sugars. Yeast and bacteria are used to convert the sugars to ethanol. Distillation removes the water from the fuel ethanol and about 2 billion gallons of fuel ethanol were produced during 2001 in the United States.

Ethanol is also produced from petroleum or from non-starchy plant components. The US government promotes the use of ethanol from biomass with a 5.4 cent per gallon subsidy on gasoline blended to contain 10% ethanol. The ethanol increases gasoline octane number and meets the oxygenated fuel requirements imposed in several cities during winter (Methyl Tertiary Butyl Ether) months. The use of ethanol is also growing to replace part of the MTBE additive that has recently been banned. MTBE was used to increase the octane number and to meet the oxygenate requirement prior to the policy to phased it out.

Cellulose is the nonstarch plant material that can also be converted to ethanol. It is the primary component of grass and wood. A cow can digest cellulose, a human cannot. Certain microorganisms can convert cellulose to sugars and then to ethanol. These cellulose feedstocks are less expensive than corn or sorghum. However, the process is more expensive and the technology does not have the support of farmers since they usually have excess production of corn and sorghum. The industry is at crossroads where economics should favor cellulose-to-ethanol production. However, there is little political support to develop cellulose-to-ethanol technology.

Table 5.1 summarizes the prospects of common alternative liquid fuels.

Table 5.1 Comparison of alternative fuels and their ability to displace petroleum

Fuel	Advantage	Disadvantage
Hydrogen	Works efficiently in fuel cells	Hydrogen stores energy and requires an energy source to make it
	Can be produced from electricity using electrolysis	It is a gas and it is difficult to store more than about 50 miles worth of hydrogen on an automobile
Ethanol	Can be produced from renewable biomass without gasification step. Oxygen content makes ethanol cleanburning	Ethanol production is primarily cost effective when produced from grain crops like corn; and the amount of ethanol that can be made from these without causing unreasonable price increases is very limited
Methanol	One of least expensive liquid fuels. Considered most compatible for spark-ignition engines and fuel cells	Unstable prices created poor perceptions with bus fleet managers. (Several managers made attempts to use methanol in the early 1990s.)
	Can be produced from natural gas, coal, municipal solid waste, and biomass	Not a good fit for distribution in current petroleum pipeline network. Transportation adds a few cents per gallon to the cost of methanol
Natural gas	Relatively inexpensive since it can be used with minimum processing	Consumers prefer liquid fuels; however, countries like Pakistan, Iran, and Argentina have shown that it can replace gasoline on a large scale

(*Continued*)

Table 5.1 (Continued)

Fuel	Advantage	Disadvantage
Propane	Alternative with good infrastructure in place and in common use	Supply is directly related to petroleum supply. Prices go up with increased consumption. Not truly an alternative fuel
Biodiesel	High conversion efficiency when produced from vegetable oils and animal fats	Feedstock oils are insufficient to give a noticeable market impact Supplies only impacts diesel fuel
Fischer–Tropsch	Can be produced from natural gas, coal, municipal solid waste, and biomass Low sulfur content meets anticipated regulations on diesel fuels	Large capital cost for large facilities (needed for economy to scale) limit growth; but will likely steadily grow through decades of twenty-first century
Electricity	Can tap into diverse and stable grid infrastructure with many low greenhouse emission alternatives	Not an energy source, only a carrier High cost of batteries and available technology for direct use limit application

Biomass Liquids

Plants produce starches, cellulose, lignin, sugars, and oils that can be converted to fuels. Starches, cellulose, and lignin are complex solid compounds that require conversion to sugars, pyrolysis, or gasification to be converted liquid fuels. Sugars can be processed into ethanol, but this process is not economical because the sugars are more valuable in the food market. The conversion of plant oils and animal fats into vehicular fuels was the most rapidly expanding alternative fuel source at the onset of the twenty-first century. However, the percentage annual increase in production represents an insignificant market share.

Soybean oil has traditionally been the largest oil crop in the world. This is changing as palm oil is now poised to exceed soybean oil production. Animal fats and greases are less expensive than vegetable oils, but the supply is limited. Waste cooking oils and fryer grease do present disposal problems for restaurants and sewage disposal systems. These feedstocks share common chemistry for conversion to a *biodiesel* fuel so fuels from waste fats and grease present a good disposal option.

The large fat and oil molecules (that contain 45–57 carbon atoms) are reacted with ethanol or methanol to form three smaller molecules containing 15–19 carbon atoms. Glycerin is produced as a by-product. The 15–19 carbon molecule liquids are less viscous than the original vegetable oils. They can be pumped and injected just like diesel fuel from petroleum. When blended with diesel these fuels freeze at an acceptable low temperature and they tend to form less deposit on diesel fuel injectors. Additional advantages include reductions in soot and hydrocarbon emissions.

Biodiesel can be mixed from 0% to 100% with petroleum diesel fuel without engine modification. Obviously, this makes it easy to take biodiesel from the laboratory to the commercial market. When vehicles convert to use biodiesel after extended service with petroleum diesel, the fuel filters may temporarily collect sediments that form. Aside from this, the transition to biodiesel is simple.

Biodiesel has three powerful factors working in favor of it: (i) it has political support because it provides a way to reduce excess vegetable oil supplies; (ii) biodiesel blends with petroleum diesel can be used without modifying the diesel engine; (iii) waste vegetable oil to biodiesel can eliminate a waste disposal problem and produce alternative fuels (limited supplies, quality issues, and large collection costs). It has been shown that biodiesel enhances the lubricating factor of the fuel, a nice feature for some applications.

ADSORBED NATURAL GAS AND SHIPPING

Natural gas can be used in spark-ignition engines requiring primarily a fuel injection system or carburetor capable of handling a gas rather than a liquid. The gas is compressed and stored under high pressures of 3000–3600 psi. Countries utilizing the most natural gas vehicles are Iran, Pakastan, Argentina, Brazil, and China. As of 2009, Argentina had 1,807,186 new generation of vehicles (NGVs) with 1851 refueling stations across the nation, or 15% of all vehicles [6].

The relatively widespread use of natural gas including up to 15% of all vehicles in Argentina has demonstrated that natural gas is available as another buffer against future oil shocks. Four problems tend to plague the further expansion of natural gas as a vehicular fuel: (i) its energy density is about one-fourth that of petroleum fuels resulting in both larger tanks and reduced ranges; (ii) many consumers prefer not to refuel using compressed natural gas; (iii) it takes considerable energy to compress gas to pressures greater than 3000 psi; and (iv) natural gas is a fossil fuel of limited supply.

If lower cost batteries became reality for automobiles, electric vehicles would be preferred over natural gas vehicles for three prominent reasons. First, it is more efficient to use natural gas in a highly efficient combined cycle or cogeneration production of electricity. Second, electricity can be produced for a diverse range of energy supplies including many having lower greenhouse gas emissions than natural gas. Finally, consumers are more comfortable with plugging into an electrical outlet than use of compressed gas refueling lines.

An alternative that would enhance both the supply of natural gas and the ease of using natural gas is adsorbed natural gas technology. This technology uses highly porous gas solids to store gas either as an adsorbed layer on the surface of the solid or as liquid-like fluids in the pores of the solids. In preferred, but not necessarily available, approaches the adsorbed natural gas tanks would hold more gas at 500 psi of pressure than compressed gas tanks hold at 3600 psi. This approach substantially reduces compression costs as well. Ultimately, the technology moderates issues associated with the cost of compressing natural gas and the low-energy density of storage.

Today, much natural gas is wastefully burned at the wells where liquid petroleum products are recovered. The absence of pipeline networks for these wells as well as energy contents less than pipeline specifications make this approach the most economical to the producers. Adsorbed natural gas technology is an approach that can both purify the natural gas (eliminating the low energy content issue) and provide for storage without the capital and operating expenses of compression.

Improved adsorbed natural gas transportation and purification is a technology waiting to happen. It will impact utilization by vehicles and large-scale supply/shipping. While electric automobiles are ultimately preferred to natural gas automobiles, effective natural gas technology is perhaps the best alternative to diesel for use in trucks, tractors, and ships.

VEHICULAR FUEL CONSERVATION AND EFFICIENCY

Defining Efficiency

Improving the efficiency of converting fuel energy to work can be as effective for reducing crude oil consumption as developing alternative fuels. It has the added advantage of reducing carbon dioxide emissions. In

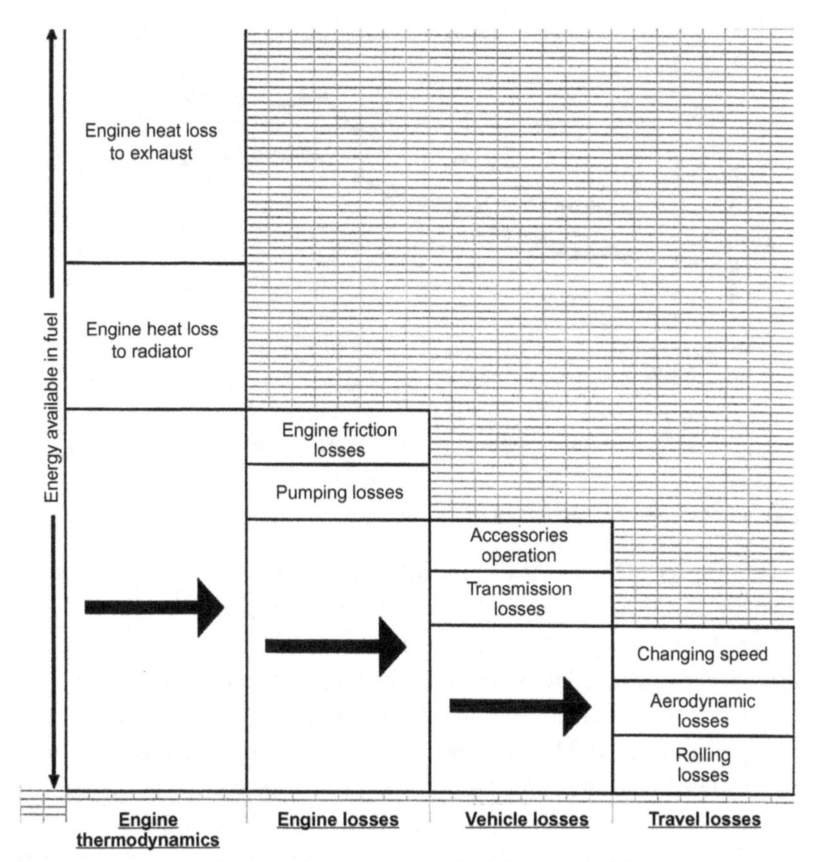

Figure 5.2 Summary of energy losses in use of fuel for automobile travel.

the 1975 Energy Policy Act, the US Congress set Corporate Average Fuel Economy (CAFE) standards in the wake of the 1973 oil crisis. Increased engine and vehicle efficiencies were identified as one of the most effective ways to reduce fuel consumption.

Figure 5.2 summarizes typical losses of energy in the process of converting the chemical energy of fuel into the performance factors of changing vehicle speed, overcoming aerodynamic drag, and highway rolling losses. While the greatest losses occur in the engine as it converts chemical energy into shaft work, improvements in the other factors will reduce fuel consumption. For example, only 25 of every 100 kW of engine power may go to travel losses, but reducing the travel losses by 10 kW would indirectly lead to reductions in all the other energy expenditures except accessories operation—totaling a 25–40 kW reduction in the total engine power demand.

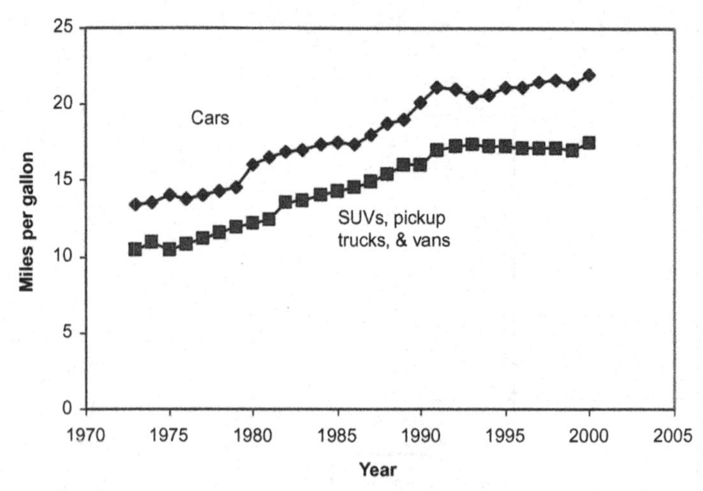

Figure 5.3 Average fuel economy of on-the-road vehicles.

The impact of the CAFE standards on the average fuel consumption is shown by the increase from 13.4 to 22 mile per gallon for passenger cars between 1973 and 2000. These improvements occurred even though additional weight was added for safety-related equipment required on the vehicles. The new car standards written in the legislation increased from 18 to 27.5 mpg between 1973 and 1985 with the 27.5 mpg standard continuing beyond 1985. As illustrated by Figure 5.3, the average fuel economy improved and leveled off shortly after 1985 as the older, less-efficient-fuel vehicles were replaced by vehicles meeting the 27.5 mpg fuel efficiency standard. The difference between the 22 mpg DOT (US Department of Transportation) report and 27.5 mpg legislated standard is partly due to the difference between public driving habits and the test drives used to certify fuel economy. Part of the difference is also due to loopholes in the law, such as getting CAFE credit for producing flexible fuel vehicles (vehicles that can run on gasoline or up to 85% ethanol in gasoline).

The fuel economy improvements of Figure 5.3 were achieved by reducing vehicle weight and improving engines. For example, a 16-valve, 4-cylinder engine wasn't even considered for small and mid-sized cars in 1973. In practice, going to four valves per cylinder reduced the air flow resistance and increased both the engine efficiency and power. In 1995, a 16-valve, 4-cylinder engine could match the performance of the average 6-cylinder engine of 1973 at a fraction of the engine weight. Reducing vehicle weight with a more aerodynamic body reduces all of the vehicle travel losses (speed, aerodynamic, and road losses).

Dependence on imported crude oil in the United States certainly is greater as we enter the twenty-first century than it was prior to the 1973 oil crisis. Any increase in vehicle fuel economy would reduce the dependence on foreign oil and reduce the production required if we moved to an alternative fuel program. Efforts to improve vehicle fuel economy focuses on four items: (i) increased use of diesel engines or new direct-injection gasoline engines; (ii) hybrid vehicles that greatly increase fuel efficiency; (iii) fuel cell replacement for the internal combustion engines; (iv) move to use lower weight vehicles.

Diesel Engines

Diesel engines are more efficient than gasoline engines for two reasons: (i) higher cylinder pressures and corresponding higher temperatures lead to improved thermal efficiency for the diesel engine; (ii) the air throttling required to control the gasoline engine power output reduces engine efficiency and is not used on diesel engines. Truck drivers and farmers are very aware of the reduced fuel costs associated with running a diesel engine as compared to a gasoline engine. The trucks and tractors include an extended range of gear ratios so the diesel engine operates at nearly optimum RPM (Rounds Per Minute) as the load changes.

The thermal efficiency of an engine is computed as the ratio the work engine delivers divided by the thermal energy available from the fuel. Table 5.2 summary lists, by engine type, the average thermal (thermodynamic) efficiencies [7]. These data show an improvement of over 30% in thermal efficiency for the diesel engine over the gasoline engine. In addition, a gallon of diesel fuel contains about 10% more energy than a gallon of gasoline, so the typical increase is about 40% more miles per gallon in favor of the diesel engine.

When the VW Beatle was reintroduced with the standard gasoline engine, it was rated at 24 mpg. With the new diesel engine upgrade, it is

Table 5.2 Typical average thermal efficiencies for diesel and gasoline engines

Engine type	Thermal efficiency
4-Stroke, spark-ignition engine	30%
2-Stroke, spark-ignition engine	22%
4-Stroke, DI compression-ignition engine	40.3%
2-Stroke, DI compression-ignition engine	43%

rated at 42 mpg. This represents an average fuel economy increase of about 75% based on energy available in the fuel.

There are obstacles to the adoption of the diesel engine in automobiles: (i) poor consumer perceptions that originated with the noise and smoke of the diesel engines that were marketed in the early 1980s; (ii) increased costs for diesel engines; (iii) the lack of catalytic converters for the diesel engine exhaust to reduce emissions. Catalytic converters are available for gasoline engines. The new diesel engines are quieter and do not have smoking problems when properly "tuned." It takes time for these perceptions to change. Lower diesel engine costs will occur with higher engine production volumes and there is a sincere desire to reduce the cost of both the engine and the drive train.

Diesel engines always run with excess air in the cylinders, more air than required to burn the fuel. Excess air reduces the hydrocarbon and particulate matter in the exhaust gas, an environmental advantage.

The diesel engine exhaust contains considerable oxygen because of the excess air. This oxygen prevents traditional catalytic converters from working with diesel exhaust gas.

Diesel fuel sold at the end of the twentieth century had high sulfur content that poisons catalytic converters. The new Tier 2 diesel requires refiners to make low-sulfur diesel fuel available. Low-sulfur diesel and a new generation of catalytic converters will then meet the new restrictive emission standards.

The low Tier 2 emission requirements are not required in all areas of the United States and adjustments may be possible for exempt regions. The new generation of small diesels should not cause the emission problems in small cities and rural communities so common in metropolitan areas.

Hybrid Vehicles

Hybrid vehicles use a small internal combustion engine and a bank of batteries to supplement engine power during acceleration. The small engine runs at optimum RPM near maximum efficiency. When the power required is more than the engine output, additional electric current is drawn from the battery pack—the wheels are either partially or totally driven by electric motors. When the vehicle is cruising, excess power that is generated charges the batteries. Hybrid vehicles can avoid idling losses since the engine continues to operate at maximum efficiency and charges the batteries when the vehicle is at a stoplight. When the batteries are fully charged and the vehicle is stopped, the engine turns off. The "on-board" computer monitors the operator's request for acceleration, braking, stopping, and waiting, and it starts and stops the engine "on demand" for power.

The fuel that would be used during engine idling and from the engine operating at nonoptimal RPM is substantially reduced. In addition, some models have electric motors that become generators when brakes are applied, to slow the vehicle, and this recovers much of the energy lost to conventional friction brakes. These features combine to boost fuel efficiency by up to 50% over conventional gasoline-powered vehicles. If a diesel engine is used in a hybrid vehicle, efficiency could be higher.

In some models, like the 2005 Honda Accord hybrid, the hybrid features are used primarily to improve vehicle acceleration. In this case, the hybrid option may have the same fuel economy as other vehicles. However, if a larger engine is required to deliver the same performance, the fuel economy of the hybrid would be better. The use of hybrid features allows a vehicle manufacturer to deliver improved performance without a new engine option for the vehicle.

In 1993, the US Department of Energy (DOE) launched the Partnership for the Next Generation Vehicle Program as described by a DOE Web site:

> The Hybrid Electric Vehicle (HEV) Program officially began in 1993. It was developed as a five-year cost-shared program that was a partnership between the *U.S. Department of Energy* (DOE) and the three largest American auto manufacturers: General Motors, Ford, and DaimlerChrysler. The "Big Three" committed to produce production-feasible HEV propulsion systems by 1998, first generation prototypes by 2000, and market-ready HEVs by 2003.
>
> The overall goal of the program was to develop production HEVs that achieved 2X fuel economy compared to similar gasoline vehicles and had comparable performance, safety, and costs compared to similar gasoline vehicles. As the program progressed, its goals began to merge with the goals of the *Partnership for a New Generation of Vehicles* (PNGV). Now DOE and its partners are striving to develop vehicles that achieve at least 80 miles per gallon. PNGV is a public/private partnership between the U.S. federal government and DaimlerChrysler, Ford, and General Motors that aims to strengthen America's competitiveness by developing technologies for a new generation of vehicles.

The PNGV program was an example of "not getting the job done." While the United States was busy talking about developing and marketing hybrid vehicles, the Japanese actually marketed these vehicles. As the 2003 deadline approached, the DOE and American auto industries essentially abandoned the program.

From about 1995 to 2010, the automobile industry was trying on fuel cells, batteries, natural gas vehicles, and hybrid vehicles. During the subsequent 5 years, hybrid (batteries and engines) and electric vehicles have tended to dominate trends in viable approaches with good upper-end potential for sustainability.

If hybrid technology increases the fuel economy from 30 mpg to 50 mpg, the fuel consumed in 100,000 miles of travel reduces from 3333 to 2000 gallons. The saving is about $2300. This fuel savings is less than the additional cost of the hybrid vehicle [8] (based on net present value). However, optimistically larger volume production of hybrid vehicles and components will bring the annualized operating cost (including vehicle depreciation) to numbers equal or less than those of conventional vehicles. With hybrid technology there is a danger of replacing imported petroleum with imported hybrid vehicle components. An advantage of the technology is the reduced dependence on foreign petroleum.

Plug-In Hybrid

PHEVs [9] are similar to hybrid vehicles with two exceptions. First, the typical 5–7 mile battery pack in a hybrid vehicle would be replaced with a 10–20 (up to 60) mile battery pack in a hybrid vehicle. The additional batteries cost about $1000 for every 10 miles of capacity. Second, the batteries can be charged from grid electricity allowing substantial operation without the engine running.

Whereas the HEV can reduce the vehicle's gasoline consumption up to 33% (50% increased fuel economy), the PHEV can reduce the vehicle's gasoline consumption by more than 80%. The PHEV displaces gasoline with both efficiency and grid electrical power.

PHEVs can displace the majority of the fuel they consume with electricity that is domestically produced and has stable prices. Half of the gasoline consumed in the United States is within the first 20 miles traveled by vehicles each day. For many owners, over 80% of the gasoline could be replaced with grid electricity. Here, a 30 mpg automobile would consume 3333 gallons of gasoline for 100,000 miles of travel. A PHEV running at 80% plug-in and 20% at 50 mpg would consume 400 gallons of gasoline.

An advantage of the PHEV is a reduction in operating costs (not including depreciation). Most of the savings is in fuel costs—while $0.46/liter ($1.75 per gallon) gasoline costs about 2.9 ¢/km (4.6 per mile), 4 cents per kWh electricity costs about 0.9 ¢/km (1.42 per mile). This example is based on a 38 mpg vehicle. The savings are greater if the conventional vehicle gets 25 mpg, the PHEV saves both by increased vehicle mileage (equivalent to 38 mpg) and use of electricity rather than gasoline.

For a PHEV, displacing gasoline with grid electricity leads a greater savings per mile of travel. Chapter 7 discusses this in greater detail along with projections of lower net present costs for PHEVs compared to conventional vehicles. In addition, the money spent on electricity stays in the United States and the hybrid vehicle component production can stay in the United States. Chapter 10 discusses in greater detail strategies for sustainability and possible ways to bring improved prosperity to more countries through energy technology.

This is where nuclear power has a major role to play in future transportation. Both HEV and PHEV options can work with gasoline or diesel engines, so these options have broad potential impact.

By plugging in at night, PHEV technology can use available off-peak electrical power (and the low 4 ¢/kWh cost previously cited). Increase demand for base load power will provide the incentive to build more efficient base load (50% efficient for new generation system power plants) rather than peak demand units (28% efficient) can provide large amounts of transportation energy without any additional expenditure for coal and without additional pollution. Substantial reductions in greenhouse gases can be realized in this approach.

ELECTRIC VEHICLE POTENTIAL

Electric cars, including plug-in hybrid electric cars, are the future of automobiles. The reason is simple; batteries are sustainable in niche car markets today and they have the potential to markedly and continuously improve both in performance and cost. But there is also a sleeping giant in transportation technology, that being direct electric vehicles (vehicles powered directly from grid electricity).

Grid Power Basis

Both battery electric and direct electric vehicles will be powered from grid electricity. History has shown that grid electricity in the United States is reliable and reasonably priced. Furthermore, generation alternatives like wind and nuclear are based on sources that are abundant, can be sustainable, and have the potential for very low greenhouse gas emissions.

Also, natural gas may not have the low greenhouse emissions of nuclear, but it is relatively inexpensive. From an energy-efficiency perspective, it is more efficient to use natural gas in a highly efficient combined cycle and to use that electricity in electric vehicles than it is to burn natural gas in less efficient combined cycles.

The combination of low cost, safe, and low greenhouse gas production is substantially inclusive of the drivers for energy commercialization today and in the near future.

Battery Electric Vehicles

An advantage of battery electric vehicles is that batteries can be charged at off-peak times which enhances options to make grid electricity less costly and more efficient. Namely, lower cost base load electrical power production can be more widely used if batteries are charged in the off-peak times.

Today's vehicular battery markets include: (i) HEVs where the increases in fuel economy justify the cost of today's batteries; (ii) upper-end vehicles such as where Tesla Motors has established profitability for battery electric cars; and (iii) confined space vehicles (e.g., warehouses) where hydrocarbon fuels cannot be used. These markets will continue to expand and the price of batteries decreases.

The two areas where battery technology must improve and is improving are battery costs and battery energy densities.

Figure 5.4 provides battery costs and price trends for applications such as electric vehicles and solar energy. The figure compares cost projections of the Energy Information Agency (EIA) with an average of multiple startup companies. At the start of 2015, the price of high performance batteries were about $600 per kWh. By 2020, these prices are expected to be cut in half to about $300–400 per kWh [10]. While the different groups making projections vary on the $300 per kWh for 2020 (Tesla's Gigafactory) will likely make this reality. This trend is projected to continue to less than $200 per kWh by 2025.

If these batteries realize 1000 full cycles (3 years at 333 cycles per year), the cost is $0.20 per kWh. Ironically, even this low price is about twice the cost many consumers pay for electricity.

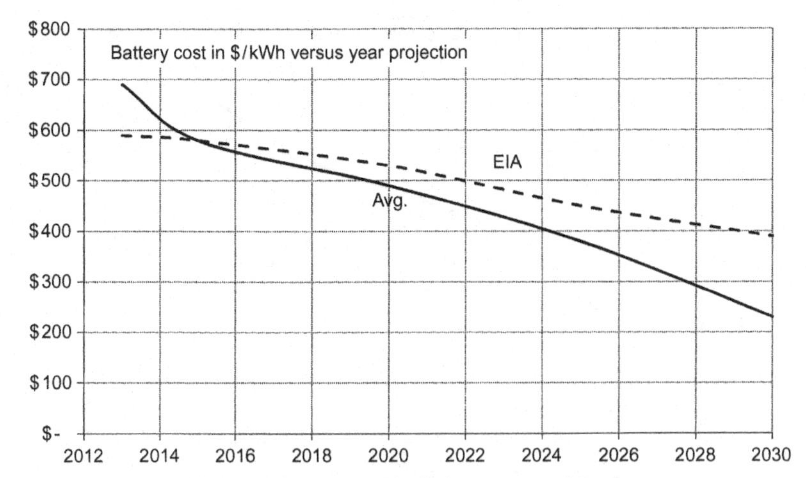

Figure 5.4 Projected prices of electric vehicle batteries in $/kWh.

The Tesla Model S is rated at 290 miles for a 85 kWh battery pack. This translates to 3.4 miles per kWh, and so $20 per kWh translates to a battery pack cost of 5.8 ¢/mile. Adding 4 ¢/mile for electricity costs about 10 ¢/mile. 10 ¢/mile is also the fuel cost for $3.00 per gallon fuel at 30 miles per gallon which does not include engine costs and oil changes.

The bottom line is that it is reasonable to expect electric vehicles to cost less per mile than gasoline-powered vehicles after 2025 and perhaps as early as 2020 if the 1000 cycle per battery is extended to 2000 cycles per battery. It is worth noting that most target specifications for electric car batteries are at least 3000 cycles.

To take advantage of lower cost electricity available at night, the maximum number of cycles per year for a plug-in vehicle is 365. However, a PHEV where the battery pack both uses grid electricity for miles and supplements the combustion engine for efficiency can realize over 500 cycles per year depending on the application. The Chevy VOLT is an example of a PHEV. If the vehicle is able to charge both at home and at work, it would be possible for batteries to achieve 1000 cycles per year.

A factor not accounted for in Figure 5.4 is the emergence of the lithium—sulfur battery. Sulfur is a very cheap material, and while lithium is not expensive, there is uncertainty on prices for large-scale demand for lithium metal. Despite the uncertainty, there is the very real possibility that lithium—sulfur batteries will be available for less than $100 per kWh after 2020; this would be a game-changing technology.

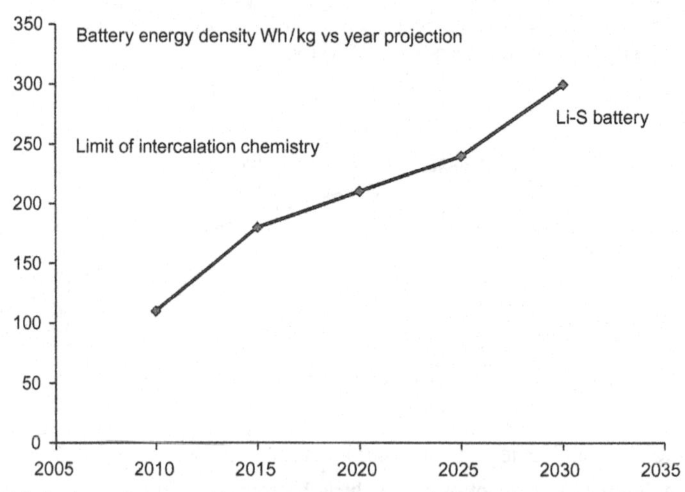

Figure 5.5 Projected energy densities of electric vehicle batteries in Wh/kg.

Reduced weights of battery packs are important for both allowing extended ranges for the vehicles and for increasing the miles/kWh. Figure 5.5 summarizes projected trends in battery weights. This graph identifies specific breakthroughs with anticipated increases in energy densities for the next couple of decades; these breakthroughs impact both the price of batteries and energy density of batteries.

An important milestone summarized by Figure 5.5 is overcoming the limit of intercalation chemistry listed as having an energy density limit of 240 Wh/kg for battery system costs by 2025 (350 Wh/kg for battery cell costs) with new battery systems overcoming this limit in 2030 having energy densities >300 Wh/kg. This convection battery (see Chapter 3) was demonstrated to overcome this barrier in 2012; more than 15 years ahead of anticipation.

Unlike the 1990s and first decade of the twenty-first century, the path is now fairly well identified for battery electric vehicles to replace liquid fuels. This path includes the ability to rapidly charge batteries in timeframes that approach liquid refueling times.

Direct Electric Vehicles

Electric trains, electric street cars, subways, and cable cars are examples of direct electric vehicles. These vehicles can have improved efficiency versus charging of batteries; and because they operate from grid electricity, they offer societal advantages of sustainability, potentially low greenhouse

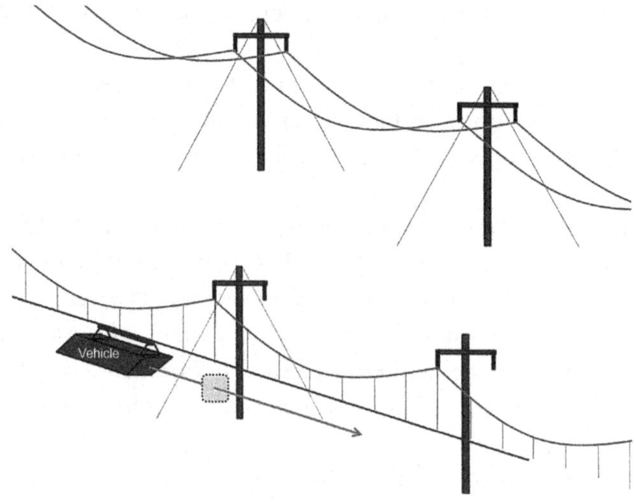

Figure 5.6 Illustration of Terreplane advanced transportation concept illustrating simple and low-cost nature of the propulsion line.

gas emissions, and security due to the diverse range of energy sources that can be used to provide electrical power.

The major problem with each of these is the high cost of the infrastructure to enable the technologies. When this issue of high cost is overcome, the second issue is the inefficiency and reduced capacity of transportation in a system that makes frequent stops.

The lower image of Figure 5.6 illustrates a configuration of the Terreplane Transit that consists of a glider (shown as container) being pulled by a propulsion carriage that runs along a horizontal propulsion line suspended (lower, straight cable) by an upper cable. A most important aspect of Terreplane vehicle is that the vehicle is a pulled flying vehicle that is attached to a propulsion carriage that moves along the lower horizontal line (propulsion line).

At higher velocities, flaps on both the vehicle and propulsion carriage are used to control lift and lateral forces so that the only primary force on the propulsion line is a pulling force. This mode of operation allows for travel along a smooth and straight (or large radius turns) path along a propulsion line that can be fabricated and installed at a small fraction of the costs of highways or rail tracks.

If the vehicle stalls, both the propulsion line may sag to the extent that both the upper cable and the propulsion line support the vehicle. Since the vehicle is not moving at high velocity, the propulsion line does not need to be straight.

When considering both the upper and lower images of Figure 5.6, it is evident that the upper cables could be used to distribute electrical power as transmission lines while the lower propulsion line is used for vehicle transit.

In at least one configuration, the propulsion cable is an upside down T that is about 3 inches wide and 3 inches high with a thickness of about 0.5 inches. Such cables could be run above existing streets and railway lines at a level that does not interfere with the traffic of those systems and minimal intrusion to the aesthetics of the environment. In the country, the ground, vegetation, and wildlife below the system would be largely undisturbed.

Since the combination of the vehicle and propulsion carriage does not have the weight burden of the engine, fuel, and suspension system of automobiles or aircraft, the weight is considerably less relative to comparable vehicles. Lower weight and nonstop service results in an energy consumption per passenger less than half of alternatives for jets, trains, or highway transit.

The system has the potential to be transformative up to and including hypersonic travel in evacuated tubes which has been characterized as the fifth mode of transportation.

REFERENCES

[1] See <http://en.wikipedia.org/wiki/Electric_car#Invention>.
[2] Owen K, Coley T. Automotive fuel reference book. 2nd ed. Warrendale, PA: Society of Automotive Engineers; 1995. p. 21.
[3] Owen K, Coley T. Automotive fuel reference book. 2nd ed. Warrendale, PA: Society of Automotive Engineers; 1995. p. 7.
[4] Owen K, Coley T. Automotive fuel reference book. 2nd ed. Warrendale, PA: Society of Automotive Engineers; 1995. p. 269.
[5] Teague JM, Koyama KK. Methanol supply issues for alternative fuels demonstration programs. SAE Paper 952771.
[6] See <http://en.wikipedia.org/wiki/Natural_gas_vehicle>.
[7] Heywood JB. Internal combustion engine fundamentals. McGraw Hill; 1988. p. 887.
[8] Duvall M. Advanced batteries for electric-drive vehicles. Reprint Report, Version 16. Palo Alto, CA: EPRI; March 25, 2003.
[9] Suppes GJ, Lopes S, Chiu CW. Plug-in fuel cell hybrids as transition technology to hydrogen infrastructure. Int J Hydrogen Energy, January, 2004.
[10] Lacy S. Report: solar paired with storage is a 'Real, Near and Present' threat to utilities. Greentechmedia. February, 2014. see <http://www.greentechmedia.com/articles/read/where-and-when-customers-may-start-leaving-the-grid>.

CHAPTER 6

Energy in Heating, Ventilation, and Air Conditioning

Contents

THE HEATING, VENTILATION, AND AIR CONDITIONING INDUSTRY

The energy used by Americans for heating, ventilation, and air conditioning (HVAC) is second only to the energy used for transportation. HVAC applications consume about the same amount of energy per year as the 130 billion gallons of gasoline consumed annually in the United States for transportation (see Table 6.1). From a greenhouse gas emission perspective, about 8.8% of carbon dioxide emissions are from space heating furnaces compared to 33.9% from electrical power production. A large fraction of the electrical power is used for HVAC.

Consumers use electricity, fuel oil, liquid petroleum gas, kerosene, and natural gas for heating. Table 6.2 shows that natural gas and fuel oil provide most of the space heating needs.

The competition created by alternative technologies and fuel choices provide relatively stable heating/cooling costs, just as with electrical power production. In the HVAC industry, energy savings result from

Sustainable Power Technologies and Infrastructure.
DOI: http://dx.doi.org/10.1016/B978-0-12-803909-0.00006-3

Table 6.1 Summary of the largest energy applications of energy

	Energy in quadrillion BTU
Amount of gasoline consumed[a]	15
Electricity produced[b]	12
Approximate energy expended on producing electricity[c]	34
Energy consumed for HVAC (including hot water heaters)[d]	14

[a]130 billion gallons times 115,000 BTU per gallon.
[b]11.6 rounded up.
[c]Assuming average efficiency of 35%.
[d]Residential [5.2 + 0.55/0.35 + 1.29 + 0.16 + 0.08 + 0.39/0.35] + commercial [1.7 + 0.35/0.35 + 0.16/0.35 0.81/0.5].

Table 6.2 Summary of energy sources for residential heating and air conditioning in quadrillion (1 with 15 zeros) BTUs

Residential air conditioning	0.42
Space heating [1]	
Electrical	0.4
Natural gas	3.6
Fuel oil	0.85
Kerosene	0.06
LPG	0.26
Total	5.2

improved building construction with better insulation and double or triple pane windows. The consumer can choose between investing in energy-efficient buildings with lower yearly energy costs or they can install less expensive HVAC equipment with higher annual energy expenses. Diversity in energy sources and competition among equipment manufacturers has made the HVAC market one of the success stories in American free enterprise, but the story does not end there.

HVAC technology can provide the flexibility needed to achieve additional reductions in greenhouse gas emissions and improve the overall efficiency of electrical power generation networks. The way to achieve these two objectives is best explained by the following examples:

Peak Load Shifting with Hot Water Heaters

In much of Europe where electrical power costs are 2–3 times those in the United States, one of the most common methods to reduce electricity costs is to run the hot water heater at night. Base load electrical power production at night can be sold at a discount to the customer to maintain the higher electrical power production efficiency operating the power plant closer to its design production capacity. The hot water heater must be large enough to supply the daily use, and some inconvenience may result in late afternoon if the system runs out of hot water. This approach benefits the customer with reduced electrical costs and reduces the greenhouse gas emissions because the electricity is produced with the most efficient power plant. Customers easily adjust to any inconvenience.

Use of Heat Pumps Instead of Fossil Fuels

Summary of Table 6.3 compares the energy required to provide 1000 BTUs to heat a building.

Of these options, use of wind energy with a 7-heating season performance factor (HSPF) heat pump would consume the least fuel, but would not be practical on a nationwide basis. However, the second-least energy consuming option is practical on a large scale. Using natural gas and a state-of-the-art technique, combined cycle power plant for electrical power production reduces the fuel consumed for heating by 20–30% compared to a gas furnace. The carbon dioxide emissions can be all but eliminated by using nuclear power rather than natural gas or coal. As heat pumps and power cycles achieve increased efficiency, less fuel is burned to produce the desired heating or electricity; this reduces the greenhouse gases released.

New generations of 50% efficient coal and nuclear power plants would provide an improved option for heating without relying on natural gas.

Table 6.3 Energies required to provide 1000 BTUs of heat to a building

BASIS (heat received)	1000 BTU
Fuel burned in 80% efficiency furnace	1250 BTU
Fuel burned in 90% efficiency furnace	1110 BTU
Electricity consumed by a 7-HSPF heat pump	447 BTU
Fuel consumed to produce 447 BTU at 50% thermal efficiency–combined cycle plant	894 BTU
Fuel consumed to produce 447 BTU at 30% thermal efficiency power plant	1490 BTU
Fuel consumed to produce 447 BTU at 45% new generation nuclear power plant	1000 BTU

Use of Thermal Energy Storage for Peak Load Shifting for Heat Pumps or Air Conditioning

The peak demand for electricity occurs during daylight hours throughout the year when people are awake and active. The peak demand for air conditioning occurs in the afternoon of a summer day, adding to the normal daytime power demand spike. An air conditioner is most efficient when the outside temperature is cool. It would be an advantage to run the air conditioner at night to cool an energy storage unit that would supply cool air during the day. During peak demand periods, peaking electric generator units typically produce electrical power at about 28% thermal efficiency. At night, only base load generators would be used such as combined cycle units operating at greater than 50% efficiency or coal-fired plants running at about 38–45% efficiency. Nuclear power plants operate at 30–33% thermal efficiency but with zero greenhouse gas emissions. The high cost of peak demand power is due to both higher fuel and higher capital costs since the peak demand units are used for only a few days per year.

A heat pump is designed to produce warm air to heat a building during the winter heating season. The coldest part of the day is at night; the time when the difference between the inside and outside temperature is greatest. Heat pumps are most efficient when this temperature difference is small so there is an advantage running the heat pump during the day to heat an energy storage unit to provide heat at night. In this case, the efficiency of the heat pump is best during daytime when the demand for electrical power is highest.

The major components of an air conditioner and a heat pump are the same. Combination units (heat pumps) are commercially available that can be used as air conditioners in summer and switched to be heat pumps in winter. These units work best in temperate climates and this is most of the United States. Development work on an efficient energy storage unit is in progress. This combination would make an energy-efficient system for homes and small commercial buildings.

Potential Impact of Thermal Energy Storage for Peak Load Shifting with Heat Pumps or Air Conditioning

The peak demand of electricity relative to base load electrical power varies during the year. Peak demand in April is typically about one and a half times the base load. In July, the peak demand can become twice the base load. It is possible to make the demand for electricity nearly constant at a higher base-line load level by using a "peak load shifting strategy." For example, a higher base load can justify the higher cost of installing a 50% efficient combined-cycle natural gas power plant operating continuously instead of the 28% gas turbine units used to supply peak power just part of the day.

During the night hours operating at a base power level, more electricity is generated than can be sold. This excess power could be "stored" (storing electricity—electrons—is not easy. It would have to be a huge "battery?"—maybe). During the hot daytime, when the demand for power exceeds the generating capacity, power would be added using the "battery."

For every 1.0 kWh that is shifted from peak demand to base load there is 2.0 kWh that benefits from the improved efficiency; 1 kWh that was shifted and used with the 1 kWh during the peak demand period. The base load has been increased to produce power for the shifted load.

Peak load shifting saves fuel and reduces costs for essentially every application. For those applications where savings are passed to consumers by the local electrical provider, the consumers can realize quick paybacks for investments in energy storage devices (see insert).

Example of Value of Peak Load Shifting

(Data from http://www.xcelenergy.com/EnergyPrices/RatesTariffsMN.asp, Rate Codes A01 and A04)

Peak demand electricity is more expensive and less efficient to generate than base load. The highest peak loads occur during the 4 months of high air conditioner use. The peak loads increase to a maximum at mid-afternoon and then decrease each day during the 4 months of summer. These peak load times represent about 14% of the full year. The generators producing this peak power operate at about 10% of annual capacity. Investment costs to build these facilities are recovered by increasing the rates for peak demand electricity or by increasing the rates for all the power produced. In addition, economics dictates that capital equipment costs be minimized and these generating units are less efficient and often use more expensive fuels.

The Northern States Power Company has programs that provide customers with incentives to reduce peak demand consumption. Their nuclear power infrastructure provides inexpensive base load availability. Two of their residential rate programs are listed below.

Standard Rate Code A01:

June–September 7.35 cents/kWh

October–May 6.35 cents/kWh

This standard rate code provides easy bookkeeping, but it does not reflect the true cost and availability of electricity. Other rate codes are used to create markets for excess winter capacity and give a reduced rate of $0.0519 during the winter for electric space heating. The time of day service option (Code A04, below) reflects the difference in costs for providing peak load versus baseline load and allows the homeowner to adjust and use pattern to reduce costs.

Standard Rate Code A04:

On-Peak June–September 13.95 cents/kWh

On-Peak October–May 11.29 cents/kWh

Off-Peak 3.27 cents/kWh

When the price of electricity reflects the cost of providing peak power versus off-peak power, the peak electricity costs 4 times the off-peak power. The definition of peak demand for this plan is 9:00 AM until 9:00 PM. This plan gives a real incentive to program use of electricity to off-peak hours.

EnergyGuide Labels

(from http://www.eren.doe.gov/buildings/consumer_information/energyguide.html)

The US government established a mandatory compliance program in the 1970s requiring that certain types of new appliances bear a label to help consumers compare the energy efficiency among similar products. In 1980, the Federal Trade Commission's Appliance Labeling Rule became effective and requireed that EnergyGuide labels be placed on all new refrigerators, freezers, water heaters, dishwashers, clothes washers, room air conditioners, heat pumps, furnaces, and boilers. These labels are bright yellow with black lettering identifying energy consumption characteristics of household appliances.

Although the labels will not tell you which appliance is the most efficient, they will tell you the annual energy consumption and operating cost for each appliance so you can compare them yourself.

EnergyGuide labels, such as illustrated by Figure 6.1, show the estimated yearly electricity consumption to operate the product along with a scale for comparison among similar products. The comparison scale shows the least and most energy used by comparable models. An arrow points to the position on that scale for the model the label is attached to. This allows consumers to compare the labeled model with other similar models. The consumption figure is

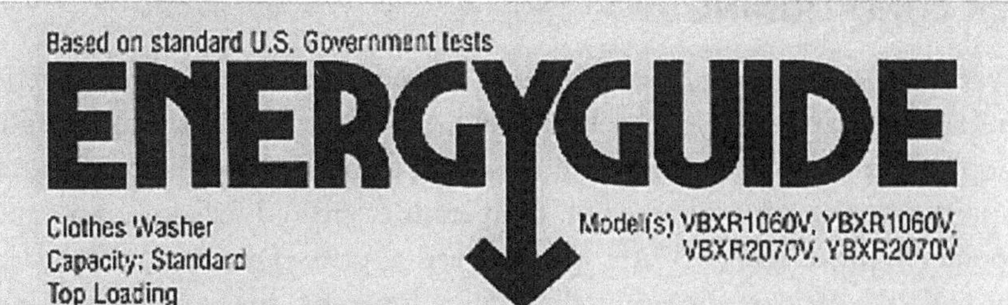

Figure 6.1 Example energy guide for a clothes dryer.

printed on EnergyGuide labels, in kWh, assuming average usage. Your actual energy consumption may vary depending on the appliance usage.

EnergyGuide labels are not required on kitchen ranges, microwave ovens, clothes dryers, on-demand water heaters, portable space heaters, and lights.

AIR CONDITIONING

Ventilation (open windows and fans), evaporative coolers, and effective building design with proper landscaping often provide air conditioning. Cooling and humidity control go hand in hand during much of the air conditioning season. The cool temperature created by an evaporative cooler is produced as water passes over a grid. Dry warm air passes through the grid evaporating the water, lowering the temperature, and cooling and humidifying the air. Modern central air conditioning systems condense water on the cooling coils out of view and it flows to a drain. It can be seen as a puddle of water under a car parked after the air conditioner is used. The energy consumed condensing the water from air can be about the same as the energy used to reduce the air temperature.

Ventilation is the least expensive of all the nonarchitectural options to keep a house cool. Evaporative coolers that cool air by evaporating water into dry, warm air are a good option in hot dry climates. Vapor-compression air conditioners are the most common air conditioners used by homeowners to cool and dehumidify the summer air. They are responsible for most of the energy consumed to cool buildings. These units use the compression and condensation of refrigerants to pump heat (see insert).

Summaries in Table 6.2 show that the energy consumed by air conditioning is about one-tenth the energy used for heating. Air conditioners use less energy because they "pump heat" out of the house. By contrast, furnaces directly convert the chemical energy in fuels to heat or they use electrical resistance heaters to convert electrical energy into heat.

The Seasonal Energy Efficiency Ratio (SEER) is defined as the BTUs of cooling provided per Watt-hour of electricity consumed. The SEER Federal efficiency standards require that heat pumps (air conditioners that provide both heating and cooling) have a SEER rating of at least 10.0 with some units providing SEER values above 14 [2]. Dividing the SEER rating by 3.41 gives the cooling and electrical consumption in the same units. Modern air conditioners remove 3–4 times more heat from the house than electrical energy consumed.

Compare the SEER heat-pumping capacity for air conditioners to the typical amount of heat added to a Midwest house during winter. Heating the house in winter takes three times as much heat as is removed in the summer. This factor of 3 seems right about considering the longer heating season (6 months heating compared to 3 months cooling) and that the outside temperatures will typically vary from about 70°F less (in winter) than the inside temperatures. In summer, the outside temperature is about 30°F higher than inside.

For commercial buildings in the Midwest, about twice as much energy is consumed for heating as for cooling. Commercial buildings are generally larger and have less outside wall area per square foot of floor space. The smaller outside surface areas provide better insulation. As a result, lighting, electrical office equipment, and heat given off by people will have a larger impact when compared to outside weather conditions. The heat from each heat-producing sources must be removed from the building during summer.

How Air Conditioners Work?

Air conditioners use the work produced by electrical motors to "pump heat" out of a house. This can be expressed mathematically as follows:

Three BTU heat pumped from house + 1 BTU of electrical work = 4 BTU heat pumped outside.

This equation shows the conservation of energy in this process and gives air conditioner a SEER rating of (3 BTU ÷ 1 BTU) × 3.41 (BTU per Wh) = 10.23.

As illustrated in Figure 6.2, a vapor compression air conditioner consists of four components: evaporator, compressor, condenser, and an expansion valve. Air inside the house circulates through the evaporator. At about 40°F, the evaporator is cooler than the inside air and cools the inside air from 80°F to about 50°F. For example, 3 BTU of heat is transferred from the house air into the refrigerant circulating in the air conditioner. The heat causes the refrigerant to boil and leave the evaporator as a vapor.

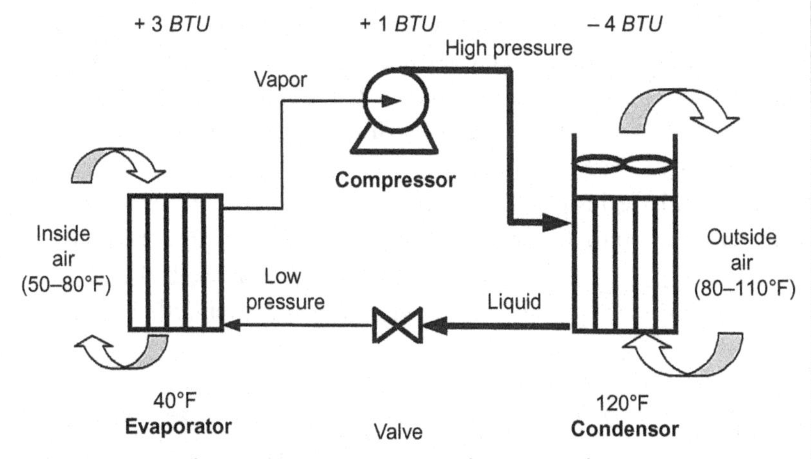

Figure 6.2 Typical vapor compression air conditioning cycle.

With the addition of about 1 BTU of work, the vapor from the evaporator is compressed to a higher temperature and pressure. The high pressure allows warm outside air to be used to condense the refrigerant vapor (high pressures increase boiling points and favor formation of liquids) in the condenser located outside the house. The 4 BTUs of heat is released from the condenser and makes the warm outside air even warmer. Liquid refrigerant leaves the condenser, proceeds through an expansion valve reducing the pressure, and causes some of the refrigerant to evaporate, which cools the liquid to 40°F in the evaporator where the cycle started.

Residential and commercial air conditioning uses 6–7%[1] of the electricity generated in the United States. While this number appears small, consider that this use occurs during the hottest one-third of the year. Summer is the warmest and the most humid in part of the United States in a year and most of the demands is during the daytime rather than night. This 6–7% rapidly increases to 50%[2] or more of the peak electrical power load during the warmest summer afternoons.

The peak load shifting of this air conditioning power demand provides an opportunity to justify more efficient base load electrical power generating capacity and for substantial reductions in greenhouse gas emissions and fossil fuel consumption.

HEATING

Air conditioning arrived with the industrial revolution at the beginning of the twentieth century. The generation of heat is as old as civilization and often considered a trivial process. The combustion of fuels converts chemical energy into heat. The electrical resistance in wires will convert electrical energy to heat.

Furnaces have historically converted wood, coal, and even cow chips (dried manure) into useful heat. The reduced soot from kerosene, fuel oil, natural gas, and liquefied petroleum gas allowed these fuels to dominate the heating fuel market into the mid-twentieth century. The

[1] (0.35 + 0.42)/11.6, see table in appendix.

[2] 6.6% × 3 (12/4 months) × 2 (warmest half of United States) × 2 (warmest half of the day)/0.8 (none-heating and none-cooling load).

historical low cost of natural gas combined with pipeline distribution to commercial and residential buildings makes natural gas the popular furnace fuel in the United States.

In the warmer parts of the United States, heating is often necessary in the winter, but the low annual fuel consumption for heating does not justify the cost of installing natural gas pipelines. For these locations as well as other locations where natural gas is not available, heat pumps are now an available alternative.

Using the same vapor compression cycle as an air conditioner, a heat pump "pumps" heat from outside air into a building. As illustrated in Figure 6.3, the four functional components of an air conditioner can be

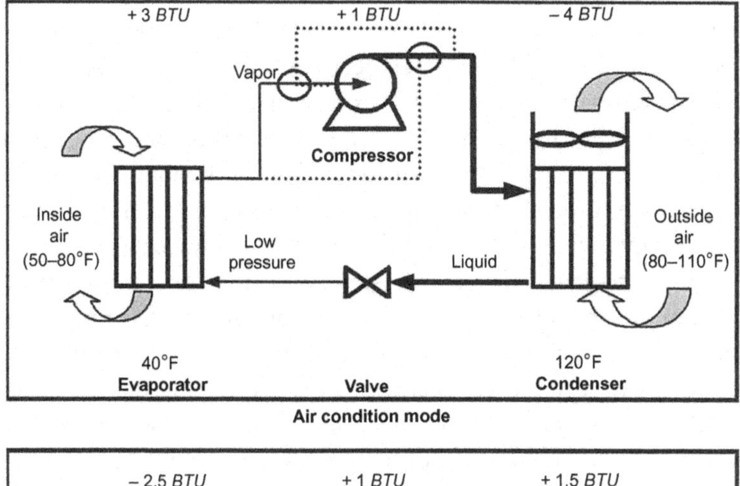

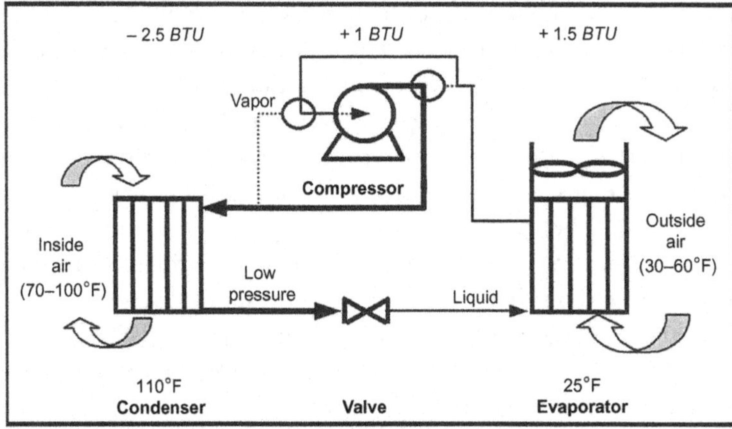

Figure 6.3 Illustration of heat pump showing operation of air conditioning versus heating modes.

configured to pump heat out of a house or pump heat into a house. This is done with the addition of two valves that reverse the flow of refrigerant (reversing the direction of the heat flow from outside to inside the house). These valves can be installed at small incremental cost when the air conditioner is manufactured, providing cooling in summer and heating in winter.

The HSPF rates a heat pump performance based on the BTUs of heat provided per Wh of electricity consumed. An HSPF rating of 6.8 or better is required by federal efficiency standards [3]. The HSPF rating is a function of outside temperatures and decreases rapidly as outside temperatures drop below 30°F. The compressor has to work harder to generate a higher pressure difference necessary to overcome higher temperature differences operating as a heat pump.

While air conditioners typically have to overcome about a 15°F temperature difference (75°F inside temperature vs a 90°F outside temperature), heat pumps often have to overcome temperature differences in excess of 40°F (75°F inside temperature vs a 35°F outside temperature). This explains why the HSPF ratings of heat pumps are lower than SEER ratings of air conditioners. As outside temperatures get lower the HSPF rating also gets lower. At temperatures lower than the freezing point of water (32°F, 0°C) the water on the evaporator coil can freeze and the heat flow slows or stops.

For an incremental increase in cost above that of a conventional air conditioner (a couple valves and minor equipment changes), the heat pump provides a significant performance advantage over electrical resistance heaters that convert electrical energy to heat. By tapping into the electrical power grid, the heat pump also uses the diversity of the electrical power infrastructure including the mix of fuels, using nuclear, wind, or hydroelectric energy. In addition, increased use of electrical power during off-peak seasons (winter) and off-peak times (night) can provide the incentive for building more efficient electrical power facilities. These improvements will become important as fossil fuel reserves are consumed and the reduction of carbon dioxide emissions becomes an important global priority.

At close to 10 billion gallons per year, liquid fuels used for residential space heating (heating oil, kerosene, and liquified petroleum gas) contribute about $10 billion of the $200 billion for annual imported crude oil (and products). These fuels along with liquid fuels used for commercial heating and hot water heaters are part of the problem and must be part of the solution for long-term energy security.

The stationary nature of space heating and hot water heating applications makes electricity particularly attractive for these applications. Heat

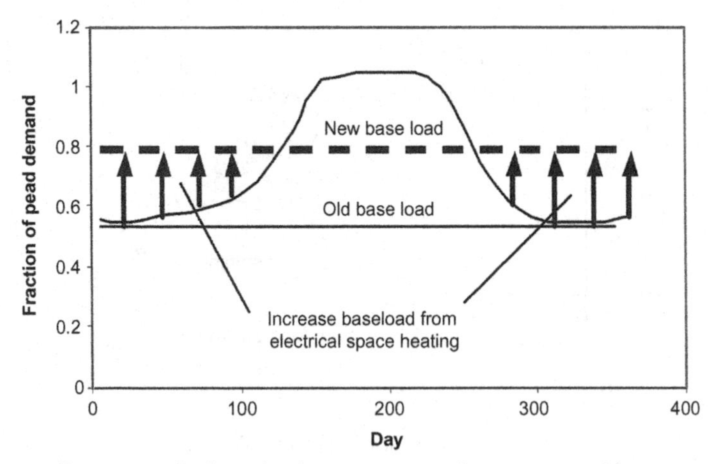

Figure 6.4 Illustration of a base load power in 365 days a year and how space heating can increase base load.

pumps used in combination with resistance heating can provide the heating demands while moving point-source fuel combustion emissions from inside the city to power plants outside the city. If the electrical load created by these applications adds to the base load during the winter, this can lead to benefits. By installing new, efficient electricity generating plants, electricity costs and greenhouse gas emission can be reduced. This will provide efficient power for space heating (during the winter) and replace less-efficient electricity generation (peak load units) during the rest of the year.

For example, creating 100 additional days of base load demand for electric heating could justify a new nuclear power plant or wind turbine farm that would provide base load power those 100 days. This generating capacity is available the other 265 days of the year. Figure 6.4 is a graphical representation of this proposal. This would replace imported liquid heating fuels and potentially replace peaking electrical power units that consume natural gas or petroleum. These changes would generally be cost-effective when all factors are considered. However, the accounting mechanisms may not be in place for local electrical providers to pass the savings on to the consumers who control when energy is used.

PEAK LOAD SHIFTING AND STORING HEAT

Peak load shifting refers to changing use of electricity from the middle of the day to night when the electricity is in low demand. Figure 6.5 illustrates how electrical demand can vary during the 24 hours of a day. This change in

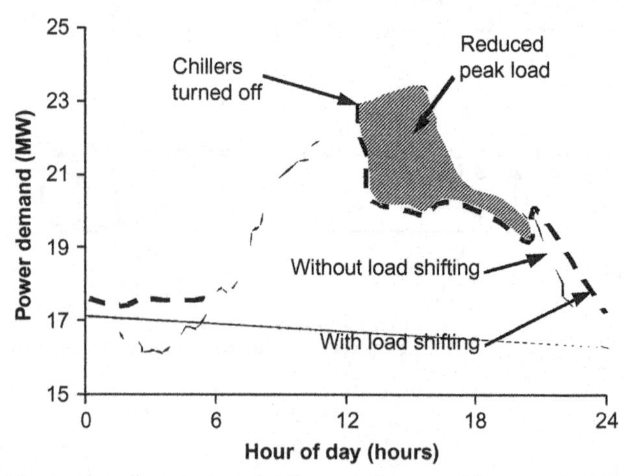

Figure 6.5 Illustration of peak demand from chillers used for air conditioning during 24 hour a day.

demand should not be a surprise since it is during daytime that air conditioners run when the outside temperature is high. It is during the day that the clothes dryer runs and hot water heaters recharge after the morning shower.

On a hot summer day, air conditioners can be responsible for over 50% of the electrical demand. Peak demand power generation is both inefficient and costly. The case study at Fort Jackson illustrates the typical costs associated with peak demand electricity and illustrates how this problem might become an asset.

For large commercial or military installations, electrical power providers offer a number of "rate plans" to pass savings to these consumers as a reward for working with the provider to reduce the cost for supplying electricity. These rate plans are typically based on the principal of reducing peak demand and purchasing predictable amounts of base load electricity. Paying premium prices for all electricity purchased above the base demand is an example of such a rate plan. For the profile of Figure 6.5 at Fort Jackson, electricity consumed beyond 19 MW is electricity at the premium price.

In 1996, the Fort Jackson Army installation paid a $5.3 million electrical bill with 51% of this ($2.7 million) as "demand" electricity. The nondemand portion of the electrical bill is referred to as the energy charge because it is intended to reflect the cost of all electricity consumed at base-load prices. During summer months, it was normal for 50% of the electrical bill to be demand charges. For most Army installations this demand bill exceeds one-third of the total electrical bill.

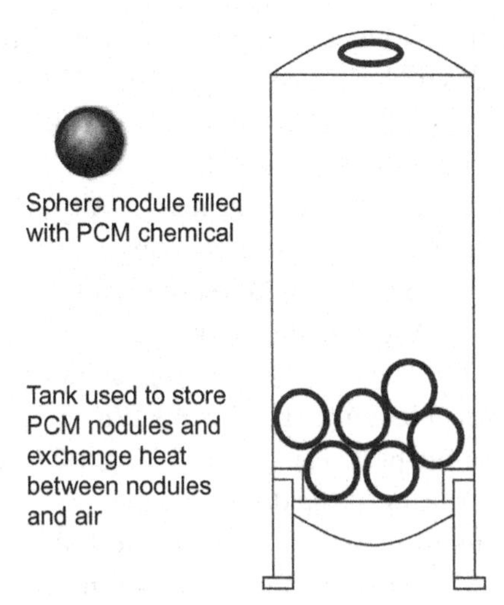

Figure 6.6 Illustration of phase-change material nodules used to store cold during the night for use during the day to shift use of electricity from day to night.

To reduce the cost of demand electricity, a chilled water storage tank was installed at Fort Jackson. During off-peak hours, air conditioners (chillers) ran to cool water to about 42°F and stored it. The chilled water is used to provide cooling during the day. The dashed line in Figure 6.5 illustrates how demand for electricity is typically reduced when the chillers are turned off at mid-day and the stored chilled water is used to provide cooling. An additional advantage of this storage system is it allows the most efficient chiller to be used at full capacity during off-peak hours while minimizing the use of less efficient units. Also, the chiller is operated more during the cooler hours of the day when the chiller operates more efficiently. A disadvantage of chilled water storage is that about 10% of the cooling energy is lost to "heat added" to the tank from the hot summer air and mixing with the circulated chilled water returned to the tank as warm water.

Ice storage is also used for peak load shifting of electricity. Ice takes in considerable energy when "thawing" back to water. Ice storage units are smaller than water storage units. Special materials, such as waxes, eutectic salts, and fat/oil derivatives have been developed to freeze at temperatures close to room temperature. These materials are referred to as phase-change materials. Figure 6.6 illustrates a configuration developed in New Zealand that uses a wax phase-change material.

In this configuration, a phase-change material (PCM) is encapsulated in a spherical nodule about the size of a golf ball. The encapsulation keeps insects and air away from the PCM and prevents the PCM from mixing with circulated air. These spherical nodules are then placed in a tank about twice the size of the building's hot water heater. Air flows through the tank and then through the house to provide cooling. Air is circulated between the air conditioner and nodule tank to freeze the PCM at night.

When used in Los Angeles, the peak load shifting reduced monthly electrical bills from $19,941–14,943. These savings were made possible because of a rate plan charging $0.11 per kWh for peak electricity and $0.061 per kWh for off-peak electricity. Using PCMs, air conditioning use went from 75% during peak demand times to 75% during off-peak times. When the tank storage is located in the building, these units can approach storage and recovery efficiencies of nearly 100%—far better than available methods for storing and recovering electrical energy.

For heating applications, the conversion of fuel to electrical power used during the winter increases base load demand at night. Shifting the electrical demand from peak load during the day to peak load during the night increases the annual base load. Even though the increase in base load may only be for a few months each summer to winter cycle, it can create the incentive to build more efficient power generation facilities by providing a higher annual base load (off-peak load) market.

Chilled water or ice storage systems are generally preferred for larger buildings or groups of buildings while PCM storage tanks are preferred for small buildings. The Ft. Jackson chilled water system was estimated to have a payback period of 5 years. A similar chilled water storage system at the Administration Center in Sonoma County, California, cut the electrical utility bills in half and saved an estimate $8000 per year due to reduced maintenance of the chillers that were sized at half capacity before installing chilled water storage [4]. Just as peak load shifting reduces the maximum peak loads for electrical power providers, peak load shifting can also reduce the peak chiller operation demands allowing use of less expensive units.

The US Department of Energy reports that federal government installations could save $50 million in electricity costs each year. Since ice or chilled water storage systems have been commercially available for over 50 years, several manufacturers and options are available.

For air conditioning, the greatest demands occur during the heat of a summer day when people are most active. The air conditioning load adds to and inflates the daytime peaking of electrical consumption that serves other activities. When heating during the winter, the coldest times are at night during off-peak hours. As a result, except for the storage of solar energy or when using ground source heat, storing thermal energy does not have the same benefits as storing chilled water.

Heat storage can eliminate fuel energy consumption when solar heat is stored. Usually, solar heating systems experience extended periods during the spring and fall when additional heat is not needed during the daylight hours, but heat may be necessary during the cool nighttime hours. For such systems, solar heating systems can be equipped with energy storage. Most solar heating systems do offer energy storage options.

THE ROLE OF ELECTRICAL POWER IN HVAC TO REDUCE GREENHOUSE GAS EMISSIONS

Energy storage and HVAC energy consumption impact strategies to provide cheap, abundant, and environmentally acceptable energy. Well-planned government programs could create incentives necessary to build the next generation of higher efficient electrical power plants and substantially reduce the amount of fossil fuels consumed for heating.

The energy consumed in the United States for HVAC applications is about the same as that burned in gasoline engines; HVAC can play a role in reducing greenhouse gas emissions if this becomes a national priority. The 14 quadrillion BTUs of energy consumed for HVAC probably underestimates the impact HVAC can have on greenhouse gas emission reduction.

The impact of near-zero-greenhouse-gas technologies is increased when electricity demand is stabilized both on the 12-month cycle through increased use of electric-based heating and on the 24-hour cycle by energy storage. Increased annual base load electrical demand should serve as incentive for building the most efficient combined cycle natural gas power production facilities (operating at over 50% thermal efficiency) and can lead to a new generation of more efficient nuclear power plants.

These technologies can be used to reduce greenhouse gas emissions to 1990 levels and provide the transportation industry with time to develop new energy technologies to attain less than 1990 levels. These technologies and conversions are cost-effective when electrical providers and consumers share the cost benefits. When electrical power providers have mechanisms to pass their peak load reduction savings to consumers and the available technology is fully implemented, the transition can occur.

We hear messages from scientists that the dangers of climate change are real; little has been proposed as cost-effective solutions. For electrical power generation and HVAC, the quantity of energy involved is big. The technology is available and the knowledge and understanding to make the transition is known.

EXAMPLE CALCULATIONS

The principle behind converting units is the mathematical axiom that any number multiplied by one is unchanged. Therefore, since the following is known to be true:

$$1000\ \text{g} = 1\ \text{kg} \quad \text{or} \quad 1000\ \text{g} = 1\ \text{kg}$$

Then

$$1 = \frac{1\ \text{kg}}{1000\ \text{g}}$$

An example application of this conversion is as follows:

$$140\ \text{g} = 140\ \text{g} \times \frac{1\ \text{kg}}{1000\ \text{g}}$$

The conversion is complete by recognizing that units, g, cancel.

$$140\ \text{g} \times \frac{1\ \text{kg}}{1000\ \text{g}} = 0.140\ \text{kg}$$

Table 6.4 summarizes commonly used conversions used in energy calculations.

Table 6.5 summarizes commonly used physical properties and abbreviations. The physical properties of gasoline and diesel fuel will vary based on the source of the petroleum, refining practices, and seasonal-specific formulations.

Table 6.4 Conversion factors and abbreviations

Conversions	Abbreviations	
1 barrel = 42 gallons	Atm	atmospheres
1 hectare = 2.47 acres	BTU	British thermal unit
	Cal	calories
1 kg = 0.001 metric tons	Cc	cubic centimeter
= 2.20462 lb_m	Cm	centimeter
= 0.00110 tons	Ft	feet
	G	gram
1 m = 3.2808 ft	Gal	gallon
= 39.37 in	GW	gigawatt
	Hp	horsepower
1 m^3 = 1000 L	In	inch
= 35.315 ft^3	J	joule
= 264.17 gal	Kg	kilogram
	l	liter
1 kJ = 0.9486 BTU	lb_m	pounds mass
= 239.01 cal	M	meter
	MW	megawatt
1 W = 1 J/s	psi	lb/in^2
= 0.001341 hp	S	second
	W	Watt
	M	milli 10^{-3}
	C	centi 10^{-2}
	D	deci 10^{-1}
	K	kilo 10^{+3}
	M	mega 10^{+6}
	G	giga 10^{+9}

Example Calculation

What is the rate of energy delivery to the driveshaft (in kW) of an automobile with a fuel economy of 30 mpg traveling 70 mph?

$$\frac{70\ \text{miles}}{3600\ \text{s}}\ \frac{\text{gal}}{30\ \text{miles}}\ \frac{115{,}000\ \text{BTU}}{\text{gal}}\ \frac{\text{kJ}}{0.9486\ \text{BTU}}\ \frac{0.3\ \text{kJ shaft}}{\text{kJ fuel}}$$

or 23.6 kJ/s. Converting to Watt, this is 23.6 kW. It would take a 23.6 kW electric motor to provide this power. Note that a typical gasoline engine efficiency of 30% was used in the calculation. This neglects frictional losses after the drive shaft.

The technology described in Chapter 6 has not changed much in the past decade.

Table 6.5 Commonly used physical properties and abbreviations
Physical properties and definitions

	Biodiesel	Ethanol	Methanol
Density (lb_m/gal)	7.35	6.63	6.64
Heating value (BTU/gal)	118,200	76,577	57,032
Gasoline (no alcohol in fuel)	115,000 – 119,000 BTU/gal		
Diesel	130,500 – 135,000 BTU/gal		
Corn	56 lb_m/Bushel, 2.5 (2.6) gal ethanol/bushel		
Soybeans	18–20% Soybean oil		
SEER	Seasonal energy efficiency rating		
SEER	BTU cooling/Wh electricity		
SEER/3.41	Wh cooling/Wh electricity		
HSPF	Heating season performance factor		
HSPF	BTU heating/Wh electricity		
Thermal efficiency	Energy delivered/energy consumed		
Gasoline engine (4-stroke)	30% Thermal efficiency, typical		
Gasoline engine (2-stroke)	22% Thermal efficiency, typical		
Diesel engine (4-stroke)	40% Thermal efficiency, typical		

REFERENCES

[1] See <http://www.eia.doe.gov/fueloverview.html>.
[2] <http://www.eere.energy.gov/buildings/appliance_standards/news_detail.html/news_id = 6781>.
[3] Rules and regulations. Fed Reg Tuesday, August 17, 2004;69(158).
[4] See <http://www.eren.doe.gov/cities_counties/thermal.html>.

CHAPTER 7

Electrical Grid Power and Strength in Diversity

Contents

PRODUCTION OF ELECTRICAL POWER

As electrical power is produced, the energy in the fuel ends up as either electrical power or waste heat. In fact, a critical step of all fuel-to-electricity conversion processes is the step where waste heat is released such as the condensers in the steam cycle that drives the generators in an electrical power plant. The thermal efficiency of the electrical power generation is defined as the electrical *energy produced* divided by the *total energy released by the fuel.*

Waste heat is released in the exhaust gases from fuel combustion exits from the power plant stacks. Waste heat is also released from the low-pressure steam in the condensers of steam cycles where the water is condensed so it can be easily pumped to higher pressures and re-used in the boiler and turbines. The steam condenser cooling water heat is released

Sustainable Power Technologies and Infrastructure.
DOI: http://dx.doi.org/10.1016/B978-0-12-803909-0.00007-5

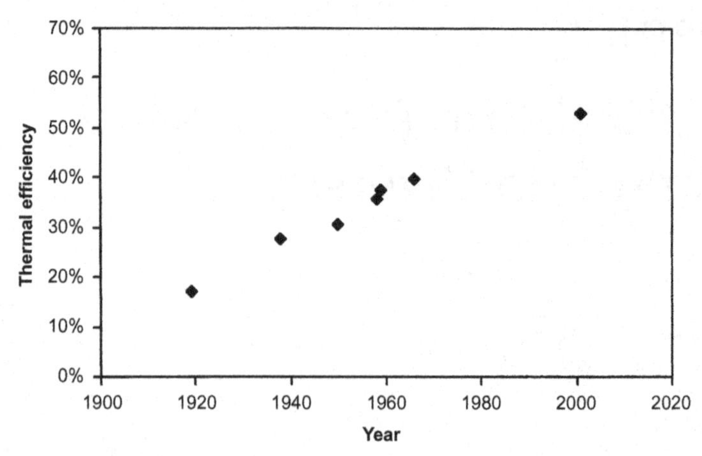

Figure 7.1 Increases in thermal efficiency electrical power generation during past century.

from cooling towers as water vapor. This is the white water-vapor cloud you see coming from all electric power plants.

The first electrical power plants were very inefficient with the thermal efficiency steadily increasing with improved power plant equipment design. Data summarized by Haywood [1] and presented in Figure 7.1 illustrates this improvement in electrical power generation efficiency over the past century. In 2002, the most efficient power plants converted 53% of the energy in natural gas to electricity. The U.S. Department of Energy was soliciting research opportunities [2] to increase the efficiency of these facilities to 60% envisioning that further increases could be achieved through incremental improvements of natural gas combined cycle technology.

Fuel cells convert the energy in chemical bonds directly to electricity. This chemical conversion to electricity could be nearly 100% efficient if the fuel cell device has no internal energy losses. Practical fuel cells today have maximum efficiencies over 50%. More advanced concepts envision natural gas fuel cells [3] using high temperature molten carbonate salts combined with a conventional steam power plant to recover heat that might achieve an overall efficiency of 70%.

Figure 7.1 summarizes the "best efficiency" for any conversion cycle over the past century. Table 7.1 lists efficiencies for different conversion options leading into the 21st century. It is clear that using natural gas for peaking power is only half as efficient as the newer combined cycle

Table 7.1 Typical thermal efficiencies for power generation options

Type of fossil fuel/atomic energy facility	Thermal efficiency
Natural Gas – Peak Power Turbine	25%
Nuclear Power [4,5]	30%–33%
Coal-Steam Cycle	38%–45%
Nuclear—Pebble Bed Modular Reactor [6]	45%
Coal—Gasification Combined Cycle	48%
Natural Gas – Combined Cycle	50%–60%
Future (2020) – Fuel Cell Combined Cycles	70%

option. These efficiencies illustrate which fuels commonly used today could benefit from technology development to improve cycle efficiency.

To make decisions about the best use of energy resources requires knowledge of the efficiency of the conversion of thermal energy to electrical power. For example, if natural gas can be converted to electrical power at efficiencies of 60%, it could be more efficient to produce electricity with natural gas and use that electricity to power automobiles than to use natural gas in an internal combustion engine operating at 30% efficiency.

Other examples exist for commercial and residential heating.

SUSTAINABILITY AND ELECTRICAL POWER

The diversity of energy sources for producing electrical power tends to provide a reliable supply with stable energy prices. For example; nuclear, wind, and biomass sources provide sustainability and near-zero greenhouse gas emissions. It is with the electrical power grid that the combination with nuclear power can provide sustainable energy for the next few centuries.

Figure 7.2 summarizes how energy is used in the U.S. Notice that 38% of the energy consumed in the U.S goes to electric power production. There is more the 28% for transportation, 22% for industry, and 12 % for commercial buildings and residence.

Electricity dominates applications like lighting, air conditioning, and appliances. The primary fuel used for transportation in the U.S. is petroleum refined to form gasoline and diesel fuel.

Industrial applications also use petroleum and natural gas as chemical feed stocks and boiler fuels to provide process heat. Electricity, petroleum

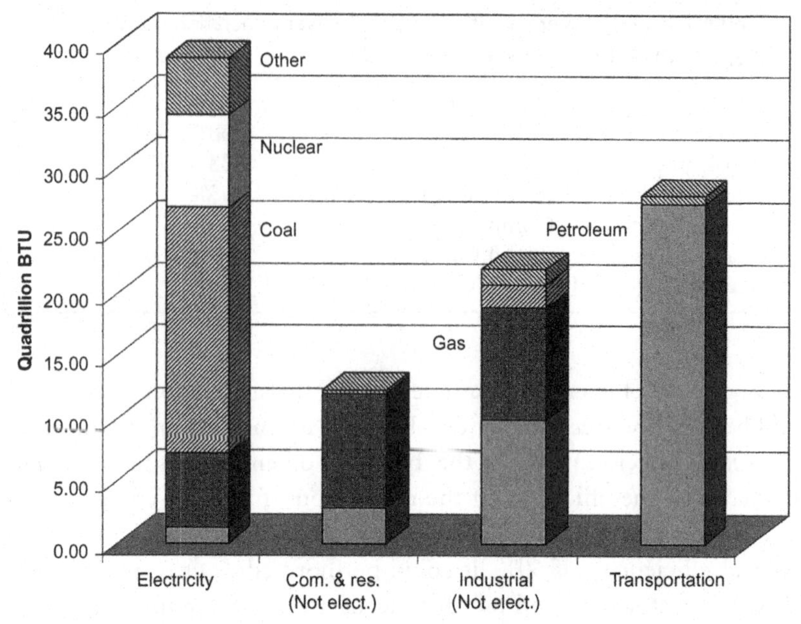

Figure 7.2 Energy consumption in the U.S. Distribution by energy source only includes sources contributing more than 2% of the energy in each category.

and natural gas provide the majority of industrial energy with some coal used in power plants on the industrial plant site.

In the Figure 7.2 presentation, rather than showing electricity in the industrial and commercial/residential energy applications, it is shown separately to demonstrate how electricity can be used to meet energy needs in these three sectors. Most of the non-electrical energy demands of commercial and residential buildings are for space and hot water heating. Natural gas is the most common hydrocarbon fuel delivered for these applications with heating oil providing a small fraction in these energy use sectors.

PEAK LOAD SHIFTING AND GRID STORAGE

The basic infrastructure of grid electricity consists of the power plant generating the electricity, wires for distribution, and transformers to adjust the voltage for individual use by customers. Problems occur when the demand by the consumers is greater than the supply provided by the power plant.

Peak load shifting is a strategy to changing the time of use of electricity from the middle of the day to nighttime when the electricity is in least demand. Figure 7.3 illustrates how electrical demand can vary during the

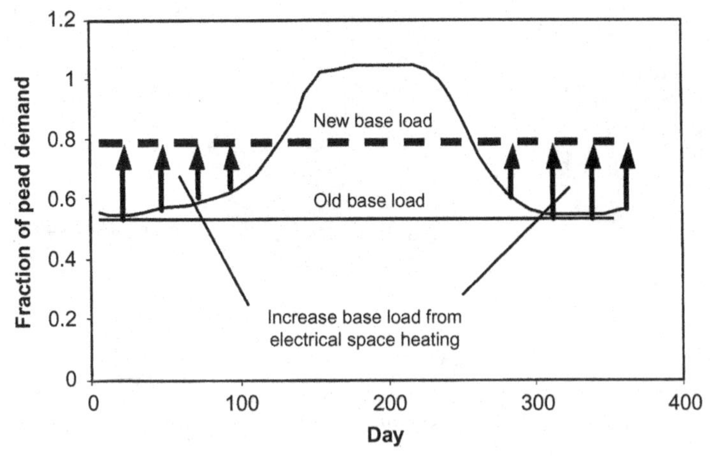

Figure 7.3 Impact of space heating on base load for electrical power generation.

24-hour day. Change in demand with time of day should not be a surprise since it is during daytime that air conditioners run when the outside temperature is high. It is also during the day that clothes dryers run and hot water heaters recharge after the morning shower.

During the heat of a summer mid-day, air conditioners can be responsible for over 50% of the electrical demand [7]. Peak demand power generation is both inefficient and costly. The primary method that electrical power companies use to match supply with demand under these conditions are to: a) charge higher prices to "big" customers during peak times to reduce demand and b) use variable load auxiliary power generation facilities that can be started and quickly assume the excess demand power load. Usually these auxiliary units are the least expensive natural gas units to purchase, but they are less efficient and cost more to operate.

The load shifting strategy can be effective for encouraging customers to use methods to store energy at night for use during the day in winter, or to store cooling capacity during the night for use during the day in summer. This is only a partial solution.

It becomes increasingly difficult to use nuclear power for more than 30% of the annual power needs (currently 19%) because nuclear power plants are not designed to start and stop as part of routine operation. Also, wind and solar power require full back up generating capacity to provide continuous electricity production.

The solution is to store electricity, which is only "technically" possible with large electrical capacitors or batteries that are very expensive.

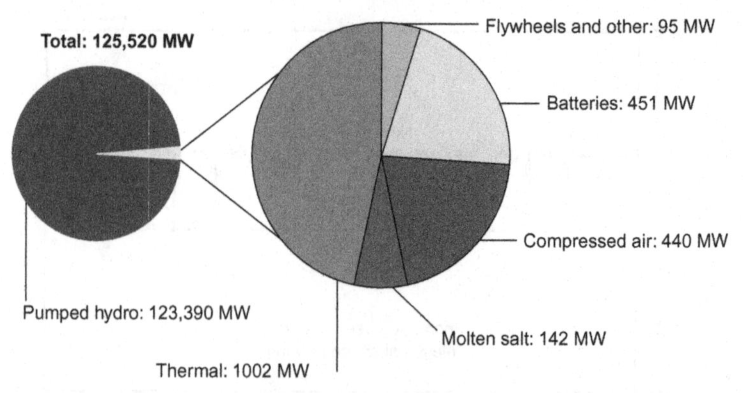

Figure 7.4 Estimated installed capacity for energy storage in global grid in 2011 [8]. *StrateGen Consulting 2011. Note: Estimates include thermal energy storage for cooling only. Figures are current as of April 2010.*

Table 7.2 Estimated installed capacity for energy storage in US grid in 2011 [9]

Storage technology type	Capacity (MW)
Pumped Hydro Power	22,000
Compressed Air	115
Lithium-ion Batteries	54
Flywheels	28
Nickel Cadmium Batteries	26
Sodium Sulfur Batteries	18
Other (Flow Batteries, Lead Acid)	10
Thermal Peak Shaving (Ice Storage)	1,000
Total:	**23,251**

Figure 7.4 summarizes the methods and amount of grid-stored energy globally. Table 7.2 provides a breakdown of the storage in the U.S.

Electric utility companies have generally exhausted available economic options for pumped hydro storage. Of the lesser options, battery storage is the one that is expected to continue to decrease in cost and increase in use.

Tesla Motors' Gigafactory was initially targeting use in electric vehicles; however, that scope has been expanded to include grid energy storage. A favorite outlet for the grid storage applications is likely to be Solarcity. The battery of choice from the Gigafactory is the lithium-ion battery.

In the U.S. technical community the favorite battery technology for grid storage tends to be flow batteries. A new battery option referred to as the convection battery is the favorite of author G. Suppes of this book. Greater technical details on battery options are discussed in Chapter 4.

INCREASED USE OF BATTERY POWER IN VEHICLES

Battery-powered automobiles have been around for over a century, and yet, only recently does this technology appear to be progressing beyond niche markets to replace petroleum. The problem is economics. Even with high-volume production of batteries, the costs are projected to be near $400 for each kWh of stored electric power. This means it costs more than $25,000 to provide 200 miles of range on a compact sedan using currently available batteries. However, new approaches in hybrid electric vehicle (HEV) technology are expected to improve the economics.

Figure 7.5 shows simplified diagrams of the series and parallel approaches to HEVs. A series design routes the engine power through the electric motor by converting the engine's mechanical energy to

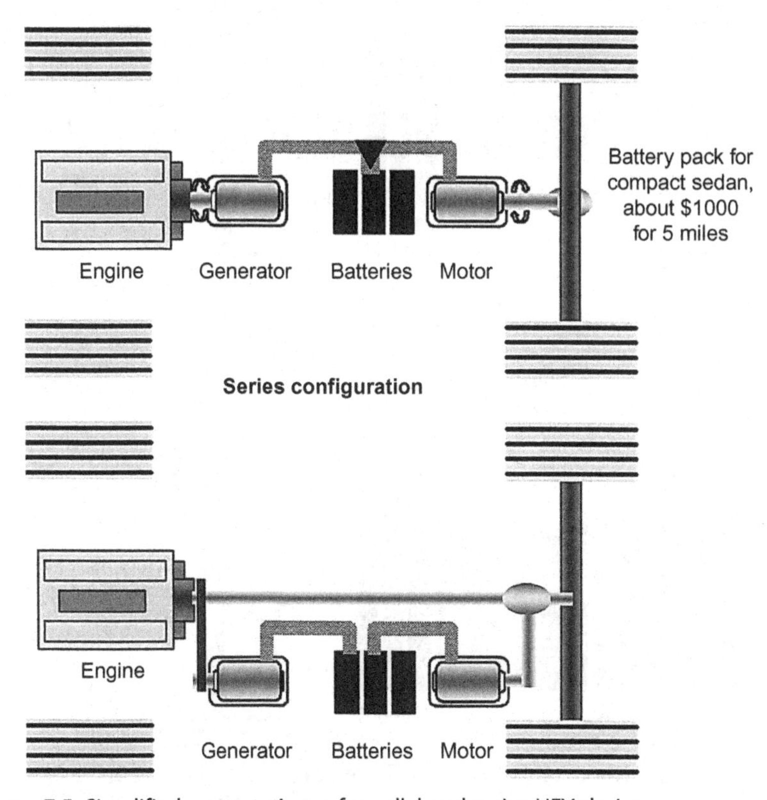

Figure 7.5 Simplified presentations of parallel and series HEV designs.

electricity. The parallel design allows the engine or the battery pack to power the wheels depending on the trip demand.

The basic concept of both designs is to use a battery pack to reduce the fluctuations in power demand placed on the engine. With a level demand on engine power, a smaller engine can be used that operates at near-optimal engine speed and efficiency. When adding to the increase in engine efficiency: a) regenerative braking and b) stopping the engine rather than idling at stop lights, the overall fuel economy can be increased by about 50%. In practice, hybrid vehicle performance today ranges from delivering improved acceleration (Honda Accord) to improving fuel economy in excess of 50% (Toyota Prius).

Figure 7.6 illustrates the addition of the "plug-in" option to the HEV. The PHEV uses a larger battery pack—extending the vehicle's battery

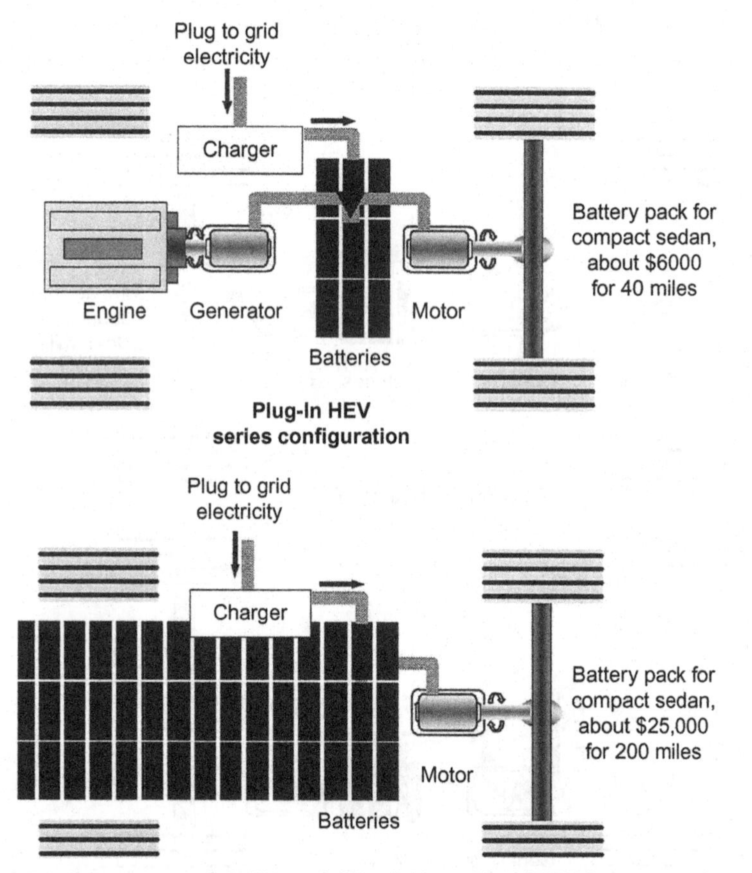

Figure 7.6 Comparison of PHEV and BEV designs. The PHEV has an engine and smaller battery pack. The BEV does not have a backup engine.

range from 3–5 miles to 20–40 or even 60 miles. The addition of a battery charger to this vehicle then allows the PHEV to provide extended travel without engine operation. Using the plug-in approach, the batteries are charged by grid electricity during the night (off peak demand) rather than from the engine generator.

For example, a 40-mile range from the battery pack used 300 days per year provides 12,000 miles per year. The average vehicle on the highway today actually travels about 20 miles per day. Depending on the travel pattern, 20–40 mile PHEVs can replace about 80% of the petroleum used by an automobile.

The widespread use of PHEVs could replace the use of all imported petroleum and reduce total liquid fuel consumption to levels that could be met with biofuels such as ethanol and biodiesel. The PHEV can succeed where the BEV fails because the PHEV matches the size of the battery pack with the application to maximize the use of the battery pack. For vehicles where the commuter only occasionally need more than 40 miles of range, it is more cost effective to provide extended range with a backup engine (about \$2,000 investment) rather than a huge reserve of batteries (e.g., batteries for 160 miles range would cost an additional \$19,000).

Figure 7.7 provides a summary of the life cycle net-present cost for operating a conventional vehicle, an HEV, and a PHEV-20.

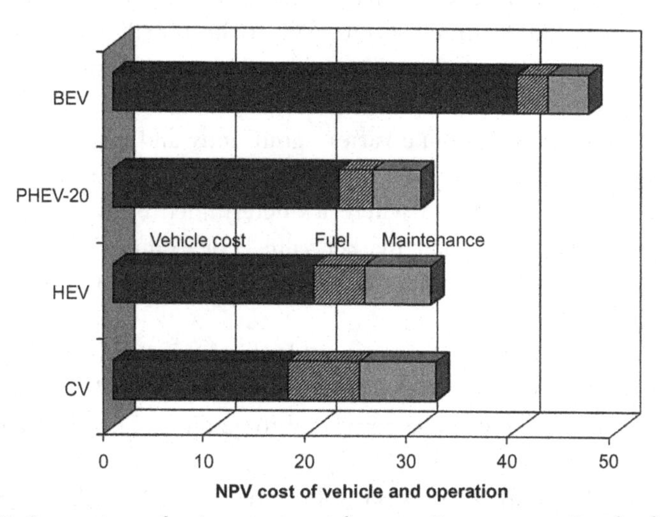

Figure 7.7 Comparison of net present cost for operating a conventional vehicle (CV), hybrid electric vehicle (HEV), plug-in HEV with a 20-mile range (PHEV-20), and a battery electric vehicle (BEV) with 200 mile range. Present values are based on a 7-year life cycle, \$1.75 per gallon gasoline, and 6 ¢/kWh electricity. Data on PHEV-20, HEV, and CV from Frank [10].

A comparison of the NPV costs of the PHEV-20 to BEV indicates that (in the right application) the PHEV can deliver the advantages of the BEV with a lower capital cost. As indicated by these cost summaries, the PHEV-20 can be more cost effective than the BEV, HEV, and conventional vehicle.

In similar analyses, Frank[135] and Duvall [11] (Electrical Power Research Institute, EPRI) qualitatively agree with those of Figure 7.7—the PHEV-20 has a lower net present cost. Results are qualitatively similar for SUVs and sedans.

The primary economic advantage of the PHEV is a lower operating cost due to electricity costs that are less per mile than gasoline costs. The downsides of the NPV cost of the PHEV are:

- the economics are based on 7 years life and many prospective consumers have a hard time seeing past the sticker price on the vehicle,
- the capital costs are higher, and
- the estimated capital costs for the PHEV are based on production of 100,000 units per year far from the handful of prototype vehicles now being produced.

The important aspect of the PHEV is the vision that economies of scale can be attained with sustainable replacement of petroleum. It is this vision and responsibility to the future that forms the basis of sustainability.

As of September 2014 global cumulative sales of plug-in hybrids totaled about 248,000 units, about 41% of the 600,000 plug-in electric cars sold worldwide [12]. This is confirmation of the vision expected by the PHEV proponents.

The recent cost of gasoline varies significantly and makes it difficult to use fuel cost as a parameter when selecting a PHEV. The price of gasoline tracks the price of crude oil where it's determined by the "commodity-trading pit" rather than a fair market value. For example, these prices for the years May 2014 to May 2015 are provided in the top part of Table 7.3.

The price of electricity is more stable, but varies with the location in the USA with representative data summarized in Table 7.3. The changes are a much smaller percentage compared to gasoline. The electricity price data are for the Transportation Sector.

The U.S. Department of Energy has prepared a New Label that provides questions and answers to 14 items related to the selection of Pug-in Hybrid Electric Vehicles [15]. This will help anyone evaluating the PHEV transportation option.

Table 7.3 Variations in gasoline, crude oil, and electricity prices [13,14]

Year	Gasoline ($/Gallon)	Crude oil ($/Barrel)
May 2014	$3.789	$107.20
May 2015	$2.780	$59.69

	Price by State in ¢/kWh		
Year	Connecticut	Missouri	California
March 2014	13.65	4.81	8.01
March 2015	14.32	6.51	8.13

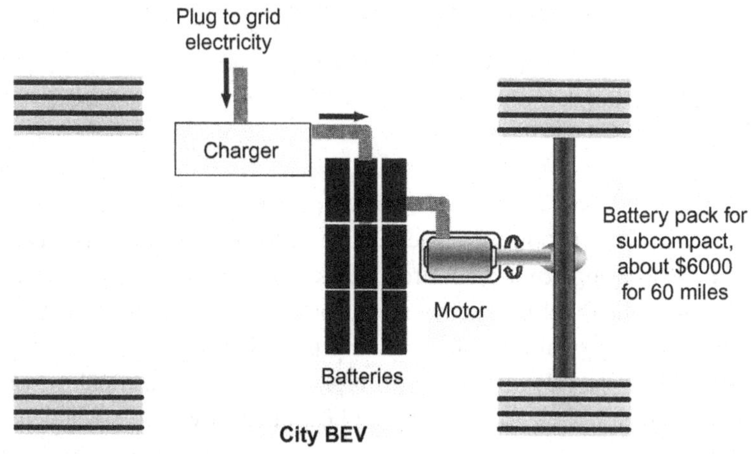

Figure 7.8 Illustration of city BEV. The city BEV is a compact vehicle that has a maximum range of 60 miles. The city BEV is a niche market vehicle that can meet the needs of "some" commuter applications.

Both the California study and the work of Professor A. Frank support the approach using PHEVs to replace petroleum with electrical power, and they document how this can be done while saving consumers money and benefiting local economies. The city BEV (see Figure 7.8) is another vehicle that can meet these objectives. The city BEV would be a low-cost vehicle designed for niche markets such as a second family car used for commuting to/from work. A range of 60 miles between charging the batteries would meet the most commuting demands.

Figure 7.8 also presents the standing of FT diesel and E85. Fischer-Tropsch (FT) diesel fared well in the California study—probably the economic advantages of producing this fuel in California from coal or natural gas. E85 fuel (85% ethanol, 15% gasoline) did not fare well in this

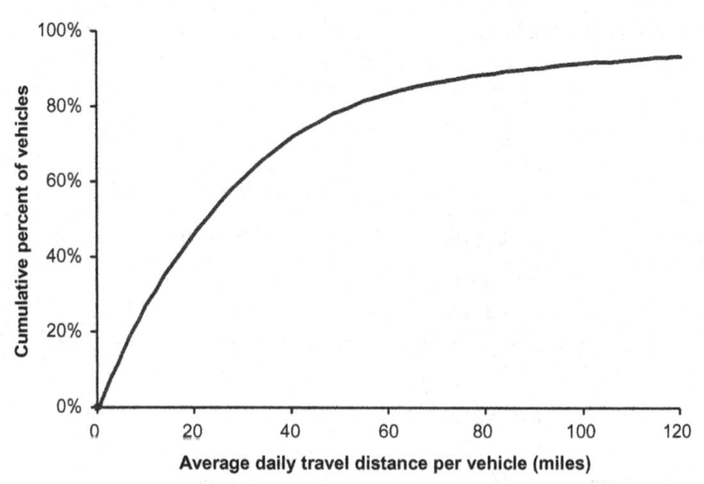

Figure 7.9 Miles traveled with typical automobile each day and implied ability for PHEVs to replace petroleum [16].

study—the production of the ethanol was assumed to be in the Midwest where the grain is produced rather than California. Transporting ethanol must be done using truck or railroad tank cars. There is always moisture in the crude oil or refined petroleum product. Fuel ethanol will pick up the water in the pipeline—contaminating the ethanol.

The PHEV is not intended to replace the conventional gasoline-powered automobile in every application; however, as depicted by Figure 7.9, most automobiles do not travel more than 30 miles in a 24-hour period. The data of Figure 7.9 implies that large sectors of the automobile market can be served with PHEV-20s (PHEV with 20 miles of range) in which the engine will be rarely used. The needs for even larger market segments would be met with PHEV-40 and PHEV-60 vehicles. Especially for two-or-more car families, there are huge potential markets for PHEVs, and these PHEVs will substantially replace the use of petroleum in the selected applications.

Additional PHEV concepts are also being studied. These include the use of fuel cells on PHEVs. Suppes documented how regenerative fuel cells could be used with batteries for energy storage at a lower cost than using either batteries or fuel cells alone. An additional advantage of this approach is that the regenerative fuel cell cost curve is much lower for mass production points than the battery pack cost curve—this means the PHEVs eventually costs less than conventional vehicles. This also points to an evolutionary path to the hydrogen economy that bypasses the need for risky investments in a hydrogen fuel distribution system [17,18]. Unfortunately, PHEVs using regenerative fuel cells are further from "technology ready for market" than PHEVs using batteries.

PHEV technology faces many of the same commercialization barriers as all new technologies. Sustainable alternatives to commercial technology are usually at odds with the momentum of established industry. Any new technology faces technical and non-technical barriers to commercialization. These barriers are at least as great as the sustainability constraints facing any new technology. New sustainable alternatives seldom present overwhelming profitability and commercialization is rarely spontaneous.

The alternative to spontaneous commercial viabilities is economic viability that includes reasonable advances in technology and presents reasonable risk. These paths are available for the transportation sector and that is about *as good as it gets*. Historically, it has taken both a path and leadership (industrial and/or political) for new applications to succeed.

Battery Energy Storage

Batteries are the preferred method of storing energy for applications ranging from watches to automobiles. They are competitive in niche automobile markets today and have the propensity for 75% to 85% reductions in price. Battery costs are depend on:

- production volume,
- type of battery (lead-acid, nickel-metal hydride, other), p
- application details: Weight and recharge time and cycle requirements, and
- how cells are combined to battery packs and systems which includes such things as heat removal.

The California Air Resources Board (ARB) has been a major driver for implementing electric vehicles. The technical assessment of batteries considered by the Battery Technical Advisory Panel (BTAP 2000). Table 7.4 summarizes key characteristics of projected battery technology based on data from these panels' investigations that included two years of surveys and site visits with industry most active in HEV technology [19].

Nickel metal hydride batteries tended to be preferred over lead in HEV applications because of life cycle and weight advantages. Life cycle advantages are especially important if the manufacturer wants to avoid creating an image that newly introduced HEVs are high-maintenance vehicles. However, the lithium iron phosphate (a lithium ion option) has

Table 7.4 Cost and capacity projections for batteries considered for HEV and PHEV applications

Property	Lead acid	Ni-metal hydride	Li ion A	Li ion B
Specific Energy (Wh/kg)	35	65	90	130
Operating Life (cycles)	400–1000	1000–2000	400–1000	800–2000
Cost @ 30,000–300,000 kWh production ($/kWh)	150–200	500–840	1000–1350	1000–1700
Cost @ >300,000 kWh production ($/kWh)	85–115	300–370	270–440	300–500
Status	mature	maturing	R&D	R&D

Table 7.5 Typical sizes of battery packs on EVs

Vehicle	Battery size (kWh)	Battery pack cost	
		$300/kWh	$840/kWh
Mid-size PHEV-20	7.0	$2,100	$5,880
Mid-size PHEV-60	19.5	$5,850	$16,380
BEV 40 city car (micro car)	9.1	$2,730	$7,644
Mid size BEV	27.0	$8,100	$22,680
Mid-size HEV 0	2.9	$870	$2,436

Cost estimates based on entire range of Ni-metal hydride batteries in Table 7.4. Battery pack sizes are averages of values reported by Duvall.[136]

been taking market share in recent years as this technology has improved. By 2020, it is anticipated that rechargeable lithium-sulfur batteries will provide costs lower than $400 per kWh as summarized in Chapter 5.

The size of battery packs varies based on vehicle size (see Table 7.5). Advocates of PHEV see the lower cost battery pack prices of Table 7.5 as an advantage for PHEVs. This has been an assist for PHEVs as they emerge with an automobile market share comparable to engine-powered automobiles. For commuting applications, the PHEVs could replace 80% of the gasoline consumption with a sustainable alternative—grid electricity.

Fuel Cell Technology

As a subject of national attention, the U.S. DOE prepared a Fuel Cell Report to Congress in February of 2003 [20]. The following statement taken from this congressional report is an official rationale for fuel cell technology and targeted performance.

Fuel cell technologies offer unique opportunities for significant reductions in both energy use and emissions for transportation and stationary power applications.

- *Efficiency improvements over conventional technologies that are inherent to fuel cells could lead to considerable energy savings and reduction in greenhouse gas emissions.*
- *The use of hydrogen in fuel cells, produced from diverse, domestic resources, could result in reduced demand for foreign oil in transportation applications.*
- *Widespread use of fuel cell technology could make a significant improvement in air quality in the United States. This would be a result of near zero emission vehicles and clean power generation systems that operate on fossil fuels, and zero emission vehicles and power plants that run on hydrogen.*

For the purposes of this report, the Department did not attempt to quantify benefits of fuel cell commercialization and compare them to the expected public and private sector costs necessary to achieve commercialization.

Significant additional fuel cell research and development (R&D) would need to be conducted to achieve cost reductions and durability improvements for stationary and transportation applications.

Additional barriers to commercialization vary by application and fuel cell type; however, cost and durability are the major challenges facing all fuel cell technologies. (See Table 7.6.)

Table 7.6 Barriers to fuel cell commercialization

Application	Barriers	Difficulty
Transportation	Cost	High
	Durability	High
	Fuel Infrastructure	High
	Hydrogen Storage	High
Stationary-Distributed Generation	Cost	High
	Durability	Medium – High
	Fuel Infrastructure	Low
	Fuel Storage (Renewable Hydrogen)	Medium
Portable	Cost	Medium
	Durability	Medium
	System Miniaturization	High
	Fuels and Fuels Packaging	Medium

For fuel cell vehicles, a hydrogen fuel infrastructure and advances in hydrogen storage technology are required to achieve the promised energy and environmental benefits.

Five types of fuel cells dominate development efforts as illustrated by the five columns of Figure 7.10. Fuel cells are expected to be suitable for a wide range of applications. Transportation applications include vehicle propulsion and on-board auxiliary power generation. Portable applications include consumer electronics, business machinery, and recreational devices. Stationary power applications include stand-alone power plants, distributed generation, cogeneration, back-up power units, and power for remote locations.

There are several fuel cell technologies being pursued. These divide into low temperature and high temperature technologies. The low temperature technology options, include phosphoric acid and polymer electrolyte membrane fuel cells (PAFCs and PEMFCs). These target transportation, portable power, and lower-capacity distributed power applications. The high temperature technologies, including molten carbonate and solid oxide fuel cells (MCFCs and SOFCs). These are for larger stationary power applications, niche stationary and distributed power, and certain mobile applications. A combination of technology developments and market forces will determine which options are successful. Currently, phosphoric acid fuel cells are the only fuel cells commercially available. More than 200 of these "first generation" power units are now operating in stationary power applications in the United States and overseas. Most are the 200-kilowatt PC25 fuel cell manufactured by UTC Fuel Cells.

A cornerstone activity of the FE fuel cell program is the Solid State Energy Conversion Alliance (SECA), a partnership between DOE, the National Laboratories, and industry. The aim of SECA is to develop and demonstrate planar solid oxide fuel cells for distributed generation applications. Performance and cost goals for the SECA Program are shown in Table 7.7.

In addition to performance goals for the solid oxide fuel cells, the Report to Congress reports the Table 7.8 performance goals for PEMFCs.

The industry was not even close to achieving the cost goal of $45/kW in 2010. One of the major problems with the energy research infrastructure in the US is the setting of performance goals that are very inflated. This setting of unrealistic goals is then followed by government funding

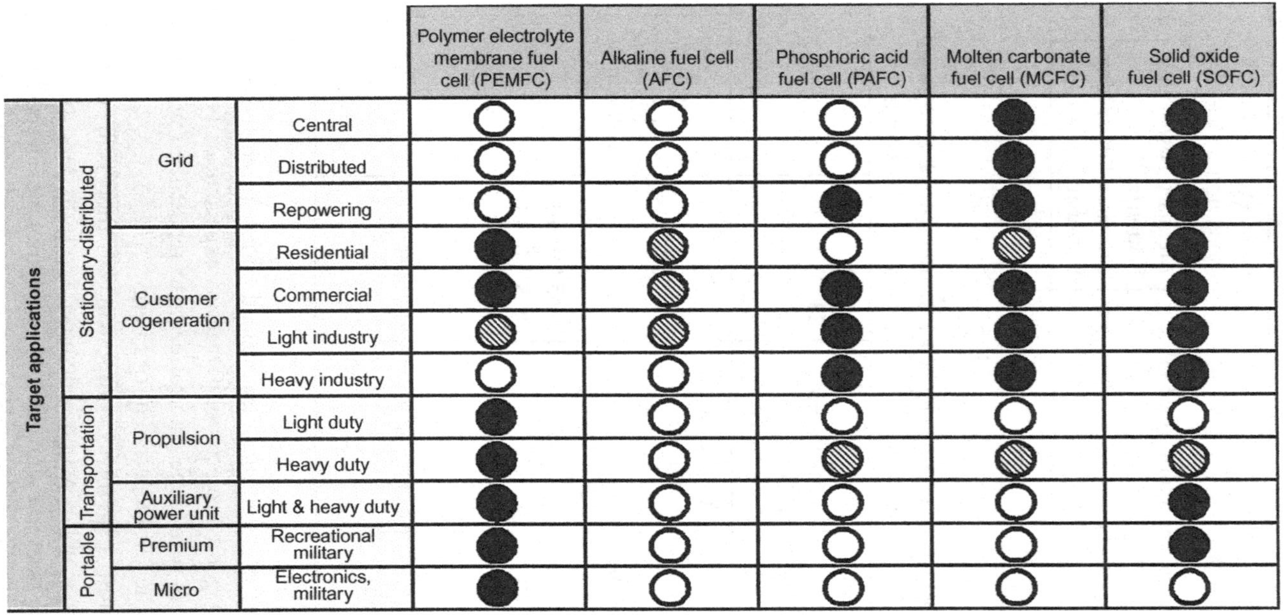

Target applications				Polymer electrolyte membrane fuel cell (PEMFC)	Alkaline fuel cell (AFC)	Phosphoric acid fuel cell (PAFC)	Molten carbonate fuel cell (MCFC)	Solid oxide fuel cell (SOFC)
	Stationary-distributed	Grid	Central	○	○	○	●	●
			Distributed	○	○	○	●	●
			Repowering	○	○	●	●	●
		Customer cogeneration	Residential	●	◍	○	◍	●
			Commercial	●	◍	●	●	●
			Light industry	◍	◍	●	●	●
			Heavy industry	○	○	●	●	●
	Transportation	Propulsion	Light duty	●	○	○	○	○
			Heavy duty	●	○	◍	◍	◍
		Auxiliary power unit	Light & heavy duty	●	○	○	○	●
	Portable	Premium	Recreational military	●	○	○	○	●
		Micro	Electronics, military	●	○	○	○	○

Figure 7.10 Fuel cell technologies and their applications.

Table 7.7 SECA performance and cost goals

	Fuel cell system	
Capital Cost	$400/kW	
Maintenance Cost	3,000 hours	
Electrical Efficiency (Full Load, LHV)	Auxiliary Power Unit	50%
	Stationary	60%
Design Life	Auxiliary Power Unit	5,000 hours
	Stationary	40,000 hours
Emissions	Near Zero	

Table 7.8 Past FreedomCAR performance & cost goals (all 2010 except as noted)

	Efficiency	Power	Energy	Cost
Fuel Cell System	60% (hydrogen) 45% (w/reformer)	325 W/kg 220 W/L		$45/kW $30/kW (2015)
Hydrogen Fuel/ Storage/ Infrastructure	70% well-to-wheel		2 kW-h/kg 1.1 kW-h/L 3.0 kW-h/kg 2.7 kW-h/L	$5/kW-h $2/kw-h $1.50/gal (gas equiv.)
Electric Propulsion		>55 kW 18 s 30 kW cont.		$12/kW peak
Electric Energy Storage		25 kW 18 s	300 W-h	$20/kW
Engine Powertrain System	45% Peak			$30/kW

primarily to those entities that claim they can reach these unrealistic goals with many good projects with realistic expectations being by-passed. To make matters worse, the pattern seems to repeat itself without accountability within the US Department of Energy.

The most expensive component of fuel cells is the membrane electrode assemblies where chemical reactions take place and ions are transported between the electrodes. Membrane electrode assemblies are basically sheets of the membrane material (polymeric, ceramic, or other) coated with the catalyst and designed with a microscopic system of channels to allow flow of fuel, oxygen, and water to/from the catalyst on the membrane surface. The electric current (the power output) is directly proportional to the area of the membrane electrode assembly, and therefore the cost of a fuel cell is proportional to power output rather than total power available. On the other hand, the cost of fuel/hydrogen storage tends to be proportional to total energy stored (more fuel, large tanks, higher cost).

The fuel cell system costs of Table 7.5 are reported in kW and the hydrogen storage is reported in kWh—consistent with fuel cell prices based on power output and fuel storage prices for the total stored energy. This means, fuel cells have performance advantages over batteries when the ratio of maximum required power output is low and the required stored energy is high. For a PEMFC fuel cell stack costing $45/kW and used continuously over an 18-hour period, the power to storage ratio is 18 hours. If a 1 kW power basis is assumed, the fuel cell cost is $45 or $2.5/kWh and the energy storage cost at $5/kWh is also $90. The overall cost of stored energy for this system is $45 + $90 for a total of 18 kWh of power, or $7.5 per kWh.

When fuel cell technology matures and these costs are realized, the cost to store energy for lower power output applications available from fuel cells could be considerably less than for batteries. In this example, a continuous supply of power for 18 hours, the fuel cell system cost is estimated at $7.5/kWh while the bottom of the cost curve for most batteries is about $300/kWh in 2028 (see Figure 5.4). It is possible that battery costs as low as about $85/kWh are possible with lithium-sulfur battery technology; but more importantly, for high energy output applications such automobiles, the $/kW for batteries are considerably lower than fuel cells.

The details are important. The $45/kW PEMFC costs are projected, will require high production volumes, and may be higher for a 1 kW system compared to a 30 kW system. Both battery options provide recharging capabilities—the recharging option on the fuel cell will increase costs. Because each application brings different details for that application, there is a future for both batteries and fuel cells.

New Solar Devices

The greatest opportunities for enhancing residential solar devices is the use of solar panels to replace other structures during new building construction. The Solar Wall [21] is an example of such a device. The solar collecting surface replaces the building siding and structural elements and collects thermal energy.

A solar device that costs $5.00 per square foot to build may cost an additional $1.00 per square foot to install. If this device was added to an existing structure it would cost $6.00 per square foot in addition to the cost of the wall on which it is placed. If during new construction, an 8 by 8 foot section of the Solar Wall were placed on the south side of the building in place of $1.00 per square foot siding and $1.00 per square foot of wall construction, the net cost for the solar addition would be $4.00 per square foot.

The cost of solar devices could approach the cost of the siding and wall structures they replace. This net cost of $4.00 could reduce to less than $2.00 with improved production efficiency. The device would pay for itself in a year or two. Solar walls on the south sides of buildings collect thermal energy and reduce heating costs while solar shingles on the southern exposures of roofs could produce electricity. Ideally, the wall and shingles could be designed to be aesthetically pleasing. Solar shingles are an example of technology that could generate electrical power cheaper than conventional power plant production technologies.

THE NEXT STEP IN TRANSPORTATION

Vehicles running on highways, trains running on tracks, ships, and jet aircraft operating between airports dominate the incredible transportation infrastructures of the world. Ironically, at a time when technologies have made remarkable advances in computers, the internet, and cell phones; the advances in transportation have been incremental. Debatably, transportation advances have been incremental since the end of World War II.

It was the comic strip character Dick Tracy that introduced the two-way wrist radio in 1946. Apple offers such a radio today (with many additional features) that gives/gets its signal from the i-phone in your pocket. This is made possible by the communication satellites first placed in space by the U.S. Military. Are there other options that can help advance sustainable energy and also reduce the time, money, and resources devoted to transportations?

What follows is a presentation of some travel vehicle options. Each requires additional preliminary design consideration, analysis of travel demands, and proof of performance studies. But most importantly, there

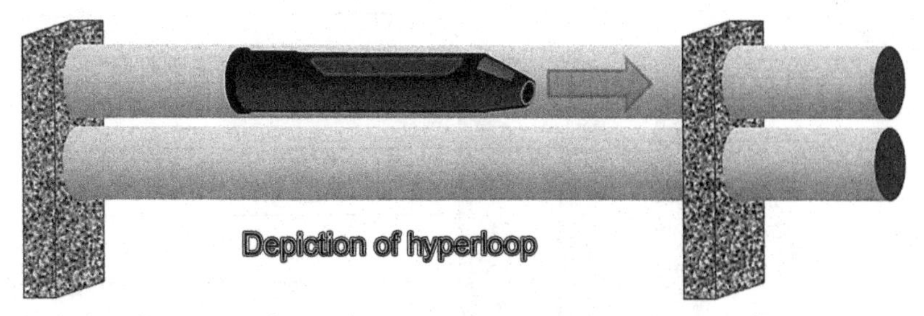

Figure 7.11 Illustration of Hyperloop transportation system.

is the possibility for advances in transportation that could meet demands of improved sustainability, lower cost, and higher performance.

It is possible that mankind can make advances in transportation by 2025 that would be as impacting as the railroads were in the 19th century.

Tesla Motors' founder, Elon Musk, has advocated the HYPERLOOP transportation system that could make a significant advance in transportation both in terms of energy efficiency and reduced transit time. The Hyperloop system as illustrated by Figure 7.11 is based on travel by vehicles in tubes that have reduced air pressures. The reduced air pressures allow higher travel speeds without the high air resistance that occurs for similar velocities at atmospheric pressure.

At the high altitudes where jets aircraft fly, the air pressures are about one fifth (20%) of the air pressure on the earths surface. Overcoming risks and/or safety issues of travel at these low pressures can be done with well-established and cost-effective technology. Supersonic (more than 768 mph at 1 atm) travel over land in tubes would be allowed even though the French Concord aircraft lost permission to provide similar passenger service over land due to the damages sonic booms caused. In evacuated tunnels, the lower the pressure the faster the speed of sound.

As illustrated by Figure 7.12, the energy cost spent per passenger for a typical trip for the Hyperloop system would be less than one tenth of that for airplanes or trains. The energy would be supplied by the electrical power grid which offers advantages of diversity and sustainability.

Two attractive aspects of the Hyperloop effort stand out. First, it has drawn public attention on how to achieve more than we have. Second, the sustained interest for multiple years may indicate that the world is ready to make such an advance.

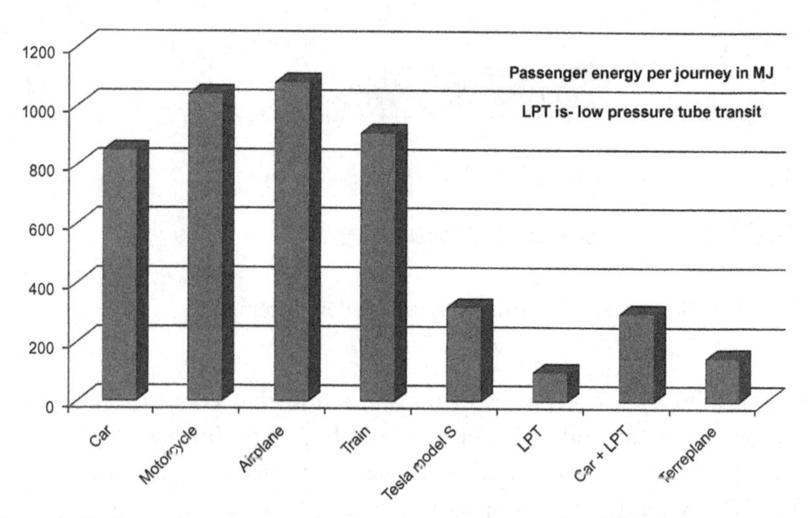

Figure 7.12 Example per passenger energy consumption for different transportation alternatives. LPT is Low Pressure Tube Transit and would apply to both Hyperloop and Terreplane vehicles operating in low-pressure tunnels.

Indeed, if the enthusiasm could coalesce on a global basis, there are more than adequate global resources to advance research, development, and demonstration.

The largest problem with the Hyperloop system is that it costs too much. This was the problem with one of the original evacuated tunnel travel concepts proposed by the Rand Institute in the 1970s and the PRT Maglev system proposed by Suppes in his 1992 patent [22] and Transportation Research Board paper of 1995 [23]. The cost problem is more than the high costs of the evacuated tube network; it is the extended time it takes to build enough infrastructures to "break even" and start paying back the investment. The issue is similar to the combination of high costs and long delays in construction of nuclear power plants. The problem is worse for the Hyperloop system.

What is required is a system that costs less and generates revenue sooner while still offering the upper-end benefits of low energy and fast travel of a mature system. The TERREPLANE Transit System will provide all the advantages of the Hyperloop while offering the additional advantage of an economically viable path to sustainability.

The Terreplane system is illustrated Figure 7.13. Upon first inspection it looks like a "ski lift" gondola that runs on a straight-line cable. However, there are two important differences. First, the horizontal line

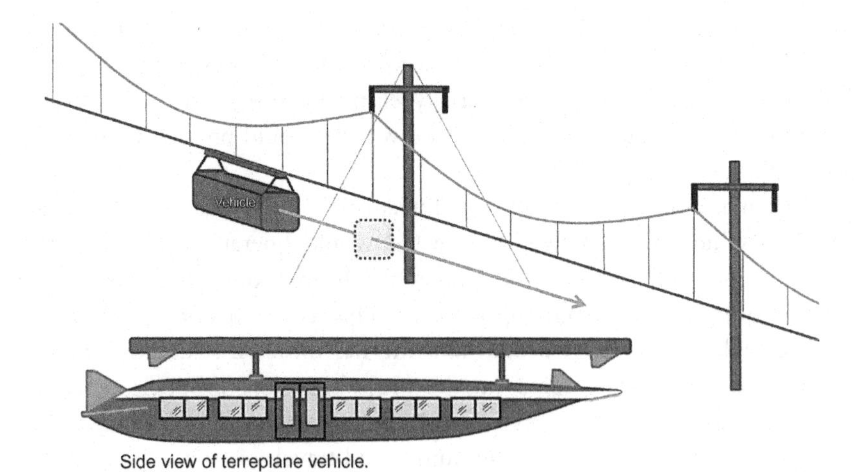

Figure 7.13 Illustration of Terreplane vehicle and nearly-straight propulsion line supported by the upper suspension cable with connecting cables between the two.

Table 7.9 Example take-off velocities of the Terreplane vehicle options

Takeoff (mph)	Wings	Atm	Lift	Vehicle
35	yes	1		Wright Brothers First Flight Aircraft
50	yes	1	1.0	TERREPLANE Personal 2-Seater
63	yes	1	1.6	--Cessna 150
75	no	1	1.0	TERREPLANE Personal 2-Seater
150	no	1	4.0	TERREPLANE Liner
200	no	1	7.1	TERREPLANE Train
300	no	1	16.0	TERREPLANE Tunneler
400	no	1	28.4	TERREPLANE Windtunneler
>500	no	<0.5	20.0	TERREPLANE Vactunneler

is referred to as a propulsion line and is stationary where wheels or linear motors on the propulsion carriage create the movement along this line. Secondly, the normal operating mode of the vehicle is to travel at velocities in excess of 150 miles per hour (mph) where the aerodynamics of the vehicle cause it to fly like a glider as it travels along the propulsion line.

Table 7.9 summarizes minimum velocities required for the Tereplane to achieve flight. For comparison, 35 and 63 mph are lift-off velocity of the Wright Brothers and Cessna 150 airplanes, respectively. Terreplane vehicle options could achieve flight at the lower velocity of 50 mph because it would be an ultra-light vehicle without the weight of an engine or fuel. However, vehicle options would have design flights speeds of 150 mph and greater to essentially eliminate the need for "obvious" wings on the vehicle. The shape of the vehicle would provide most of the needed lift.

When at operating velocity, the Terreplane vehicle and propulsion carriage to which the vehicle is connected would operate in a "controlled flight." The primary force exerted on the propulsion line is the pulling force required to maintain the velocity. That is the major advantages of the Terreplane system: A propulsion line has minimal structural requirements and, the pulling force on the propulsion line produces a natural "straightening" of the propulsion line during travel. Extremely smooth surfaces are required to guide the high-speed travel.

The Terreplane System does not have a traditional "track" like a train or the Hyperloop; it has a propulsion line and no track. Figure 7.13 illustrates a suspension cable and connecting cables that support the weight of the propulsion line and keep the propulsion line relatively horizontal and straight. If the vehicle gets stalled on the propulsion line, the propulsion would sag and both the suspension cable and the propulsion line would support the weight of the vehicle.

Figure 7.14 illustrates how the propulsion carriage is connected to vehicle with a connecting arm. The connecting arm allows the vehicle to move, within limits, relative to the propulsion carriage and thereby allows

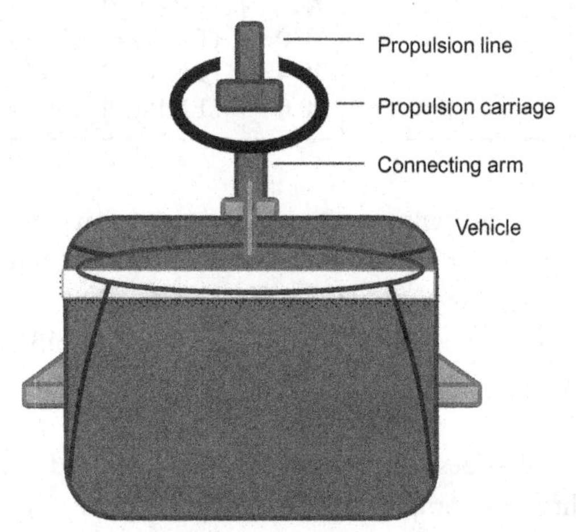

Figure 7.14 Illustration of Terreplane propulsion line, propulsion carriage, and connecting arm.

the flaps to compensate for disturbances such as wind gusts or earth movement (e.g., earthquakes).

The Terreplane System has the ultimate advantages of the very low costs for the land-based propulsion line. Estimates place the costs of the open-air transit corridor of the Terreplane system at less than one third the cost of a paved highway, train track, or hyperloop track. If the transit corridor were to "piggy-back" on electrical power lines, the incremental cost would be less than 10% of the least expensive land-based alternative.

Furthermore, the suspension lines and propulsion lines could be installed above existing streets or railway lines, in the city or in the country. Unlike the "Chicago L" (elevated mass transit train line) that is intrusive and a general eye-sore for downtown Chicago, propulsion line of the Tereplane system would have little to no impact on downtown aesthetics.

Contrary to being intrusive to city environments, a suspension cable and connecting cables down the middle of a city street could support street lights, stoplights, signs, and electrical power distribution that otherwise obstruct the sidewalks.

Initial applications would be for commuter traffic where it could reduce daily roundtrip transit times for two hours to less than twenty minutes. A network of systems would have switching capabilities and allow non-stop service as the standard between dozens of location.

Compared to a transit system with large numbers of small aircraft in independent flight, the Terreplane offers improved safety by connection to a propulsion line and stations that assist with landing, takeoff, and turbulence. In bullet format, the advantages over personal aircraft transit system are:

- Safer propulsion-line assisted takeoff, landing, and flight control.
- Safer by removal of fuel maintenance and fuel fire hazard from the vehicle.
- Safer by elimination of engine maintenance.
- Safer travel limited to the propulsion line pathway.
- Improved energy efficient due to absence of fuel/engine weight from vehicle.
- More sustainable due to electrical from the power grid.
- Direct use of grid electricity without the losses of charging and discharging of batteries.

Comparing the Hyperloop system to the Terreplane system, this is a partial list of advantages:

- Replacement of an expensive guide way with a less expensive propulsion line.
- A propulsion line/guide way that straightens during travel makes high-speed travel possible.
- Low cost regional transit outside tunnels with optional corridors in tunnels for hypersonic travel.
- Direct use of the electrical power grid energy for most operation.

This system is practical for travel distances from one mile to greater than 3000 miles. It provides a transportation infrastructure that could start generating revenue within a year of initial construction and be sustainable following initial investments less than 1% of what would be required for the Hyperloop.

Another advantage is the difference between the "air jet" propulsion of Hyperloop and the linear motor propulsion of the Terreplane. For the Hyperloop system, jets of air exit one vehicle ultimately blowing against the next vehicle in the tube. The Terreplane has no exhaust blast air; rather, the Terreplane tends to push air down the tunnel in the direction of travel. The wind created in the tunnels could top 100 mph; this means at 500 mph travel in the tunnel would have the equivalent air resistance of 400 mph.

This "wind tunnel phenomena" increases energy efficiency and allows faster travel without resort to reduced pressures in the tunnels. It allows local commuter lines to feed directly into inter-city lines. A local commuter line might operate at 150 mph and reduce a 50-minute current-day transit to 10 minutes. That same 150 mph local commuter line could connect to a 500 mph inter-city line offering non-stop service for dozens of locations in Los Angeles to dozens of locations in San Francisco at travel times of about an hour.

In the country, the towers supporting the suspension cables could also support wind turbines. Such systems would consist of these support towers at ⅜ to ½ mile intervals in rural areas and would provide both electrical power generation and electrical power transmission.

The Terreplane system is a new transportation system that provides a path to economic and energy sustainability while providing major reductions in both travel time and energy use for travel. It could evolve to travel for certain corridors in low-pressure tunnels like the Hyperloop. It is an available option.

INCREASED USE OF ELECTRICAL POWER IN SPACE HEATING

Traditional Electrical Space-Heating Markets

For residential and commercial buildings, electricity can be used as an alternative to natural gas, propane, and heating oil to provide space and water heating. As none of these fossil fuels qualify as "long term" sustainable alternatives, the increased use of electricity can provide a sustainable alternative.

Historically, the use of electricity for space or water heating was an available option when natural gas was not available. Since natural gas became available (before 2002) at about $3 per million Btus (or 1.02 ¢/kWh) and local electricity prices as low as 6 ¢/kWh, the choice was always "use available natural gas." Natural gas has returned to lower prices with introduction of Fracking providing "abundant shale gas."

For locations without natural gas, heating oil ($1.65/gallon) and propane ($1.30/gallon) have been available at about $12.6 per million Btu (4.3 ¢/kWh). In these locations the use of electrical resistance heating could be attractive in warmer climates—the higher cost of electricity over fuel could be less than the annualized cost of installing separate propane or heating oil furnaces.

Today and in the future, these economics change due to advances in **heat pumps** and changing natural gas prices. In 2005 the price of natural gas exceeded $12 per MBtu (4.1 ¢/kWh) which translates to 4.55 ¢/kWh in a 90% efficient furnace. At locations where temperatures rarely go below 20°F, heat pumps can reduce the electrical resistance heating costs by 50–66% (COPs of 2 to 3) resulting in delivered heat at an average price of about 2.4 ¢/kWh. Often these heat pumps cost little more than central air conditioning units. Commercially available air conditioner (summer)/heat pump (winter) units are now available.

Figure 7.15 summarizes the ratios of heating costs from fuels (90% efficiency) over heating costs using a heat pump with a COP of 2.0. The shaded numbers are those over 1.0 and represent combinations where a heat pump is cost effective.

In Southern and Southwestern states the heat pump has become an economically sustainable alternative for space heating. The difficulty interpreting the data of Figure 7.15 is identifying the fuel and electricity costs for your region. If pipeline natural gas is compared to retail price of electricity today we find, $4.83 (March) and $3.16 (April, 2015) per MBtu for natural gas [24]. Electricity costs double depending on location

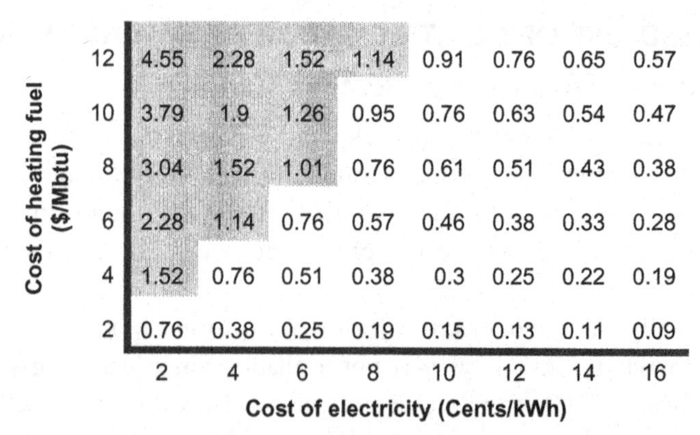

Cost of heating fuel ($/Mbtu) \ Cost of electricity (Cents/kWh)	2	4	6	8	10	12	14	16
12	4.55	2.28	1.52	1.14	0.91	0.76	0.65	0.57
10	3.79	1.9	1.26	0.95	0.76	0.63	0.54	0.47
8	3.04	1.52	1.01	0.76	0.61	0.51	0.43	0.38
6	2.28	1.14	0.76	0.57	0.46	0.38	0.33	0.28
4	1.52	0.76	0.51	0.38	0.3	0.25	0.22	0.19
2	0.76	0.38	0.25	0.19	0.15	0.13	0.11	0.09

Figure 7.15 Ratios of costs for heating with fuel versus heating with electrical heat pump. Ratios based on COP of 2.0 and heater efficiency of 90%. Heated regions show where heat pump is more cost effective.

7.93 ¢/kWh (Iowa) or 14.27 ¢/kWh (California) [25]. The cost of natural gas has been volatile and the price of electricity can double but is fairly stable in different regions of the U. S. This makes it very difficult to use a "cost analysis" to decide whether a heat pump is a good option.

With this economic environment it might be best to explore the heat pump option at your local appliance dealer. A reputable dealer can quote the price for the heat pump equipment. There will be an installer to bid on the installation. A decision to install a heat pump will place the price you pay on the usually more stable price of electricity.

Emerging Electrical Space-Heating Markets

Plans for *expanding* the use of heat pumps to cooler regions depend more on heat pump technology.

Figure 7.16 shows a typical performance curve for a heat pump. At temperatures above 32°F, the COP (ratio of delivered heat to consumed electricity, both in kWh) for a heat pump can be sustained above 2.0. However, the performance falls off quickly at about 32°F because ice can build up on the evaporator that interferes with airflow through the coils and creates an increased resistance to heat flow. This happens because the evaporator coils will be about 10°F cooler than surrounding air, and these temperatures are below the dew point (or ice point) of the moist, outdoor air. The increased thermal resistance requires even cooler evaporator temperatures to achieve the desired heat transfer and this decreases the

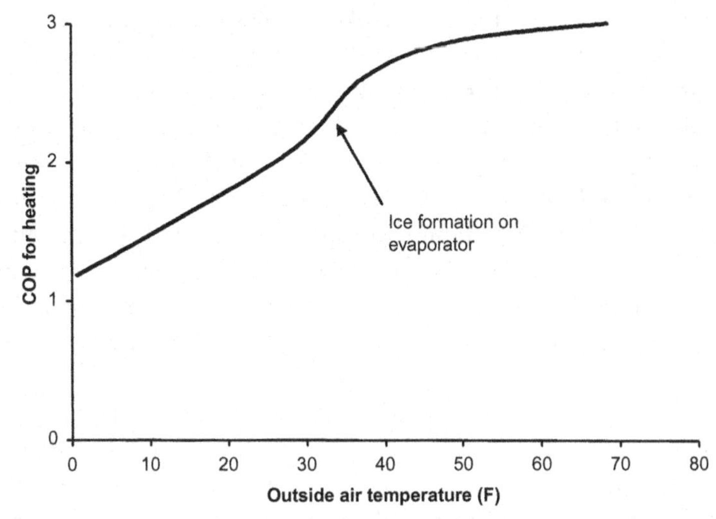

Figure 7.16 Dependence of a typical heat pump performance on outside temperature.

COP. Heat must be periodically delivered to the evaporator coils to defrost them. This pause in service further decreases the effective COP.

In view of the performance curve of Figure 7.12, a good option in the Midwest, West Coast, and the southern two thirds of the East Cost is to install both a heat pump and a furnace. As temperatures go below about 20°F, programmable thermostats are available that automatically switch from the heat pump to the furnace. The impact of this approach is an incremental substitution of fossil fuels to electrical power. At locations where air conditioners are common, the incremental cost for upgrading a central air conditioning unit to a heat pump can be small. The combination of a heat pump, an 80% efficient furnace can be less costly and more sustainable than an air conditioner and a 90% efficient furnace.

Hinrichs and Kleinbach [26] report that reducing the thermostat setting from a constant 72°F to 68°F during the day and 55°F at night can reduce heating fuel consumption (costs) 25–50% (Dallas versus Minneapolis). An extension of this approach to systems that use both heat pumps and furnaces can provide further reductions in natural gas use.

Temperatures often fluctuate by 20°F or more between day and night, and this leads to the COP increasing from 1.8 (20°F) to 2.7 (40°F). The combination of the thermal mass of the house (i.e., 13°F decrease in temperature during the night) and effective insulation can reduce periods of inefficient heat pump operation.

Programming of thermostats to take advantage of fluctuations in temperatures between day and night can also extend the useful range of heat pumps to more-northerly regions. By heating the house during the day (when the outside temperature is 20°F) rather than at night (when the outside temperature may be 0°F) the heat pump goes from an unacceptable COP of less than 1.5 to a daytime COP of 1.7 or higher.

The use of lower nighttime thermostat settings with heat pumps is a cost-effective approach to reduce heating costs while converting to more-sustainable energy options. For new construction or when replacing an air conditioning unit, the savings start at once and extend to the life of the new system.

The down side of this approach is that heat must be built up in the thermal mass of the house during the warmest part of the day (after about 10:00 AM) which means that the house will be cool during the morning. To avoid taking a shower in a 55°F bathroom, resistance heaters can be used to heat the bathrooms without significant compromise of savings. Alternatively, phase change materials in the bathrooms could substantially reduce temperature fluctuations in these rooms—a phase change material approach that is not cost effective for the entire house may be cost effective for a couple rooms.

Ground Source Heat Pumps

As indicated by the performance curve Figure 7.16, heat pump performance improves at warmer evaporator temperatures. Warmer evaporator temperatures can be attained even in Northern states using ground source heat pumps.

Ground source heat pumps employ ground, groundwater, or surface water as the heat sink rather than outside air. Water circulation from the ground source over the evaporator coils can be from open or closed water cycles. The use of closed cycle ground system is more-widely used and debatably has the least impact on the environment. Only the use of the ground system will be discussed.

Figure 7.17 compares a traditional air source of heat pump energy to a closed ground source heat pump. By placing the water circulation-piping network well below the frost line of the ground. This provides a circulating water supply at the ground temperature well above 32°F,. The system can typically supply the heat pump evaporator through the winter, even with sustained outside air temperatures below 0°F.

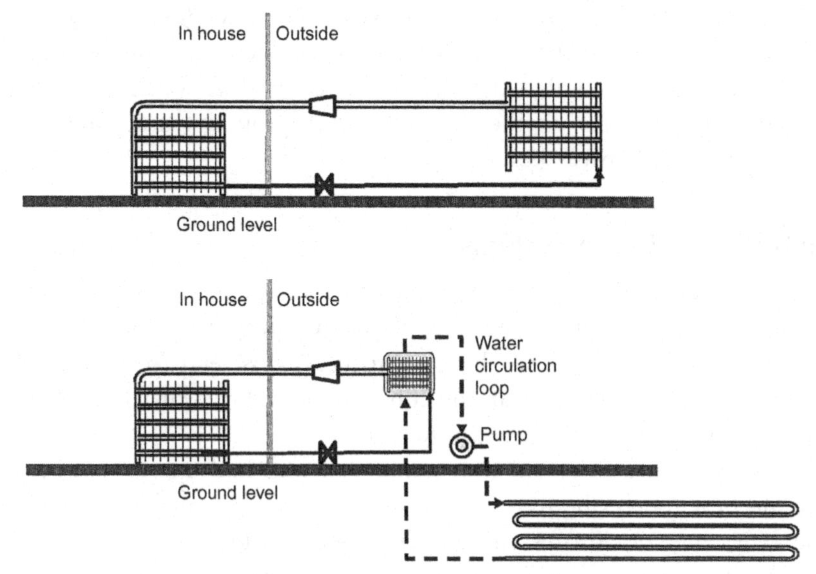

Figure 7.17 Comparison of heat pump using air heat sink to ground source unit.

Properly designed ground source heat pumps eliminate the need for backup furnaces and increase the efficiency of both heating and air conditioning. The major drawback of ground source heat pumps is their costs. A typical cost for a conventional heat pump is about $1,000 per ton of air conditioning capacity ($2,000 for a 2000 square foot home, and it will vary by construction and location). The cost for a comparable ground source system is about $3,000 per ton of capacity.[1] Polyethylene U-tube pipes are usually used for heat exchange with the ground. Horizontal systems (as illustrated in Figure 7.13) are typically less expensive but require larger land area for the layout than vertical systems (that usually extend 150–250 feet into the ground).

The payback period for most ground source systems is about 15 years or longer. The high costs of ground source heat units tend to limit their use in the Midwest to federal buildings and schools large enough to negotiate reduced electrical costs during the heating season (e.g., programs that avoid use of peaking electrical power in the summer). They usually qualify for federal programs that cover much of the up front costs.

[1] Based on DOE EERE example for installation at Fort Polk, LA where 6,600 tons of cooling capacity for 4,000 homes were supplied at a cost of $19 million.

In the Great Lakes region, open cycle lake water (or ground water connected to the lakes) is available and can reduce the costs of ground source units—the combination of lower costs (lake water) and high heating requirements (northern states) makes ground source unit more common in the Great Lakes region.

Hybrid Heat Pump Systems

The commercialization of HEV vehicles is successful because the use of engines and batteries to propel vehicles has advantages beyond the use of either engines or batteries. The combined systems tend to have smaller engines and battery packs than the conventional vehicles or BEVs. Similar opportunities exist in the heat pump market.

Could a $1,500 per ton heat pump that uses both above and below ground heat sources be more cost-effective than the non-hybrid unit? In the wide range of climates, the answer depends on location.

In summary:

- Heat pumps are in use today with applications ranging from moderate heating demands in Southern states to intensive ground source units in the North.
- With unstable natural gas and heating oil prices, larger regions of the U.S. can benefit from the lower relative annualized costs using heat pumps.
- There are regions where a heat pump can be used in combination with furnaces to provide space heating at a lower cost than furnaces alone—especially where air conditioners are already used (the heat pump option on the air conditioner system costs little more than the air conditioner system alone).
- Programming thermostats to operate the heat pump during the warmer daytime hours can extend their use for consumers willing to put up with cooler houses during the morning hours. This works best if there is a demand for electric power at night (e.g., charging of PHEVs) that does not produce new peak demands for electricity.
- Hybrid heat pump systems have the potential to improve the heat pump economics.

Switching from fossil fuels to electrical power can bring sustainability and increased cash flows to local economies. The PHEV and heat pump technologies are available and can give consumers savings.

INCREASED USE OF ELECTRICAL POWER FOR HOT WATER HEATING

Energy used to heat water can be a larger fraction of the energy demand in Southern states and a smaller fraction of space heating energy in northern states. Hot water heating is often the highest or second-highest nonelectrical energy cost in residences and commercial buildings.

Heat pumps can efficiently pump heat from ambient (80°F) temperatures to 160–180°F hot water and are available. Water often enters a house at temperatures below 70°F, even heating to an intermediate temperature of 120°F with a heat pump might reduce fossil fuel heating of water by half.

Heating water with heat from the condensing coil of an air conditioner also has potential, but it would likely require a second heat exchanger. A conventional outside air heat exchanger must also be in place to keep the air conditioner running when the water is hot. The cost of buying and maintaining two heat exchangers will not be justified in northern state climates where there are 30 days or less when air conditioners operate.

EXAMPLE CALCULATIONS

Automobile Cruising kWh Calculation:

Estimate the maximum sustained energy requirement of an automobile cruising at 80 mpg with a fuel economy of 38 mpg. Assume 30% of the energy in the fuel is delivered to the wheels and the gasoline has an energy density of 115,000 Btu/gallon. This 30% efficiency is referred to as the powertrain efficiency (also known as drivetrain efficiency); it includes the engine, transmission, drive shafts, differentials, and turning the wheels.

Solution:

Part 1 – calculate the rate of fuel consumption in kW.

$$\begin{aligned}\text{Fuel Consumption} &= 80\ \text{mph} \div 38\ \text{mpg} \times 115{,}000\ \text{Btu/gallon}\\ &= 242{,}105\ \text{Btu/h}\end{aligned}$$

Conversions:

$$1\ \text{h}/3600\ \text{sec}$$
$$1\ \text{kJ}/0.9486\ \text{Btu}$$
$$1\ \text{kW} = 1\ \text{kJ/sec}$$

Therefore:

$$\begin{aligned}\text{Fuel Consumption} &= 242{,}105\ \text{Btu/h} \div 3600\ \text{sec/h}\\ &\quad \times 1\ \text{kJ}/0.9486\ \text{Btu} \times 1\ \text{kW}/(\text{kJ/sec})\\ &= 70.9\ \text{kW}\end{aligned}$$

Part 2: Assume that 30% of the fuel's energy makes it to the wheels.

$$\text{Power to Wheels} = 70.9\ \text{kW} \times 0.30\ \text{Wheel:Fuel Power}$$
$$= 21.3\ \text{kW}$$

Alternative Automobile Cruising kWh Calculation:
For the previous example, repeat the estimate for 30 mpg, and at 30 mpg, estimate the energy economy in kWh wheel energy requirement per mile. Assume 30% of the fuel's energy is delivered to the wheels.

Solution:

$$\text{Power to Wheels} = 21.3\ \text{kW} \times 38/30(\text{mpg/mpg})$$
$$= 27\ \text{kW}$$

Conversions:

$$0.002778\ \text{kWh}/9.486\ \text{Btu}$$
$$\text{Energy Economy} = 115{,}000\ \text{Btu/gallon} \div 30\ \text{mpg}$$
$$\times 0.002778\ \text{kWh}/9.486\ \text{Btu}$$
$$\times 0.30\ \text{kWh to wheel/kWh in fuel}$$
$$= 0.337\ \text{kWh/mile}$$

Battery Pack Sizing Calculation:
For the previous example, estimate the size of a battery pack for a 60-mile range and compare this to numbers reported in this chapter.

Solution:

Assume the battery is specified in delivered power and the electric motor is 90% efficient:

$$\text{Ideal Motor Battery Pack} = 0.337\ \text{kWh/mile} \times 60\ \text{miles}$$
$$= 20.2\ \text{kWh}$$
$$\text{Batter Pack (90\% motor)} = 20.2\ \text{kWh} \div 0.9\ \text{Wheel:Battery Power}$$
$$= 22.45\ \text{kWh}$$

Comparison: Table 7.5 estimates 19.2 kWh for a mid-sized PHEV-60. Since a PHEV is both an HEV and a PHEV it is reasonable that 30 mpg is consistent with a large sedan rather than a mid-sized sedan. 30 mpg is low for a mid-sized HEV sedan. In view of this, the calculation agrees with the sizing reported in Table 7.5.

REFERENCES

[1] Haywood RW. Analysis of engineering cycles. 4th ed. Oxford, England: Pergamon Press; 1991.

[2] Development of technologies and capabilities for developing coal, oil, and gas energy sources. Solicitation for financial assistance applications no. DE-PS26-02NT41613, U.S. Department of Energy National Energy Technology Laboratory, Pittsburgh, PA, August 21, 2002.

[3] See <http://www.pnl.gov/fta/5_nat.htm>.

[4] See <http://www.uic.com.au/nip57.htm> reports 33% thermal efficiency for nuclear power as reasonable assumption. Also reports high energy output as compared to energy input.

[5] See <http://www.nucleartourist.com/world/koeberg.htm> confirms 33% efficiency—after improvements.

[6] See <http://www.worldandi.com/public/2001/April/nuclear.html>.

[7] See <http://www.cecer.army.mil/techreports/soh_stor/Soh_Stor-03.htm>, Fort Hood, TX example.

[8] See <http://www.energy.gov/sites/prod/files/oeprod/DocumentsandMedia/FINAL_DOE_Report-Storage_Activities_5-1-11.pdf>.

[9] See <http://www.energy.gov/sites/prod/files/oeprod/DocumentsandMedia/FINAL_DOE_Report-Storage_Activities_5-1-11.pdf>.

[10] Frank A. 30 Years of HEV Research Leading to Plug-in HEVs. PHEV Workshop, 2003.

[11] Duvall M. Advanced batteries for electric drive vehicles. EPRI Rep 1009299. May, 2004.

[12] <http://en.wikipedia.org/wiki/Plug-in_hybrid#Sales_and_main_markets>.

[13] <http://www.eia.gov/petroleum/weekly/gasoline.cfmq>.

[14] <http://www.eia.gov/electricity/monthly/epm_table_grapher.cfm?t=epmt_5_6_a>.

[15] <http://www.epa.gov/carlabel/phevlabelreadmore.htm>.

[16] Suppes GJ, Lopes SM, Chiu CW. Plug-in fuel cell hybrids as transition technology to hydrogen infrastructure. Int J Hydrogen Energy January, 2004;29:369–74.

[17] Suppes GJ. Plug-in hybrid electric vehicle roadmap to hydrogen economy. SAE Paper 2005-01-3830.

[18] Suppes GJ. Plug-in hybrid with fuel cell battery charger. J Hydrogen Energy 2005;30:113–21.

[19] Anderman M. BTAP June 2000 report. See <www.arb.ca.gov/msprog/zevprog/2003rule/03board/anderman.pdf>.

[20] Fuel Cell Report to Congress. Report ESECS EE-1973, U.S. Department of Energy, February, 2003.

[21] <http://www.solarwall.com/>.

[22] Suppes G. Compact magnetic levitation transportation system. US Patent 5,146,853. 1992.

[23] See <https://faculty.washington.edu/jbs/itrans/suppes.htm>.

[24] See <http://www.eia.gov/dnav/ng/hist/n9102us3m.htm>.

[25] See <http://www.eia.gov/electricity/monthly/epm_table_grapher.cfm?t=epmt_5_06_a>.

[26] Hinrichs RA, Kleinbach M. Energy its use and the environment. 3rd ed. New York: Brooks/Cole; 2002.

CHAPTER 8

The Future in Nuclear Power

Contents

Sustainable Power Technologies and Infrastructure.
DOI: http://dx.doi.org/10.1016/B978-0-12-803909-0.00008-7

Shortly before World War II the physics research community learned that the uranium-235 isotope would fission when exposed to a beam of neutrons. When a uranium nucleus split a huge surge of energy and two or three neutrons were released. The potential use of these data indicated it would be possible to assemble a powerful explosive weapon. All of the research that led to the production of the two nuclear bombs that exploded over Hiroshima and Nagasaki Japan was labeled "top secret." This ended World War II with the surrender of Japan.

There were lots of freight shipped during World War II and the German submarines were a plague—sinking a high percentage of the Allied surface vessels. Submarines used diesel electric generators to charge their batteries that allowed them to cruise underwater using stored electric power. A submarine had to come close to the surface so that the diesel engines could "breathe" to charge the batteries. This signaled their location. Admiral Hyman Rickover was assigned the task of "taming" the nuclear fission process to produce energy to charge the submarine batteries while they were under water. This made the Nuclear Navy possible with submarines cruising under water—undetected for 90 of more days.

Civilian contractors building submarines made their living building coal-fired electric power plants. There was a Federal "Atoms for Peace Initiative" that made a perfect fit for using the submarine nuclear power plant as the staring model for civilian nuclear power plants. The secrets for this application were suspended and civilian nuclear power was launched. This chapter presents some details of this effort and includes some proposals pointing to the future of civilian nuclear energy.

ENERGIES OF NUCLEAR PROCESSES

The huge quantities of energy liberated in nuclear power plants come from the nuclei of atoms. In the fission process, relatively stable nuclei are induced into excited states that fission and release energy as they form new smaller stable atom nuclei. The heat produced in the nuclear reactor is converted to work through a heat engine power cycle.

Atomic Nature of Matter—Terms

Following is a short summary of the terms used in the physical theory of matter to describe the nature of chemical compounds and the processes involving the nucleus of atoms that occur in a nuclear reactor.

Atoms consist of three basic subatomic particles. These particles are the proton, the neutron, and the electron.

Protons are particles that have a positive charge, have mass about the same as the mass of a hydrogen atom. Protons exist in the nucleus of an atom. The nucleus of the hydrogen atoms is one proton and the mass of the hydrogen atom defines the atomic mass unit (amu) used in nuclear calculations.

Neutrons are particles that have no electrical charge and have a mass about the same as a hydrogen atom (approximately 1 amu). Neutrons exist in the nucleus of an atom.

Electrons are particles with a negative charge and have a mass about 1/1837 the mass of a hydrogen atom. Each electron exists in a well-defined, unique orbital shell around the nucleus of an atom.

The **atomic number** of an atom is the number of protons in the nucleus.

Nuclides are atoms that contain a particular number of protons and neutrons.

Isotopes are nuclides that have the same atomic number of protons (and electrons, therefore the same chemical properties) but differ in the number of neutrons.

The **mass number** of an atom is the total mass number of nucleons (protons and neutrons) in the nucleus.

The **stability** of a nucleus is determined by the forces acting within it. There is the long-range repulsive electrostatic force that acts between the protons, very strong at the very close distances in the nucleus. The gravitational force between the nucleons in the nucleus is negligible. The nuclear force is a very short-range strong attractive force, independent of the charges, acting between all of the nucleons holding the nucleus together.

The **radius** of a nucleus ranges from $1.25-7.74 \times 10^{-13}$ cm (for hydrogen and uranium-238). The average diameter of an atom, except for a few very light atoms, is about 2×10^{8} cm, making the atom more than 25,000 times as large as the nucleus. The nucleus is very small, very dense, and contains nearly all of the mass of the atom.

Radioactive nuclides are atoms that disintegrate by the emission of a particle or electromagnetic radiation; most commonly an alpha or beta particle, or gamma radiation. There are three classes of radio nuclides:

1. Primary—with half-lives greater than 10^{8} years. These may be alpha or beta emitters.
2. Secondary—formed by the radioactive transformation of uranium-235, uranium-238, or thorium-232.
3. Induced—these radio nuclides have geologically short half-lives and are formed by induced nuclear reactions. All of these reactions result in transmutation with a new (radioactive or nonradioactive) nuclide formed.

Nuclear fission is a nuclear reaction that splits the atom nucleus forming two new atoms each with about half of the original mass. There is the release of a great quantity of energy since the mass of the new atoms is slightly less than the mass of parent atom; the mass loss is converted to energy by the Einstein equation.

The natural stability of atoms is characterized by their half-life. The half-life of U-238 is 4.5×10^{9} years. If 1 pound of pure U-238 were flying through outer space today, in 4.5 billion years that meteor would have a total mass slightly less than 1 pound—one half pound would be U-238 and the other half would be fission products, mostly lead. Since the earth is about 5 billion years old, the U-238 present on earth today is about half of that present when earth was formed.

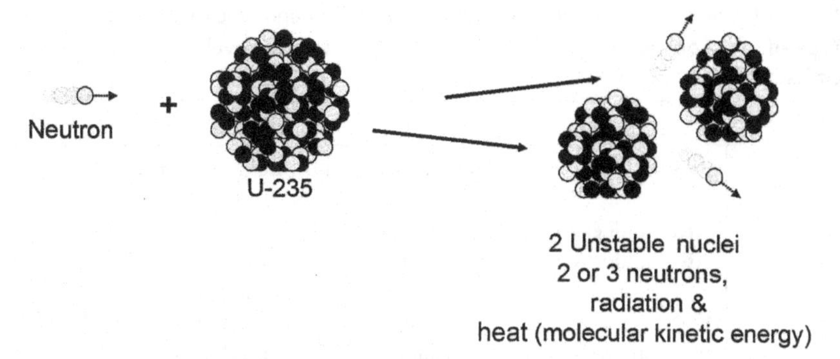

Figure 8.1 Illustration of neutron-induced fission of U-235.

While the atomic stability is typically discussed in relative rather than absolute terms, atoms with half-life greater than 4.5 billion years are generally considered stable. Lead (Pb-206) is stable; the change in concentration of a 1-pound lead meteor would be negligible over a 4.5-billion-year period.

The decay of U-238 is a natural process (specifically, an α-decay process[1]). An unnatural decay of a nucleus, nuclear fission can be induced by collision with a neutron. A nuclear reactor environment is designed to sustain a critical concentration of free neutrons that gives a constant and controlled source of heat from induced fission. U-235 is the isotope that provides most of the energy nuclear reactor. Figure 8.1 illustrates the overall process by which U-235 releases heat through neutron-induced fission.

The energies released from the excited nuclei are not common in nature. However, analogous electron processes are regularly observed. For example, an incandescent light bulb operates on the principle of using electric power to increase the energy of the metal filament (high temperature)—some of this energy produces excited states of the electrons that surround the metal nuclei. These *excited* electron states emit visible radiation (light) as they return to lower energy, more stable states or *ground* states.

Table 8.1 provides several example emissions that occur when electrons and nuclei go from *excited* states to *ground* states (referred to as stable states for nuclei)—electrons and nuclei have multiple excited states and one or two stable/ground states. The energies are reported in electron volts; one electron volt is equivalent to 1.602×10^{-19} joules or 1.18×10^{-19} foot pounds.

[1] The α-decay half-life for U-238 is 4.5×10^9 years. The fission decay half life is 8.0×10^{15} years.

Table 8.1 Examples of different emissions from nuclei and electrons

Type of emission	Source	Energy level
Nuclear		
Beta	Atomic decay in nuclear reactor	Disintegration energy of S-38—2.94 MeV
Alpha	Atomic decay in nuclear reactor	—
Neutron	Atomic decay in nuclear reactor	Fission release—~2 MeV Fast neutron—>1 MeV Thermal neutron—0.025 eV
γ-ray	Nuclear transition from excited state to lower energy state	Relaxing of excited states of Ni-60—1.174, 2.158, and 1.332 MeV U-234 decay to Th-230—0.068 MeV
Electron		
X-ray		Typically from 5–100,000 eV

Table 8.2 Examples of energy levels in electron volts for different processes

Other processes	
U-235 fission to Rb-93 + Cs-140	200 MeV
Ionization	Remove outer electron from lead—7.38 eV Remove inner electron from lead—88,000 eV
Mass defect	Mass defect of Li-7—931.5 MeV
Binding energy	Binding energy of Li-7—1784 MeV

Among the lowest energy emissions from electrons is visible light resulting from electricity flowing through an incandescent light bulb. Among the highest energy emissions are neutrons emitted as part of atomic decay (2,000,000 eV). As illustrated by the comparison of Table 8.2, the energies associated with atomic processes are much larger than electron processes; atomic processes tend to be useful in large power plants while electronic processes tend to have applications in homes and offices.

Electron emissions tend to be photonic (light, energetic x-rays); nuclear emissions may be photonic (γ-rays) or have white particle mass (α, β, and neutron) as energy components.

Table 8.3 While electronic emissions dissipate, nuclear emissions do not dissipate

Type of emission	Product of emission
Nuclear	
Beta	Is newly produced, excited electron
Alpha	Is newly produced, excited helium atom nucleus
Neutron	Is newly produced, excited atomic hydrogen (but remains a neutron if incorporated into another nucleus)

Of the emissions in Table 8.1, only the neutrons can collide and combine with an atom nucleus—often leading to an unstable state of the nucleus. In some physics laboratories atomic accelerators are able to increase the energy of particles that collide to produce excited nuclei or new elements.

If neutron loses enough energy through collisions, it will at sufficiently low energy convert to atomic hydrogen (one proton and one electron). Beta (β) particles become electrons, and alpha (α) particles become helium (He-4). These transitions are summarized in Table 8.3.

The US Department of Energy publication entitled *DOE Fundamentals Handbook, Nuclear Physics and Reactor Theory, Volume 1 of 2* [1] provides a detailed summary of key aspects of nuclear physics including the following excerpts (italics) that describe nuclides, nuclear stability, and conventions for reporting the atomic information.

CHART OF THE NUCLIDES

A tabulated chart called the Chart of the Nuclides lists the stable and unstable nuclides in addition to pertinent information about each one. Figure 8.2 *shows a small portion of a typical chart. This chart plots a box for each individual nuclide, with the number of protons (Z) on the vertical axis and the number of neutrons ($N = A - Z$) on the horizontal axis.*

The completely gray squares indicate stable isotopes. Those in white squares are artificially radioactive, meaning that they are produced by artificial techniques and do not occur naturally. By consulting a complete chart, other types of isotopes can be found, such as naturally occurring radioactive types (but none are found in the region of the chart that is illustrated in Figure 8.2).

Located in the box on the far left of each horizontal row is general information about the element. The box contains the chemical symbol of the element in addition to the average atomic weight of the naturally occurring substance and the

Z \ N	Element	0	1	2	3	4	5	6	7	8	9	10
6	C 12.011 CARBON			C8 2.0E−21s	C9 127 ms	C10 19.3 s	C11 20.3 m	C12 98.90	C13 1.10	C14 5730 a	C15 2.45 s	C16 0.75 s
5	B 10.811 BORON			B7 4E−22 s	B8 770 ms	B9 8E−19 s	B10 19.9	B11 80.1	B12 20.20 ms	B13 17.4 ms	B14 14 ms	B15 9 ms
4	Be 9.012182 BERYLLIUN			Be6 5.0E−21 s	Be7 53.28 d	Be8 ~7E−17 s	Be9 100	Be10 1.6E6 a	Be11 13.8 s	Be12 24 ms		Be14 4 ms
3	Li 6.941 LITHIUM			Li5 ~3E−22 s	Li6 7.5	Li7 92.5	Li8 0.84 s	Li9 177 ms		Li11 8.7 ms		
2	He 4.002602 HELIUM		He3 0.000138	He4 9.999862	He5 7.6E−22 s	He6 807 ms	He7 3E−21 s	He8 119 ms	He9 VERY SHORT			
1	H 1.0079 HYDROGEN	H1 99.985	H2 0.015	H3 12.3 a								
0												

Figure 8.2 Excerpt from chart of nuclides.

average thermal neutron absorption cross section, which will be discussed in a later module. The known isotopes (elements with the same atomic number Z, but different mass number A) of each element are listed to the right.

Information for Stable Nuclides

For the stable isotopes, in addition to the symbol and the atomic mass number, the number percentage of each isotope in the naturally occurring element is listed, as well as the thermal neutron activation cross section and the mass in atomic mass units (amu). A typical block for a stable nuclide from the Chart of the Nuclides is shown in Figure 8.3.

Information for Unstable Nuclides

For unstable isotopes the additional information includes the half-life, the mode of decay (for example, β^-, α), the total disintegration energy in MeVTable (million electron volts), and the mass in amu when available. A typical block for an unstable nuclide from the Chart of the Nuclides is shown in Figure 8.4.

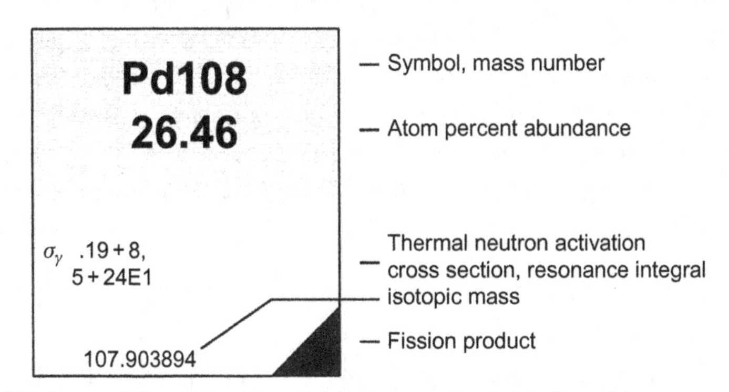

Figure 8.3 Presentation format for stable isotopes in chart of nuclides.

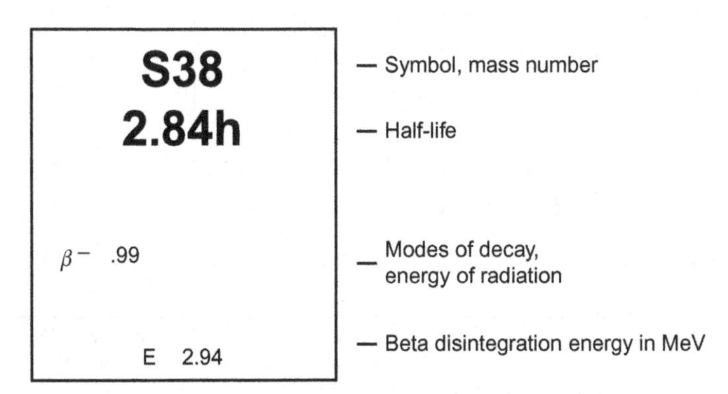

Figure 8.4 Presentation format for unstable isotopes in chart of nuclides.

Neutron–Proton Ratios

Figure 8.5 *shows the distribution of the stable nuclides plotted on the same axes as the Chart of the Nuclides—it provides the skeleton of the complete Chart of Nuclides. As the mass numbers become higher, the ratio of neutrons to protons in the nucleus becomes larger. For helium-4 (2 protons and 2 neutrons) and oxygen-16 (8 protons and 8 neutrons) this ratio is unity. For indium-115 (49 protons and 66 neutrons) the ratio of neutrons to protons has increased to 1.35, and for uranium-238 (92 protons and 146 neutrons) the neutron-to-proton ratio is 1.59.*

If a heavy nucleus were to split into two fragments, each fragment would form a nucleus that would have approximately the same neutron-to-proton ratio as the heavy nucleus. This high neutron-to-proton ratio places the fragments below and to the right of the stability curve displayed by Figure 8.5. *The instability caused by the excess of neutrons is generally rectified by successive* ***beta emissions****, each of which converts a neutron to a proton and moves the nucleus toward a more stable neutron-to-proton ratio.*

Careful measurements have shown that the mass of a particular atom is always slightly less than the sum of the masses of the individual neutrons, protons, and electrons of which the atom consists. The difference between the mass of the atom and the sum of the masses of its parts is called the ***mass defect*** *(Δm).*

The loss in mass, or mass defect, is due to the conversion of mass to binding energy when the nucleus is formed. Binding energy is defined as the amount of

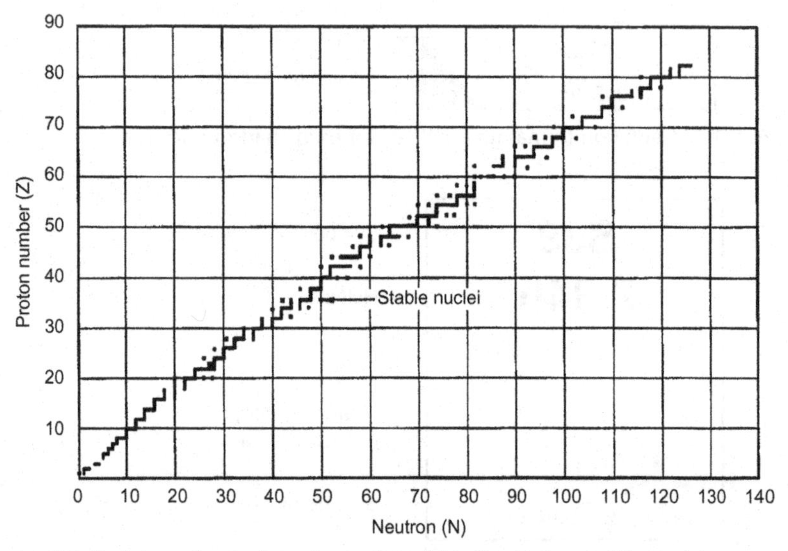

Figure 8.5 Skeleton of complete chart of nuclides illustrating stable nuclei.

energy that must be supplied to a nucleus to completely separate its nuclear particles (nucleons). It can also be understood as the amount of energy that would be released if the nucleus was formed from the separate particles. ***Binding energy*** *is the energy equivalent of the mass defect. Since the mass defect was converted to binding energy (BE) when the nucleus was formed, it is possible to calculate the binding energy using a conversion factor derived by the mass-energy relationship from Einstein's Theory of Relativity.*

Energy Levels of Atoms

The electrons that circle the nucleus move in fairly well defined orbits. Some of these electrons are more tightly bound in the atom than others. For example, only 7.38 eV is required to remove the outermost electron from a lead atom, while 88,000 eV is required to remove the innermost electron. The process of removing an electron from an atom is called ionization, and the energy required to remove the electron is called the ionization energy.

In a neutral atom (number of electrons = Z) it is possible for the electrons to be in a variety of different orbits, each with a different energy level. The state of lowest energy is the one in which the atom is normally found and is called the ground state. When the atom possesses more energy than its ground state energy, it is said to be in an excited state.

An atom cannot stay in the excited state for an indefinite period of time. An excited atom will eventually transition to either a lower-energy excited state, or directly to its ground state, by emitting a discrete bundle of electromagnetic energy ***called an x-ray****. The energy of the x-ray will be equal to the difference between the energy levels of the atom and will typically range from several eV to 100,000 eV in magnitude.*

Energy Levels of the Nucleus

The nucleons in the nucleus of an atom, like the electrons that circle the nucleus, exist in shells that correspond to energy states. The energy shells of the nucleus are less defined and less understood than those of the electrons. There is a state of lowest energy (the ground state) and discrete possible excited states for a nucleus. Where the discrete energy states for the electrons of an atom are measured in eV or keV, the (k = 1000) energy levels of the nucleus are considerably greater and typically measured in MeV (M = 1,000,000).

A nucleus that is in the excited state will not remain at that energy level for an indefinite period. Like the electrons in an excited atom, the nucleons in an excited nucleus will transition towards their lowest energy configuration and in

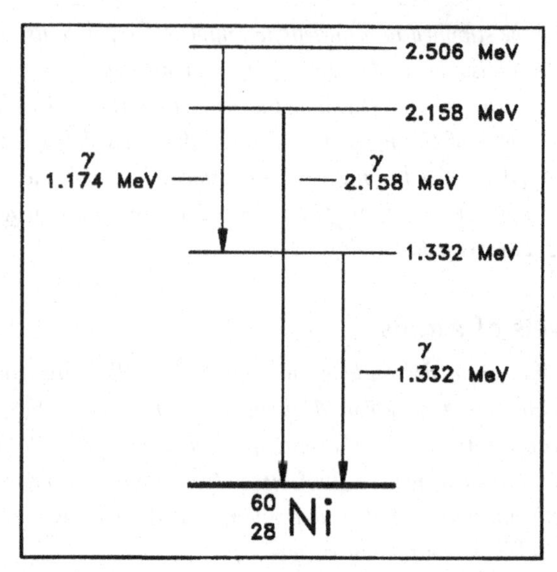

Figure 8.6 Energy-level diagram for Nickel-60.

doing so emit a discrete bundle of electromagnetic radiation called a gamma ray (γ-ray). The only differences between x-rays and γ-rays are their energy levels and whether they are emitted from the electron shell or from the nucleus. The ground state and the excited states of a nucleus can be depicted in a nuclear energy-level diagram. The nuclear energy-level diagram consists of a stack of horizontal bars, one bar for each of the excited states of the nucleus. The vertical distance between the bar representing an excited state and the bar representing the ground state is proportional to the energy level of the excited state with respect to the ground state. This difference in energy between the ground state and the excited state is called the excitation energy of the excited state. The ground state of a nuclide has zero excitation energy. The bars for the excited states are labeled with their respective energy levels. Figure 8.6 *is the energy level diagram for nickel-60.*

Stability of Nuclei

As mass numbers become larger, the ratio of neutrons to protons in the nucleus becomes larger for the stable nuclei. Non-stable nuclei may have an excess or deficiency of neutrons and undergo a transformation process known as beta (β) decay. Non-stable nuclei can also undergo a variety of other processes such as alpha (α) or neutron (n) decay. As a result of these decay processes, the final nucleus is in a more stable or more tightly bound configuration.

Natural Radioactivity

In 1896, the French physicist Becquerel discovered that crystals of a uranium salt emitted rays that were similar to x-rays in that they were highly penetrating, could affect a photographic plate, and induced electrical conductivity in gases. Becquerel's discovery was followed in 1898 by the identification of two other radioactive elements, polonium and radium, by Pierre and Marie Curie.

Heavy elements, such as uranium or thorium, and their unstable decay chain elements emit radiation in their naturally occurring state. Uranium and thorium, present since their creation at the beginning of geological time, have an extremely slow rate of decay. All naturally occurring nuclides with atomic numbers greater than 82 are radioactive. X

NUCLEAR DECAY

Table 8.4 provides examples of different types of nuclear transitions. These transitions can be from a highly unstable nucleus or from a relatively stable nucleus. A highly unstable nucleus has a short half-life (time in which the concentration of that isotope is reduced by 50% due to atomic transition) while stable molecules have long half-lives.

During and immediately after the burn of a nuclear fuel rod, radiation levels are very high due to the large number of radio nuclides with short half-lives. By definition, these short half-life nuclei rapidly undergo nuclear transitions. For a given nuclei, this process continues in a **decay chain** until a molecule with a stable or long-half-life nucleus is formed. The following decay of rubidium-91 to zirconium 91 illustrates a decay chain. The numbers under the arrows indicate half-lives in seconds, hours, days, and years.

Table 8.4 Example notations of nuclear processes

Process	Formula	Description [1]
Alpha decay	${}^{234}_{92}U \rightarrow {}^{230}_{90}Th + {}^{4}_{2}\alpha + \gamma + KE$	KE is kinetic energy of the α-particle is helium nucleus
Beta decay	${}^{239}_{93}Np \rightarrow {}^{239}_{94}Pu + {}^{0}_{-1}\beta + {}^{0}_{0}\bar{\nu}$	Neutron converted to proton ν represents a neutrino—interacts little with atoms and escapes at speed of light
Beta decay	${}^{13}_{7}N \rightarrow {}^{13}_{6}C + {}^{0}_{+1}\beta + {}^{0}_{0}\nu$	Proton converted to neutron through positron formation
Electron capture	${}^{7}_{4}Be + {}^{0}_{-1}e \rightarrow {}^{7}_{3}Li + {}^{0}_{0}\nu$	Proton is converted to neutron by electron capture

$$ {}^{91}_{37}Rb \xrightarrow[58.0\ s]{\beta^-} {}^{91}_{38}Sr \xrightarrow[9.5\ hrs]{\beta^-} {}^{91}_{39}Y \xrightarrow[58.5\ d]{\beta^-} {}^{91}_{40}Zr $$

These decay chains are important when treating fission products. The short-lived products will rapidly decay. If spent fuel is stored for 30 years at the nuclear power plant, this time will reduce the concentration of all nuclides with half-lives less than 3 years to a concentration less than 0.1% of the initial concentration.

Fortunately, the majority of the short-lived isotopes decay to stable nuclides. About 10% of the fission products remains as high-level waste after 30 years of storage—the remainder have decayed to stable nuclides.

CONDITIONS FOR SUCCESSFUL NUCLEAR FISSION

For nuclides to successfully undergo neutron-induced fission, a number of conditions must be met that are analogous to a chemical reaction. Table 8.5 summarizes and compares the factors that lead to fission with those conditions that promote chemical reactions.

These topics are discussed in the following four sections.

Table 8.5 Factors impacting the rate of nuclear fission versus analogous factors for chemical reaction

Factor	Nuclide	Chemical reagent
Materials must have a propensity to react	A low critical energy that corresponds to classifications as fissile or fissionable	A low activation energy
Materials must have ability to go to lower energy state	Products must have a higher binding energy	Products must have a lower Gibbs free energy
Degree of molecular excitement should be optimal	The energy of the neutron must be correct—high (fast) or low (thermal) energy level may be optimal	Temperature must be high enough to react but low enough to stabilize the products
Events must be concentrated rather than disperse	Concentrations of reacting materials (e.g., U-235) must be high enough to sustain reaction but not so high to run away (explode)	High concentrations are needed for reasonable reactor size, or a solvent must be used to avoid run away

(a) Uranium and Other Fertile Materials

A nuclear reactor is designed to provide a flux of neutrons with the right energy to provide a constant, steady rate of nuclear fission. Each U-235 yields about 200 MeV per atom of uranium that undergoes fission.

U-235 is referred to as a **fissile** material because U-235 will absorb a neutron with very low kinetic energy (referred to as **thermal neutrons**) and this produces fission. Table 8.6 summarizes the three types of materials that are of interest in nuclear fission fuels. Fissile atoms undergo fission because a neutron of low kinetic energy can induce fission.

When a neutron combines with a stable nucleus, a binding energy (BE) corresponding to that neutron addition is released. When that BE is greater than a **critical energy** (specific to the nuclide before addition of the neutron), the nuclide can undergo fission. Table 8.7 provides the binding energies (MeV/nucleon) and critical energies of the five **fissile** and **fissionable** materials. Th-232 and U-238 are fissionable, but not fissile because it takes higher energy neutrons to bring sufficient kinetic energy so that the sum of kinetic and binding energies exceeds the "critical energy" producing fission.

When U-238 or Th-232 absorb neutrons and fission does **not** occur, they can undergo the decay chain summarized in Figure 8.7 resulting in

Table 8.6 Definitions and examples of nuclear fission fuels

Material category	Definition	Examples
Fissile	Nuclides for which fission is possible with neutrons of *any energy level*	U-235, U-233, and Pu-239
Fissionable	Nuclides for which fission is possible with neutron collision	U-235, U-233, Pu-239
Fertile	Materials that can absorb a neutron and become fissile materials	U-238 and Th-232

Table 8.7 Critical energy versus energy released with absorption of additional neutron [1]

Target nucleus	Critical energy E_{crit}	Binding energy of last neutron BE_n	$BE_n - E_{crit}$
Th-232	7.5 MeV	5.4 MeV	−2.1 MeV
U-238	7.0 MeV	5.5 MeV	−1.5 MeV
U-235	6.5 MeV	6.8 MeV	±0.3 MeV
U-233	6.0 MeV	7.0 MeV	+1.0 MeV
Pu-239	5.0 MeV	6.6 MeV	+1.6 MeV

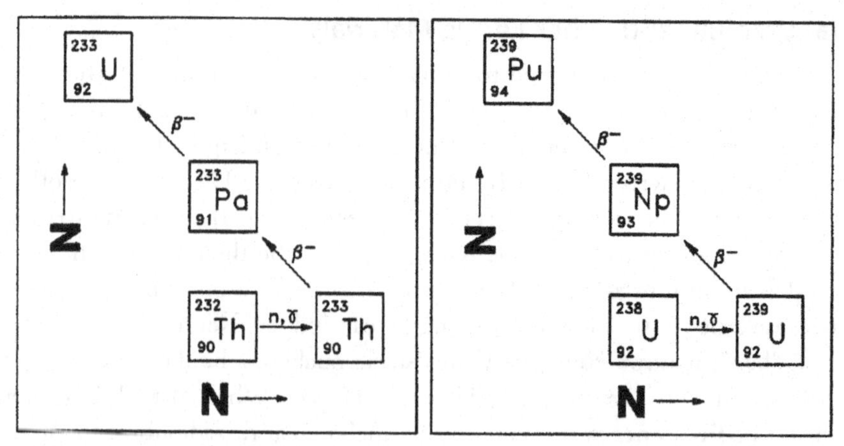

Figure 8.7 Decay chain for fertile collisions with Th-232 and U-238.

the formation of U-233 and Pu-239. U-238 and Th-232 are referred to as **fertile materials** because absorption of a neutron can produce a fissile material. Because of these nuclear processes, it is possible for a nuclear reactor to produce more fuel than is consumed—reactors designed to do this are called **breeder reactors**. In light water **converter reactors** (also referred to as burner reactors), that consume more fuel than is produced, about one-third of the energy produced is a result of Pu-239 production with subsequent Pu-239 fission. At the end of the nuclear fuel burn in a light water reactor about 0.9% Pu-239 remains in the fuel. The new fuel initially contained 3.4% U-235 (0% Pu-239).

The decay chains in Figure 8.7 (absorption without fission) are broadly referred to as transmutation processes. Transmutation is important for converting fertile fuel to fissile fuel. The susceptibility of materials to transmutation is covered in the next section under the topic of absorption cross section.

Transmutation is important for creating fissile materials and for converting problem radioactive wastes into more benign materials. Not all nuclides in nuclear waste present the same degree of waste handling problems. For example, nuclides with short half-lives (less than about 5 years) can be stored until the radioactive decay reaches benign levels. Wastes with very long half-lives tend to be less hazardous than the uranium mined to create the nuclear fuel. However, wastes with intermediate half-lives are more hazardous than natural ores. They take too long to decay in 30–60 years used for temporary storage. Transmutation can transmute some of these waste materials into new nuclides that decay quickly or that are stable.

(b) Binding Energy Constraints

Available technology limits sustainable fission power to fissionable materials originating from natural uranium and thorium. For fission to occur, the nuclei produced from the nuclear transformation must have a higher BE than the nuclei undergoing fission (see definition of BE). BE is defined so that higher binding energies represent more permanent nuclei. The most stable nuclei, like iron, have the highest binding energies.

The BE trends in Figure 8.8 illustrate that those nuclei with atomic weights greater than about 60 can undergo fission to produce more tightly bound nuclei. Nuclei with atomic weights less than about 60 can undergo fusion to produce more tightly bound nuclei.

The total energy release from the fission of U-235 is about 200 MeV. About 187 MeV of the energy is immediately released in the form of kinetic energy of the fission fragments, kinetic energy of the fission neutrons, and γ−rays. The excited product nuclei will release the remaining 13 MeV in the form of kinetic energy of delay beta particles and decay γ−rays. Table 8.8 reports average quantities of instantaneous and

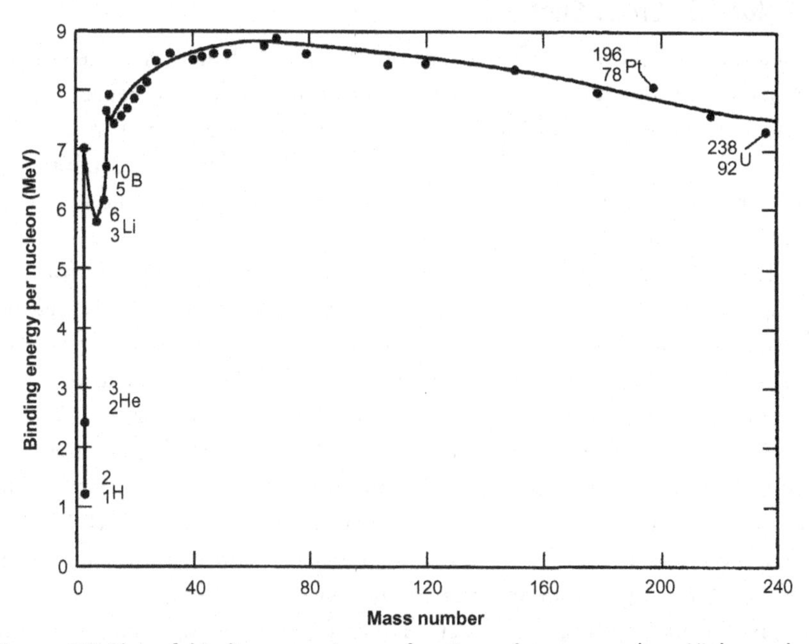

Figure 8.8 Plot of binding energies as function of mass number. Higher values reflect more stable compounds. The values are the binding energy per-nuclei release of energy if free protons, neutrons, and electrons combined to form the most stable nuclei for that atomic number.

Table 8.8 Instantaneous and delayed energy from fission [1]

Instantaneous	
Kinetic energy of fission products	167 MeV
Energy of fission neutrons	5 MeV
Instantaneous γ-ray energy	5 MeV
Capture γ-ray energy	10 MeV
Total	187 MeV
Delayed	
β-Particles form fission products	7 MeV
γ-rays from fission products	6 MeV
Neutrinos	10 MeV
Total	23 MeV

delayed energy release from U-235 fission by a thermal neutron. Of these emissions, the 10 MeV of energy from the neutrinos escape the reactor system.

(c) Nuclear Cross Sections

Nuclear cross sections are tabulated for atoms and characterize the susceptibility of the nuclide to interact with a neutron. Different representative cross sections are reported for different types of interaction. While fissile, fissionable, and fertile classifications indicate what happens if a neutron is absorbed by a nuclide, the cross section indicates the size of the target for neutron capture.

The cross sections are dependent on the energy of the neutron and the properties of the nuclide. These microscopic cross sections may be viewed as the area available for a neutron to hit to induce reaction. A larger cross section provides an increased probability for reaction.

Table 8.9 provides example nuclear cross sections for U-235 and U-238. Cross sections (reported in barns, 1 barn $= 10^{-24}$ cm^2) for both fission and capture are provided. The thermal neutrons (<1 eV) typically have cross section 20–30 times larger than fast neutrons (1–2 MeV). It is this large cross section for U-235 and thermal neutrons that made it the fuel of choice for commercial nuclear reactors.

Fission and capture cross sections are two of the four cross sections that dominate nuclear reactor behavior. Table 8.10 illustrates these and includes elastic and inelastic cross sections.

Table 8.9 Example cross section areas [2]

Nuclide	Kinetic energy of neutron (eV)	Fission cross section (barns)	Capture cross section (barns)
U-235	0.5	50	7
U-235	1,000,000	2	0.15
U-238	0.5	0.6	N/A
U-238	1,000,000	0.1	0.02

Table 8.10 Illustration of prominent cross sections in nuclear reactors

Process	Description [1]
Fission	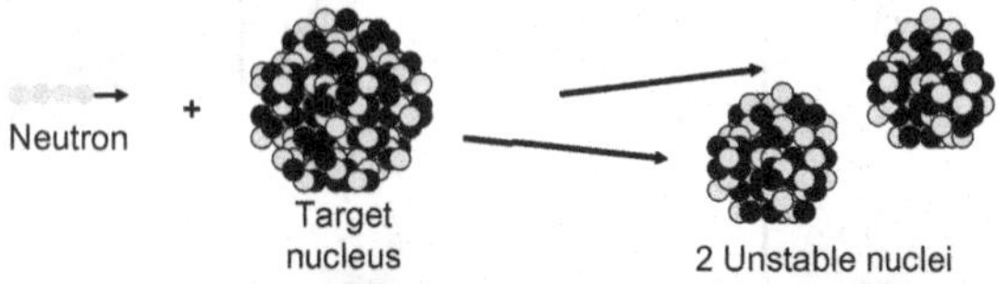 Neutron + Target nucleus → 2 Unstable nuclei
Transmutation	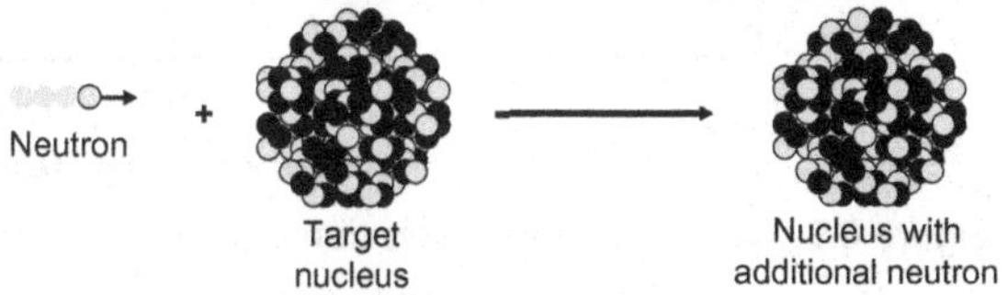 Neutron + Target nucleus → Nucleus with additional neutron
Scattering (elastic)	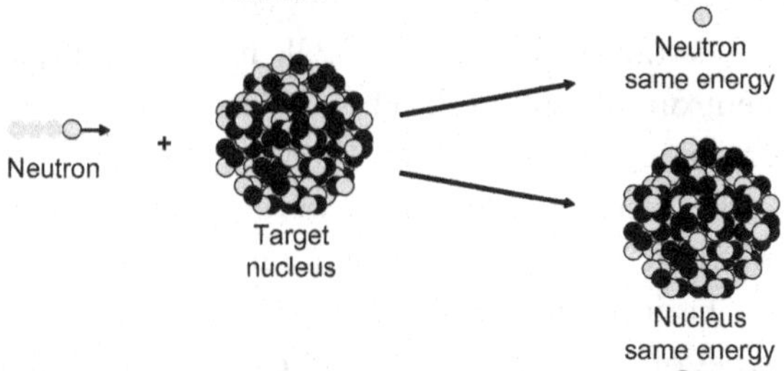 Neutron + Target nucleus → Neutron same energy; Nucleus same energy
Scattering (inelastic)	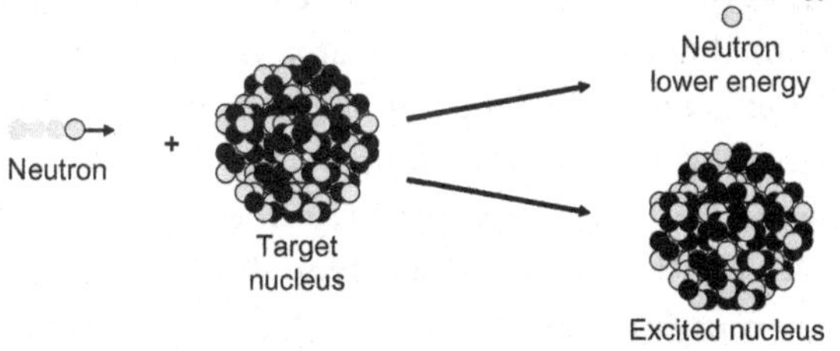 Neutron + Target nucleus → Neutron lower energy; Excited nucleus

The fission cross sections for U-235 is about 50 barns for the thermal neutron versus 2 barns for the fast neutron. For U-235 (fissile), the fission cross section is greater than the capture cross section, fission will occur more often than capture.

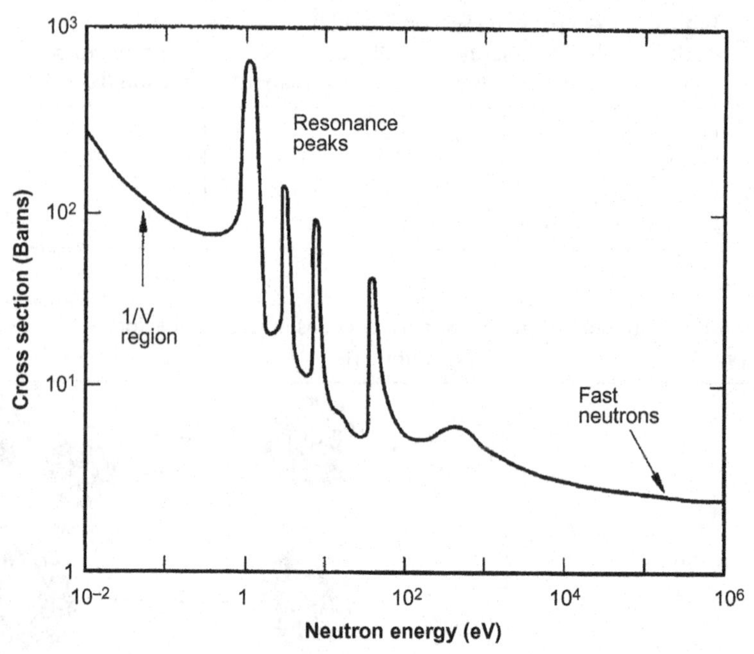

Figure 8.9 Typical neutron absorption cross section versus neutron energy.

Fertile nuclides like U-238 have critical fission cross sections below a neutron energy level. The fission cross section for U-238 is equal to the capture cross section at 1.3 MeV. Higher energy neutrons will tend to cause fission while lower energy neutrons will tend to cause transmutation, the pathway to forming plutonium.

Fission, capture, scatter, and total cross sections are a few of the different types of cross sections that are characterized. Figure 8.9 shows a typical plot of total nuclear cross section area versus the energy level of the neutron. The complex nature of the free neutron interaction with nuclei goes beyond the scope of this text with much yet to be learned. Key points have been presented; especially important is the distinction between thermal neutrons (<1 eV) and fast neutrons (typically >1 MeV). The thermal neutron is key in propagating reactions in the Generation II nuclear reactors including current commercial light water reactors. For fast-spectrum reactors (the Generation IV designs) fast neutrons are key to the performance. Fast neutrons can directly induce fission in U-238 and can fission actinides.

Actinides are nuclides with atomic numbers between 89 and 104 (at an atomic number of 92, uranium is an actinide). Actinides such as plutonium (Pu), neptunium (Np), americium (Am), and curium (Cm)

Table 8.11 Transuranic elements of primary interest to AFCI program including uranium as reference

92	93	94	95	96
U	Np	Pu	Am	Cm
Uranium	Neptunium	Plutonium	Americium	Curium

Why does transuranic matter? [3]

- Transuranic elements affect repository performance by dominating long-term heat load and long-term radiotoxicity.
- Transuranic elements and enriched uranium are the only materials of concern for proliferation.
- Transuranic elements can be destroyed while producing extra energy if recycled in (fast spectrum) nuclear reactors.

are formed in nuclear reactors (see Table 8.11). Once formed, they can continue absorbing thermal neutrons, eventually reducing the number of neutrons available to promote fission. Fast neutrons tend to produce fission. So, fast neutrons cause actinides to release fission energy (rather than inhibit other fission processes).

Fast-spectrum reactors are important for sustainable nuclear power. Fast-spectrum reactors eliminate the need to separate the actinides when reprocessing nuclear fuel. This reduces the cost and promotes sustainable economics. Using all the actinides as fuel removes them from the waste stream and eliminates the long-term storage problem.

(d) Concentrated Events

Fissile materials U-235 and Pu-239, meet the constraints of fission and, release energy as they form smaller, more stable nuclei. The chain reaction is maintained by the neutron flux. The final components of the controlled release of the nuclear energy are the initiation of the neutron flux and maintaining the neutron flux. The neutron flux is the number of neutrons passing through an area of 1 sq. cm per second. Since the neutrons tend to be moving through solids (stopped or scattered only by the dense nuclei of the atoms in the solid), the energy of the neutrons decreases (they slow down) as they travel through the reactor core.

A discussion of materials for initiating the neutron flux is beyond the scope of this text. There are such materials that are used to start the reactor by initiating fission.

Sustaining the neutron flux is the most important criteria in nuclear reactor design. The neutron flux is depleted by neutron capture and by scattering out of the reactor core volume. The neutron poisons (boron as boric acid in the reactor cooling water) are used to maintain the neutron flux for constant energy production.

In a controlled reactor environment, the flow of neutrons (the neutron flux) achieves a steady state consistent with the desired heat (energy) release. This is achieved with the right concentration of U-235 or Pu-239 present—achieved by concentrating them in the fuel rods and proper spacing of the fuel rods. The right fuel rod concentration is typically between 2.6% and 4.0% U-235 in a light water reactor. Some of the proposed Generation IV designs may use concentrations up to 20%. The spacing of the fuel rods in the reactor and the fissile isotope concentration in the fuel provide the controlled release of energy. Since the medium (water) between the fuel rods changes the kinetic energy of the neutrons, it is important to match the medium with fuel composition and spacing.

Light water reactors are designed for controlled delivery of thermal neutrons (<1 eV) to the fuel rods. Liquid water (not water vapor) between the fuel rods provides an average of 12 scattering collisions with water to produce the thermal neutrons that will successfully fission another U-235 nucleus.

If water is absent, the energy level of the neutrons is too high, the lower nuclear cross section leads to fewer successful absorption processes—and to the escape of the neutrons from the reactor core. In light water reactors, this happens if water vapor is present between the fuel rods and this will lead to a "passive" shutdown of the reactor. The flow of cooling water must be maintained when the reactor is shut down to remove the decay heat from the fission products in the fuel that continue spontaneous decay and energy release.

In fast flux Generation IV reactors, the reactor configurations and fuel isotope concentrations are such that the system relies on the collisions of fast neutrons to propagate the nuclear fission process. The higher energy neutrons allow direct use of fissionable materials (both fissile and fertile isotopes) to propagate the nuclear fission process.

The neutron absorption by U-238 leads to formation of all the transuranic elements formed in reactor fuel since each element formed is promoted by absorbing a neutron. In a fast flux reactor these actinide nuclei accept fast neutrons and they undergo fission. In thermal flux reactors, these higher actinides accumulate and contribute to the radioactive "waste" problem. The excitation and fission (energy release) of all actinides in Generation IV reactors represent an important step toward sustainable nuclear energy because this process reduces waste, makes fuel recycling easier. It allows total use of the uranium fuel. This includes the vast stockpiles of depleted uranium left from producing military highly enriched U-235 and the low enriched domestic fuel for domestic electric power plants.

TRANSMUTATION

For every 100 kg of fuel introduced into a light water reactor, about 3.4 kg of fission products are produced at refueling. Of these fission products, about 0.4 kg remain as high level radioactive waste after about 30 years of storage at the nuclear power plant. This 0.4 kg can be placed is a repository as "high-level" waste for about 1000 years to become stable, or it could be transmuted. The transmutation of I-125 (see insert) is considered viable with existing methods. Iodine is about 0.1 kg of the high-level waste in a metric ton of spent fuel. Other techniques could be developed for the other 0.3 kg. Following is from a Department of Energy (DOE) Report to Congress [4]. This provides a summary of transmutation possibilities.

What Is Transmutation?

Transmutation refers to the ability to transform one atom into another by changing its nuclear structure. This is accomplished by bombarding the atoms of interest with neutrons either in an accelerator or a nuclear reactor. In the context of spent nuclear fuel, transmutation can convert plutonium and other actinides into isotopes with more favorable characteristics.

β (release)

Neutron + I-129 → I-130 → Xe-130 (non-radioactive)

While plutonium-based fuels have been manufactured on a commercial basis, almost no work has been done on making or irradiating fuels that contain neptunium, americium, or curium.

Transmutation fuels that can significantly destroy the higher actinides should be capable of very high burnups to minimize the number of recycles required to reduce material losses during separations and fabrication steps. They should be easily fabricated in hot cells or some other remote environment due to the high radiation levels from the minor actinides. If these advanced fuels are to be useful candidates for potential deployment with Generation IV systems, research, development, and testing would be needed beyond Phase II. Advanced Fuel Cycle Initiative (AFCI), an internationally supported program, Series Two would apply considerable effort to evaluating the various fuel types that could serve as an optimum fuel for fast spectrum reactor or accelerator-driven transmutation systems.

The determination of the optimum fuel form for transmutation—a fuel that may be easily fabricated using remote handling technologies contributes to the safe operation of the reactor and results in a final waste form acceptable for a repository to be designated—is a major research objective of the program.

Oxide, nitride, metallic, dispersion, ceramic, and coated particle fuel forms are currently under investigation. Fabrication of several test fuel specimens of these fuel forms containing plutonium mixed with minor actinides is underway. The Department (DOE) plans to irradiate these fuels in the Advanced Test Reactor (ATR) in Idaho with a more ambitious follow-on irradiation program to be carried out in France by other European partners. A consortium of institutions is planning the construction of an experimental assembly containing minor actinide fuels that would come from several countries; this assembly would be irradiated in a French fast spectrum reactor (PHENIX).

Successful testing in the ATR and initiation of the French PHENIX tests during Phase II would permit DOE to select the most promising path forward for AFCI Series Two transmutation fuels including planning for potential Phase III scaled-up fast spectrum irradiations in foreign facilities.

Fast spectrum systems can be either fast reactors (which employ critical reactor cores that operate 12–18 months between refueling cycles) or accelerator-driven systems that employ reactor cores that are subcritical by nature (i.e., they need a constant source of neutrons to maintain a normal operating state). The external source of neutrons is produced by an accelerator and a target system. Both systems employ fast neutrons; however, the accelerator system has the advantage that it can transmute all radioactive elements without producing any plutonium in the process.

Accelerator systems are more expensive than fast reactors, and require significantly more research and development, although the fuel technology is basically the same.

While the Department, based on the systems analysis carried out in Phase I of this research, does not expect accelerator transmutation systems to be used as the primary transmuter of the long-lived toxic materials present in spent fuel, they may have an important role assuring the very low levels of toxicity that serve as the technology goals of this activity. The relatively high construction and operating costs of accelerator-based systems make them unsatisfactory for widespread application as commercial-scale transmuters. Fast reactor systems, however, may prove sufficiently economic to justify their eventual deployment—this is a key element of evaluation in the multinational Generation IV Nuclear Energy Systems Initiative. (The Generation IV International Forum: Update, October 2002 is included as Appendix B.)

Accelerator-Driven Systems Physics and Materials Research and Development

Many countries are considering accelerator-driven systems (ADSs) as a viable approach to transmutation because these systems may be capable of destroying long-lived radioactive isotopes of all types without making plutonium. An ADS consists of an accelerator that produces high-energy protons that strike a heavy metal target to produce high-energy (fast) neutrons through a spallation process to drive a subcritical reactor assembly.

Accelerator-driven transmutation (see insert) has been an important part of nuclear physics research for decades.

NUCLEAR FUSION

It is difficult to predict what energy options will be available in 30, 100, or 200 years. Nuclear fusion is the primary source of energy in the universe and may be an option that can be made available. The following is a summary of nuclear fusion energy as prepared by the Congressional Research Service (CRS) of the Library of Congress [5].

The potential benefits of controlled fusion are great. Successful development of a fusion power plant, however, is proving to be a very difficult scientific and technological challenge. Although progress has been steady, it may be at least 35–50 years before an operating demonstration plant is built.

Fusion occurs when the nuclei of light atoms, such as isotopes of the element hydrogen (deuterium and tritium), collide with sufficient energy to overcome the natural repulsive forces that exist between the protons of these nuclei (see Figure 8.10). When this collision takes place, a D-T reaction is said to have occurred. If the two nuclei fuse, an alpha particle, the nucleus of helium is formed with release of a neutron and lots of energy. For the fusion reaction to take place, the nuclei must be heated to a very high temperature and forced together. In a hydrogen bomb, this is done by exploding a fission bomb, uranium, or plutonium, producing the

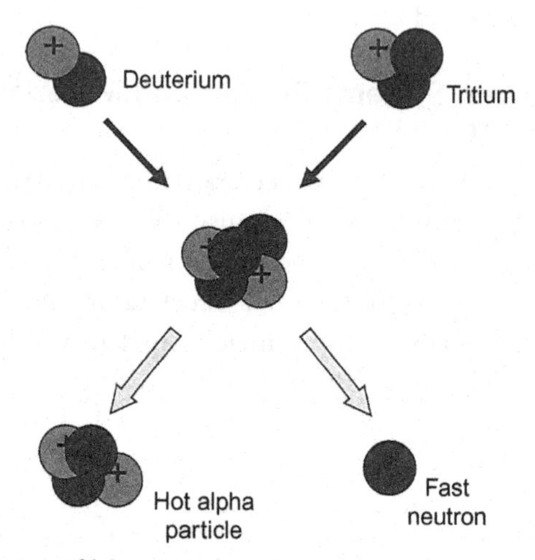

Figure 8.10 Illustration of laboratory fusion.

high temperature and pressure causing deuterium and tritium to fuse releasing a more powerful explosion.

Fusion reactions are possible between a number of light atoms, including deuterium alone (a D-D reaction); deuterium and helium-3, an isotope of the element helium (a D-3He reaction, see Chapter 3); and hydrogen and the element lithium, a light metal. All of these reactions occur much less frequently at a given temperature than the D-T reaction. For instance, the fusion energy produced from D-T reactions in a mixture of deuterium and tritium will be about 300 times greater than that from D-D reactions in a mixture of deuterium alone. For this reason, research into controlled fusion has concentrated on developing deuterium–tritium fueled fusion reactors.

Potential Benefits of Magnetic Fusion Energy Fuel Resources

The potential benefits of controlled fusion are many. Foremost is that in principle the fuel for a fusion power plant is essentially inexhaustible. One out of every 6670 water molecules is a deuterium atom. There are no technical barriers to extracting deuterium from water. Tritium, however, does not occur in nature. It can be produced from the element lithium, which is also abundant, although much less abundant than deuterium. To achieve the full resource potential of fusion energy will require reaching the conditions of plasma density, temperature, and confinement time necessary for energy production from reactions involving deuterium alone. As described below, these conditions are much harder to reach than for deuterium and tritium which has proved difficult enough.

Fusion researchers, however, note that even if success is reached with the D-T reaction, research will need to continue to reach power production from the D-D reaction.

Environmental and Safety Considerations

There also could be important environmental benefits from fusion. First, a controlled fusion power plant would be inherently safe. A reaction that became "uncontrolled" in such a plant would extinguish itself almost instantly with no part of the system melting and with no significant release of radioactive material. Even major accidents that could occur such as failure of the structure of a fusion power plant would not result in any radiation release. Of course, such an accident could result in significant cost because of severe reactor damage.

A second environmental benefit is that the radioactive waste products produced in a fusion plant would be less of a problem than those produced in a fission plant. Because of the nature of controlled fusion, it would be possible to reduce the long-term buildup of radioactive waste products by a factor up to a million times less that of a fission system of comparable size. The quantity of radioactive material produced in a power plant of a given size may be comparable for the two types of reactions (at least for the first generation, deuterium–tritium fusion plants), the half-life of the radioactive products from such a fusion plant would be on the order of 100 years or less. This compares to tens of thousands of years for those from a fission plant. Radioactive products from fusion plants, therefore, would decay much faster than those from fission plants, resulting in the large differences cited above. The counter-argument to this advantage of fusion is that the path to utilizing/destroying even the most dangerous radioactive products of a fission reactor is not only attainable, it has already been demonstrated as a viable technology.

More advanced fusion systems using fuel combinations that produce few or no neutrons, such as the D-3He reaction, would produce substantially less radioactive waste.

CRS-3 Paths to Fusion Energy Production

Two paths are being taken in attempts to attain controlled fusion. The first is to confine the light nuclei by a magnetic field and to heat them with an external source of electromagnetic energy. In this case, the deuterium and tritium are in a gas-like condition called "plasma." This process is called magnetic fusion energy (MFE). The other way is to heat clusters of very small spheres containing deuterium and tritium by compressing these clusters with powerful lasers or beams of particles. Such a process is called inertial confinement fusion (ICF) and simulates—on a very small scale—the process of a hydrogen bomb. Once the reaction starts in either case, it is possible that the heat generated by the fusion reactions would be sufficient to cause other light nuclei to collide sustaining the reaction without an external energy source. Such a condition, called ignition, has not yet been reached in practice. While substantial progress has been made over the last several years in both ICF and MFE, even the least stringent condition of break-even—the point where power produced by the fusion reactions equals the power supplied by the external energy source—is still to be achieved. A fusion power plant would operate between break-even and ignition. The ratio of power out to heating power supplied would be significantly greater than break-even, but external energy would still be supplied to control the reaction rate.

By way of comparison, stars operate by using their enormous mass and gravitational force to confine the colliding nuclei. Enough heat is generated by the fusion reactions to force other nuclei to collide and undergo fusion so the reaction is sustained. Because of the large gravitational forces, these nuclei are unable to escape to the stellar region before they gain the necessary energy to fuse with one another.

Achieving break-even and power amplification would be only the first steps in the process of producing useful power. The energy from the nuclear reactions would have to be converted to another form that could be used to do work. Energy is carried away from the fusion reactions in the form of neutrons moving at high speed. Because neutrons do not have an electrical charge, they are not confined by the magnetic field and will leave the plasma region. The neutrons will give their energy up if they collide with atoms of another material, causing that substance to heat. A prime candidate for this material for future fusion power plants is the liquid metal lithium. Lithium that is heated by colliding neutrons could then transfer that heat to water, producing steam. The steam, in turn, would drive a steam turbine and generator, producing electricity. While there are no fundamental scientific barriers to this process, putting it into practice will be a complicated engineering task requiring substantial development. A second reason for using lithium is that reaction between the lithium atoms and the neutrons would produce the tritium necessary for the reactor fuel.

It is also true that the water used to transfer heat from a PWR (a fission reactor) must be stored to allow the tritium to decay before the water is released. This is in the operating manual at the Callaway Nuclear Plant.

Magnetic Fusion Energy Research

Both MFE and ICF research activities have been funded by the U.S. DOE ([http://www.doe.gov]). The ICF program currently is primarily oriented to defense applications, for simulation of nuclear weapons, although energy applications are an important part of the research effort. Nearly all of the funds for ICF research come from DOE's Defense Programs IB91039 01-15-02 (http://www.dp.doe.gov). A major initiative of the DOE ICF program is the National Ignition Facility (NIF) (http://www-lasers.llnl.gov/lasers/nif.html) at DOE's Lawrence Livermore National Laboratory which is currently entering the detailed engineering design stage. The NIF is primarily for weapons applications, but it will also carry out important research for potential energy production from inertial fusion.

MFE research is within DOE's civilian programs and is located in the Office of Energy Research. Although funding for ICF research now exceeds that for magnetic fusion, the latter has been and continues to be the major fusion energy focus in the United States.

While it is difficult to predict what energy options will be available in 30, 100, or 200 years, it is with certainty that there are hundreds of years of energy available from nuclear fission using known technology and available fuel.

It may be possible for the first Generation IV reactors to be in operation by 2040—25 years from now. Several of the Generation IV reactors have been demonstrated. The CRS report estimates fusion reactors in "*at least 35 to 50 years before an operating power plant is built*"—fusion power production has yet to be demonstrated as viable.

There has always been a difference between what is practiced and what is known to be practical in nuclear technology. This also goes for what is being optimistically projected (for fusion) relative to what the data show is practical.

RADIOLOGICAL TOXICOLOGY

The radioactivity of uranium ore is often considered a threshold level of acceptable radiation. In practice, a concentrated uranium ingot can be handled with little concern of radioactive toxicology. Handling fuel pins need not be performed remotely when preparing fuel for nuclear reactors (one of the downsides of reprocessing methods is that the fuel will be radioactive and will require remote handling).

A brief introduction to radiation poisoning is necessary to understand the risks of radiation and methods for reducing risks. Both the US Environmental Protection Agency and US Nuclear Regulatory Commission (NRC) have Web sites that detail how one can be exposed to radiation poisoning and the impact of that exposer.

The following is an EPA summary on sources of radiation and radiation poisoning [6]:

What is radiation?—*Radiation is energy that travels in the form of waves or high-speed particles.*

When we hear the word 'radiation,' we generally think of nuclear power plants, nuclear weapons, or radiation treatments for cancer. We would also be correct to add 'microwaves, radar, electrical power lines, cellular phones, and sunshine' to the list. There are many different types of radiation that have a range of energy forming an electromagnetic spectrum. However, when you see the word 'radiation' on this Website, we are referring to the types of radiation used in nuclear power, nuclear weapons, and medicine. These types of radiation have enough energy to break chemical bonds in molecules or remove tightly bound electrons from atoms, thus creating charged molecules or atoms (ions). These types of radiation are referred to as 'ionizing radiation.'

What's the difference between radiation and radioactivity?—*Radiation is the energy that is released as particles or rays, during radioactive decay. Radioactivity is the property of an atom that describes spontaneous changes in its nucleus that create a different element. These changes usually happen as emissions of alpha or beta particles and often gamma rays. The rate of emission is referred to as a material's "activity."*

Each occurrence of a nucleus throwing off particles or energy is referred to as disintegration. The number of disintegrations per unit time (minutes, seconds, or hours) is called the activity of a sample. Activity is expressed in curies. One curie equals 37 billion disintegrations per second. (Since each disintegration transforms the atom to a new nuclide, transformation is often substituted for disintegration in talking about radioactive decay and activity.)

Exposure from radiation can occur by direct exposure, inhalation, and indigestion.

Direct (External) Exposure—*The concern about exposure to different kinds of radiation varies:*

- *Limited concern about alpha particles. They cannot penetrate the outer layer of skin, but if you have any open wounds you may be at risk. (Note: prolonged exposure should be avoided)*
- *Greater concern about beta particles. They can burn the skin in some cases, or damage eyes.*
- *Greatest concern is about gamma radiation. Different radionuclides emit gamma rays of different strength, but gamma rays can travel long distances and penetrate entirely through the body.*

Gamma rays can be slowed by dense material (shielding), such as lead, and can be stopped if the material is thick enough. Examples of shielding are containers; protective clothing, such as a lead apron; and soil covering buried radioactive materials.

Inhalation—*Exposure by the inhalation pathway occurs when people breathe radioactive materials into the lungs. The chief concerns are radioactively contaminated dust, smoke, or gaseous radionuclides such as* radon.

Radioactive particles can lodge in the lungs and remain for a long time. As long as it remains and continues to decay, the exposure continues. For radionuclides that decay slowly, the exposure continues over a very long time. Inhalation is of most concern for radionuclides that are alpha *or* beta *particle emitters. Alpha and beta particles can transfer large amounts of energy to surrounding tissue, damaging DNA or other cellular material. This damage can eventually lead to cancer or other diseases and mutations.*

Ingestion—*Exposure by the ingestion pathway occurs when someone swallows radioactive materials. Alpha and beta emitting radionuclides are of most concern for ingested radioactive materials. They release large amounts of energy directly to tissue, causing DNA and other cell damage.*

Ingested radionuclides can expose the entire digestive system. Some radionuclides can also be absorbed and expose the kidneys and other organs, as well as the bones. Radionuclides that are eliminated by the body fairly quickly are of limited concern. These radionuclides have a short biological half-life.

Shielding and the distance between the radiation emitting source and the person achieve minimizing direct exposure to radiation. Reduce the time in the presence of the radiation-emitting object. Minimizing inhalation and indigestion is achieved by keeping radioactive isotopes out of the environment. Once radiation is in the environment, the materials can be removed or isolated so the isotopes do not get into water, air, or vegetation.

The following NRC summary describes health effects upon radiation exposure [7].

Biological Effects of Radiation—*We tend to think of biological effects of radiation in terms of their effect on living cells. For low levels of radiation exposure, the biological effects are so small they may not be detected. The body has repair mechanisms against damage induced by radiation as well as by chemical carcinogens. Consequently, biological effects of radiation on living cells may result in three outcomes: (1) injured or damaged cells repair themselves, resulting in no residual damage; (2) cells die, much like millions of body cells do every day, being replaced through normal biological processes; or (3) cells incorrectly repair themselves resulting in a biophysical change.*

The associations between radiation exposure and the development of cancer are mostly based on populations exposed to relatively high levels of ionizing radiation (e.g., Japanese atomic bomb survivors, and recipients of selected diagnostic or therapeutic medical procedures). Cancers associated with high dose exposure (greater than 50,000 mrem) include leukemia, breast, bladder, colon, liver, lung, esophagus, ovarian, multiple myeloma, and stomach cancers. Department of Health and Human Services literature also suggests a possible association between ionizing radiation exposure and prostate, nasal cavity/sinuses, pharyngeal and laryngeal, and pancreatic cancer.

The period of time between radiation exposure and the detection of cancer is known as the latent period and can be many years. Those cancers that may develop as a result of radiation exposure are indistinguishable from those that occur naturally or as a result of exposure to other chemical carcinogens. Furthermore, National Cancer Institute literature indicates that other chemical and physical hazards and lifestyle factors (e.g., smoking, alcohol consumption, and diet) significantly contribute to many of these same diseases.

Although radiation may cause cancers at high doses and high dose rates, currently there are no data to unequivocally establish the occurrence of cancer following exposure to low doses and dose rates—below about 10,000 mrem (100 mSv). Those people living in areas having high levels of background radiation—above 1000 mrem (10 mSv) per year—such as Denver, Colorado have shown no adverse biological effects.

Even so, the radiation protection community conservatively assumes that any amount of radiation may pose some risk for causing cancer and hereditary effect, and that the risk is higher for higher radiation exposures. A linear, no-threshold (LNT) dose response relationship is used to describe the relationship between radiation dose and the occurrence of cancer. This dose-response model suggests that any increase in dose, no matter how small, results in an incremental increase in risk. The LNT hypothesis is accepted by the NRC as a conservative model for determining radiation dose standards recognizing that the model may over estimate radiation risk.

High radiation doses tend to kill cells, while low doses tend to damage or alter the genetic code (DNA) of irradiated cells. High doses can kill so many cells that tissues and organs are damaged immediately. This in turn may cause a rapid body response often called Acute Radiation Syndrome. The higher the radiation dose, the sooner the effects of radiation will appear, and the higher the probability of death. This syndrome was observed in many atomic bomb survivors in 1945 and emergency workers responding to the 1986 Chernobyl nuclear power plant accident. Approximately 134 plant workers and firefighters battling the fire at the

Chernobyl power plant received high radiation doses—80,000 to 1,600,000 mrem (800 to 16,000 mSv)—and suffered from acute radiation sickness. Of these, 28 died within the first three months from their radiation injuries. Two more patients died during the first days as a result of combined injuries from the fire and radiation.

Because radiation affects different people in different ways, it is not possible to indicate what dose is needed to be fatal. However, it is believed that 50% of a population would die within thirty days after receiving a dose to the whole body, over a period ranging from a few minutes to a few hours, between 350,000 to 500,000 mrem (3500 to 5000 mSv). This would vary depending on the health of the individuals before the exposure and the medical care received after the exposure. These doses expose the whole body to radiation in a very short period of time (minutes to hours). Similar exposure of only parts of the body will likely lead to more localized effects, such as skin burns.

Conversely, low doses—less than 10,000 mrem (100 mSv)—spread out over long periods of time (years to decades) don't cause an immediate problem to any body organ. The effects of low doses of radiation, if any, would occur at the level of the cell, and thus changes may not be observed for many years (usually 5–20 years) after exposure.

Genetic effects and the development of cancer are the primary health concerns attributed to radiation exposure. The likelihood of cancer occurring after radiation exposure is about five times greater than a genetic effect (e.g., increased still births, congenital abnormalities, infant mortality, childhood mortality, and decreased birth weight). Genetic effects are the result of a mutation produced in the reproductive cells of an exposed individual that are passed on to their offspring. These effects may appear in the exposed person's direct offspring, or may appear several generations later, depending on whether the altered genes are dominant or recessive.

Although radiation-induced genetic effects have been observed in laboratory animals (given very high doses of radiation), no evidence of genetic effects has been observed among the children born to atomic bomb survivors from Hiroshima and Nagasaki.

ENERGY EFFICIENCY IN THE NUCLEAR ENERGY INDUSTRY

Providing a sustainable source of energy is a universal goal for all nations. The world population continues to grow and an adequate energy supply must be in the plans to maintain or improve the quality of life. Sustainability is a strategy to meet the energy needs of the present generation and increase our ability to serve the demands of future generations.

There are more than 437 nuclear power plants that currently provide about 11% of the world's electricity. In the United States, about 19.5% of the electricity is produced with 99 nuclear power plants [8,9]. There are 66 plants under construction in the world—five of them in the United States. Most of these new plants will be Generation III designs modified to improved energy efficiency, safety, and proliferation security.

Today, nuclear energy represents a commitment to the production of electricity. This recommends there be a plan for the future. Such a collaborative was formed in 2006 called the Global Nuclear Energy Partnership charged to develop and deploy advanced nuclear fuel cycle technologies. This program faltered and has been replaced by the International Framework for Nuclear Energy Cooperation in June 2010 [10].

Under this plan, recycling spent nuclear fuel will greatly reduce the amount of "nuclear waste" destined for disposal. This program will require advanced Generation IV fast neutron flux reactors that use the transuranic (elements beyond uranium in the periodic table—this includes plutonium) as fuel. Experimental fast flux reactors have demonstrated long lived fission products can be transmuted (change their atomic number and reduce or eliminate their radioactivity) by exposure to fast neutrons.

The early nuclear reactor technology has been proven reliable and economic on a commercial scale without the environmental impacts of fossil fuel power plants. Fossil fuel power plants (burning coal and natural gas) contribute the major fraction of our electric power but they are also major sources of the increasing concentration of greenhouse gases in the atmosphere. The sustainable future of nuclear power depends on improving the technology for new energy systems to replace old nuclear plants as they are retired from service.

New generation nuclear power plants will need to meet the performance standards on safety, low environmental impact, and competitive prices. Since all of these standards are measured per kWh of electrical power produced, improved thermal efficiency is a win−win situation and will be discussed first. After a discussion of thermal efficiency, the Generation IV designs will be reviewed. Finally, the lessons history offers will be discussed.

As the name "heat engine" implies, heat engines are based on methods of converting high-temperature energy (heat) into work and discharging low-temperature heat. The illustrative example of Figure 8.11 is for a steam cycle operating at 33% thermal efficiency. For every

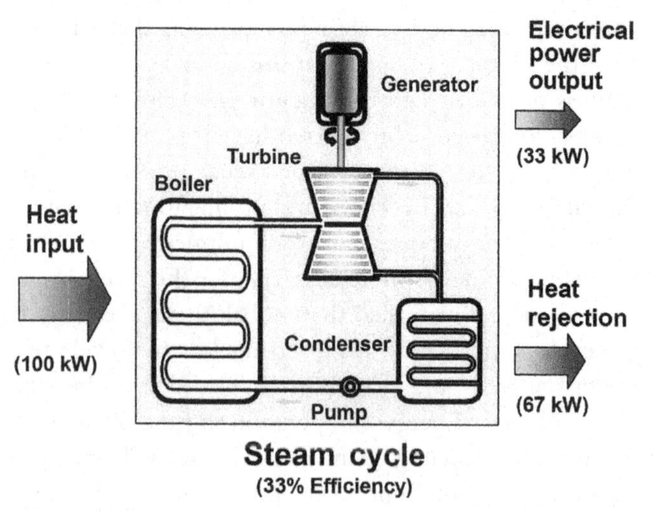

Figure 8.11 Illustration of steam cycle operating at 33% thermal efficiency.

100 kW of high temperature heat going into the engine, 33 kW of work is produced and 67 kW of waste heat is rejected into the environment. As illustrated by the following equation, thermal efficiency is defined as the net work produced divided by the heat input from the high-temperature reservoir. An energy balance on a power plant also allows the thermal efficiency to be written in terms of the heat at high temperature and heat rejection at low temperature.

$$\begin{aligned} \text{Eff}_{\text{thermal}} &= \text{Net work produced}/\text{Heat input} \\ &= (Q_H - Q_L)/Q_H = 1 - Q_L/Q_H \end{aligned} \tag{8.1}$$

The majority of advances in heat engine technology targets increasing the thermal efficiency. A French engineer Nicolas Leonard Sadi Carnot (1796–1832) recognized that the thermal efficiency of a heat engine increased with increasing temperatures of the heat input and decreasing temperatures of the heat rejection. The best possible efficiency for a given source of heat and reservoir for rejecting heat is the Carnot cycle. The Carnot cycle is a reversible heat engine operating from hot and cold reservoirs at constant temperatures of T_H and T_L, respectively. Equation 8.2 provides the thermal efficiency of the Carnot cycle where the temperatures are in degrees Kelvin, the absolute temperature scale.

$$\text{Eff}_{\text{Carnot}} = 1 - T_L/T_H \tag{8.2}$$

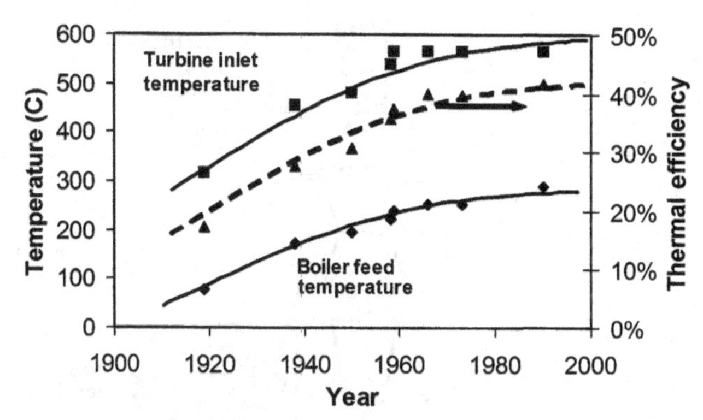

Figure 8.12 Evolution of thermal efficiency in steam cycle. Higher temperature steam turbine operation was a key improvement.

Equation 8.2 indicates that as the temperature of heat input in a power cycle increases, the thermal efficiency increases. This trend is verified by the historic data of Figure 8.12 that shows the evolution of the steam power cycle (typically coal fired).

For large commercial power plants, T_L is fixed by the environment, (often a river, lake, or a cooling tower) because it is the only place large enough to take in vast amounts of heat without an increase in temperature. Heat rejection during warm summers increases T_L in Equation (8.2), and this lowers the efficiency. A major component of power plants are the cooling towers that circulate and evaporate water to provide a practical low temperature place to reject the low temperature heat from the steam turbine.

Most locations have climates that allow the cooling water heat rejection at 40°C or less year round. Because the heat rejection temperature is controlled by local climate, the only degree of freedom in the Carnot cycle equation for increasing efficiency is to increase the high-temperature energy source (T_H).

Figure 8.13 illustrates the basic steam cycle. The generator is driven by the turbine and produces electrical power. Steam is condensed by cooling it in the condenser—the flow of cooling water through the condenser results in a heat rejection to the surroundings. The heat input occurs in a boiler (not shown). The boiler produces steam that is directed to the turbine(s), and this steam is returned to the boiler in the form of liquid water with a pump to overcome the pressure drop through the turbine(s). The work put into the pump is small compared to the power produced by the turbines resulting in a net production of electrical power.

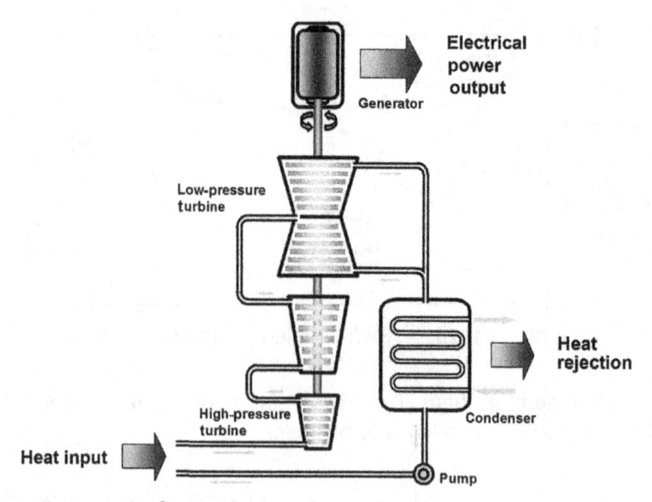

Figure 8.13 Illustration of staged expansion.

In practice, T_H is the highest temperature that a working fluid reaches in a power cycle. This is usually the temperature of the working fluid as it leaves the heat exchanger producing the steam just prior to expansion in the turbine. In a nuclear reactor, the maximum temperature occurs when the working fluid is in contact with the fuel rods; in a natural gas power plant it is the combustion temperature; and in a pulverized-coal-fired power plant it is the temperature of steam as it exits the steam super heater.

In practice, heat input is not at a constant temperature since the working fluid increases in temperature as it is heated (or as combustion takes place with a gas turbine). The Joule efficiency is defined to take into account that practical engines do not operate with all heat input at a constant temperature. Equation 8.3 defines the Joule efficiency.

$$\text{Eff Joule} = 1 - T_{\text{Lavg}}/T_{\text{Havg}} \tag{8.3}$$

Because the Joule efficiency does not account for process irreversibility, further modification is needed to correlate with actual processes. A correlation effective for the historic data of Figure 8.12 applies an overall reversibility factor (f) that indicates that the low-temperature heat rejection increases with increasing irreversibility. This empirical formula is provided by Equation 8.4.

$$\text{Eff}_{\text{modified Joule}} = 1 - T_{\text{Lavg}}/[f\ T_{\text{Havg}}] \tag{8.4}$$

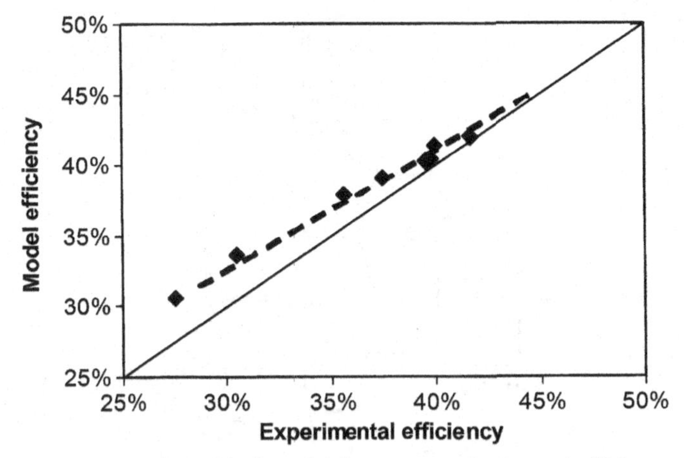

Figure 8.14 Accuracy of empirical model for power cycle thermal efficiency.

Figure 8.14 compares the historic data with Equation 8.4 and a reversibility factor of $f = 0.77$ represents the performance of the steam power cycle.

In the correlation of Figure 8.14, T_{Lavg} was taken as 313 K (40°C) and T_{Havg} was the arithmetic average of the boiler feed temperature and the turbine inlet temperature. Based on this correlation, the efficiency increases with the average temperature at which heat is received by the working fluid. Implicit in this correlation is that good design practices and efficient turbines/pumps are used. The reversibility factor of 0.77 is obtained with state-of-the-art turbines and pumps as well as designs where the minimum approach temperatures for heat exchangers is low, about 10°C.

Practical Brayton power cycles fueled with natural gas depend on materials development increasing the temperatures at which the metals of turbines can operate and large heat exchangers can be economically manufactured. As illustrated by the trends in boiler feed temperature of Figure 8.12, regenerative heating of the working fluid is just as important as increasing turbine operating temperatures to increase T_{Havg} and the thermal efficiency of the power cycle.

After partial expansion of the steam, some of it is diverted to feed water heaters. This feed water heating uses the lower quality energy of partially expanded steam rather than heat provided by combustion or the nuclear reactor. Moving from 17% to 42% thermal efficiency, the number of heaters increased from 2 to 8 or more. Higher pressures are necessary to increase the boiler feed temperature above 290°C—the higher

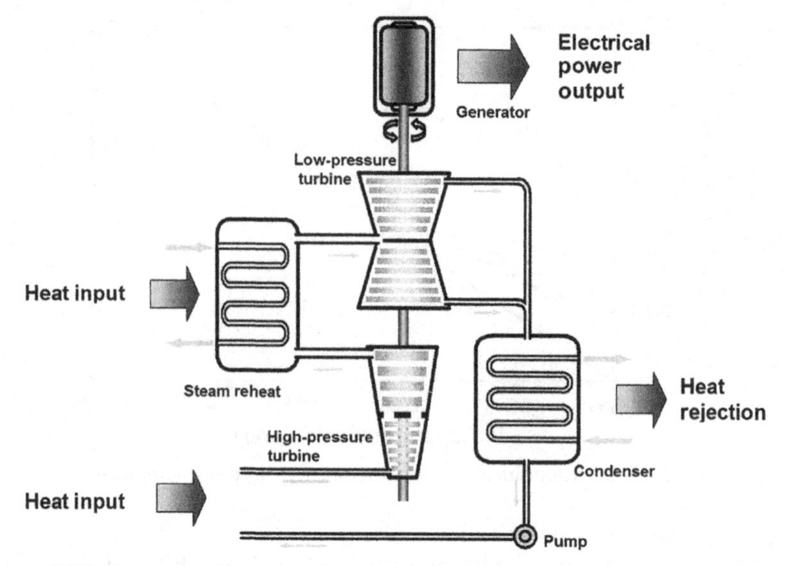

Figure 8.15 Illustration of steam reheat in power cycle.

pressures required reheat of the steam after expansion through the high pressure turbine. Moving from 17% to 42% thermal efficiency required reheating the steam two times as it moved from the inlet high pressure to the turbine exit pressure.

Figure 8.15 shows a steam cycle with one steam reheat and one feed water heater. When steam is produced at higher pressure, a steam reheat is used to keep excessive water from condensing in the turbine. Excessive condensation leads to erosion of the turbine blades and failure of the turbine. Reheating the steam before completing expansion in the low-pressure turbine avoids the excessive condensation problem and provides additional high-temperature heat input that increases the thermal efficiency.

Both open and closed feed water heaters can be used to preheat the boil feed water. A small amount of condensing steam heats the feed water to the temperature of the steam. Higher steam pressures, repeated steam reheat, and multiple feed water heaters were all needed in the evolution of the steam cycle to achieve the higher T_{Havg} and converting more heat to work.

Figure 8.16 superimposes the increases in Carnot, Joule ($T_{Lavg} = 313$ K), and modified Joule ($f = 0.77$) efficiencies as the working fluid (steam) temperature increases. Nuclear and coal-fired power plants closely follow the modified Joule curve since the reversibility factor of 0.77 is characteristic of current turbine and regenerative heat-transfer efficiencies. Natural gas combined cycles do not follow the modified Joule

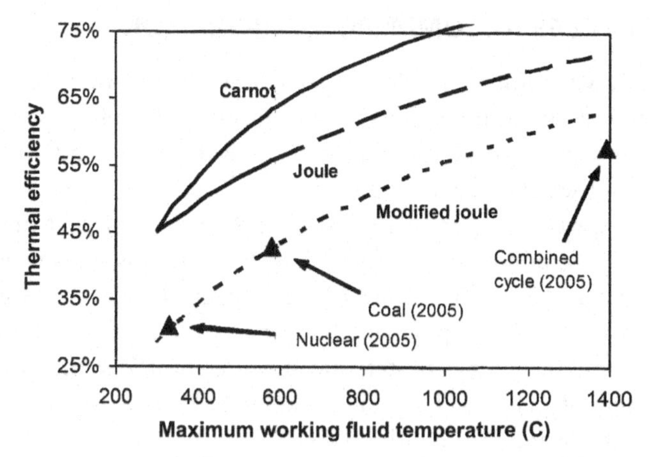

Figure 8.16 Comparison of efficiency projections of different models. The Joule and Modified Joule models assume a feed temperature of 313 K.

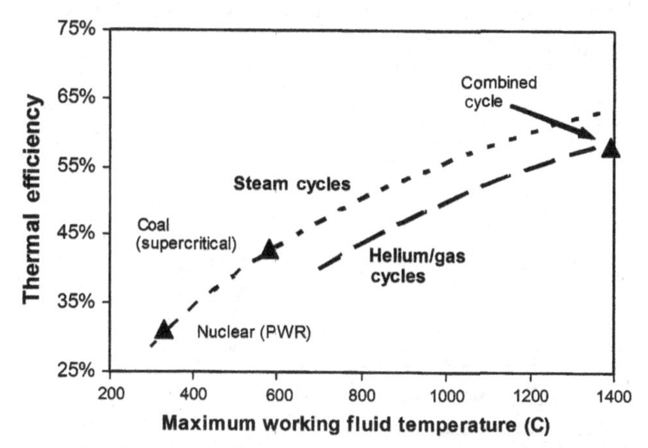

Figure 8.17 Projected thermal efficiencies as a function of maximum steam temperature and a low temperature of 313 K.

curve because the lost work associated with air compression and heat transfer to the low temperatures cycle are inefficiencies of the combined cycle that are not included in the inefficiencies of the steam cycle.

Figure 8.17 shows the (based on the modified Joule equation) performance potential for the steam cycle and combined cycles. The correlation in Figure 8.17 represents the goals for the new Generation IV reactors. The higher nuclear reactor temperatures produce higher thermal efficiency producing electricity from nuclear power. Higher thermal efficiencies yield reductions in both capital and fuel costs for the nuclear power system.

STEAM CYCLES IN COMMERCIAL OPERATION

The concepts for improved efficiency of heat engines are well known. It is the practical design limitations of current nuclear boiling water reactors (BWRs) and pressurized water reactors (PWRs) that limit the thermal efficiency of these nuclear power plants.

For comparison, Figure 8.18 illustrates a pulverized coal-fired power plant. Coal is ground into powder so that when it is introduced into a flame it burns similar to a spray of liquid fuel. The hot flue gases rise from the flames to steel pipes in the upper section of the fire box that comprise the boiler, super heater, and steam re-heater of the power plant steam generator. The steel piping contains the liquid, vapor, and super-critical fluids passing through the boiler.

The materials used to fabricate the pipes limit the high temperature and pressure of the steam generator. Multiple steam reheats can be placed between the partially expanded steam flow as it passes between the stages of the steam turbine.

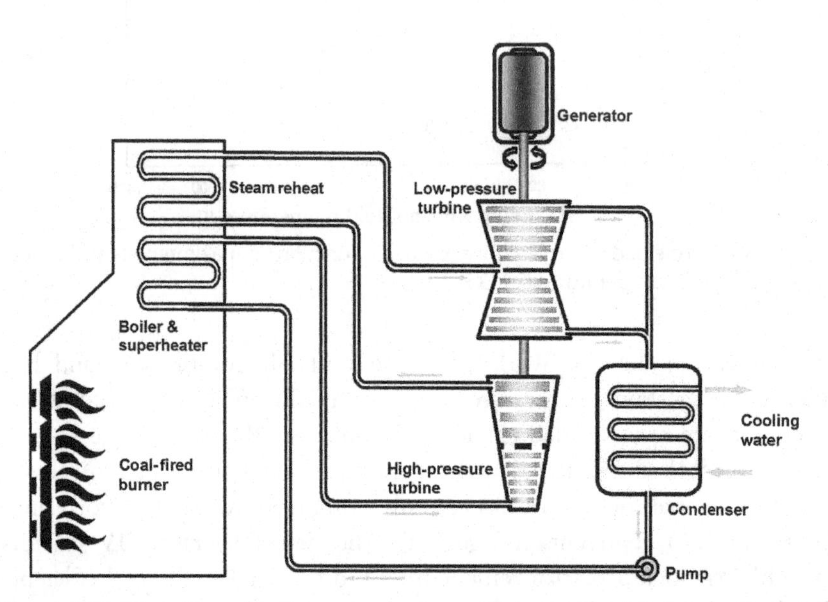

Figure 8.18 Illustration of boiler, super heater, and steam reheat in a pulverized coal power plant.

Boiling Water Reactors

Figure 8.19 is a diagram of a nuclear BWR. Water enters the reactor, pre-heated by the feed water heaters (to about 150°C, not shown). Both the pressure and temperature in the reactor are maintained below the critical points of water (374°C, 221.MPa). The operating temperature is set near 286°C and the pressure near 70 MPa.

The water surrounding the fuel rods in the core of the BWR must be maintained as liquid because the core is designed for water to serve as the neutron moderator (slow the neutrons). The core is designed to operate with the neutrons dissipating most of their energy (velocity) through collisions with water molecules before colliding with nuclear fuel. If water vapor bubbles are present an insufficient number of collisions with water occur and the neutrons have too high an energy to produce fission, and the nuclear fission reaction will not be self-sustaining. While this is a desirable feature in case of pump failure, normal operation requires that liquid water surround the fuel rods. In a BWR, the fraction of vapor in the core can be adjusted by changing the circulation rate of water through the core, the water circulation rate works to control the nuclear fission rate.

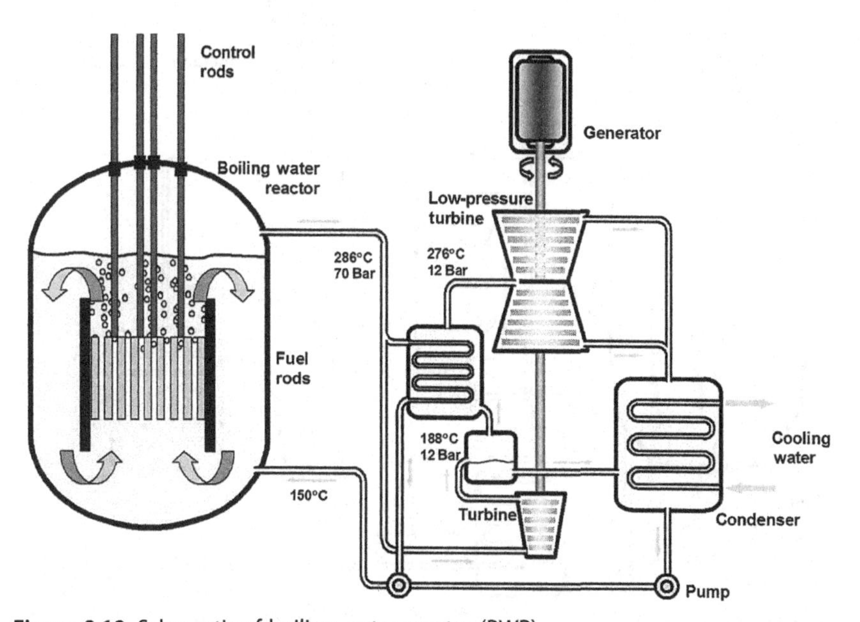

Figure 8.19 Schematic of boiling water reactor (BWR).

BWR systems employ high-volume jet pumps (not shown) to assist the circulation of water through the reactor core. Steam is formed, but the high water circulation rate rapidly carries the steam to the top of the reactor vessel where it is separated from the water and flows to the steam turbines.

Steam leaving the BWR is usually saturated. At saturated steam near 286°C, expansion through a (condensing) turbine produces liquid water. This water must be removed when about 10% in the steam condenses. As illustrated in Figure 8.19, the BWR power cycle uses staged expansion to remove the condensed steam in a separator (~188°C) rather than as a means to superheat the steam to attain higher efficiencies. Heating this 188°C saturated steam with 286°C primary steam produces a superheated steam that can then be expanded through the turbine.

The T_{Havg} of the BWR is quite low at about 218°C; however, the thermal efficiencies are about 33%.

Pressurized Water Reactors

The PWR uses a closed cycle with water in a isolated, pressurized water loop circulated between the reactor core and heat exchangers that produce steam for the steam turbine power cycle. Figure 8.20 is a schematic

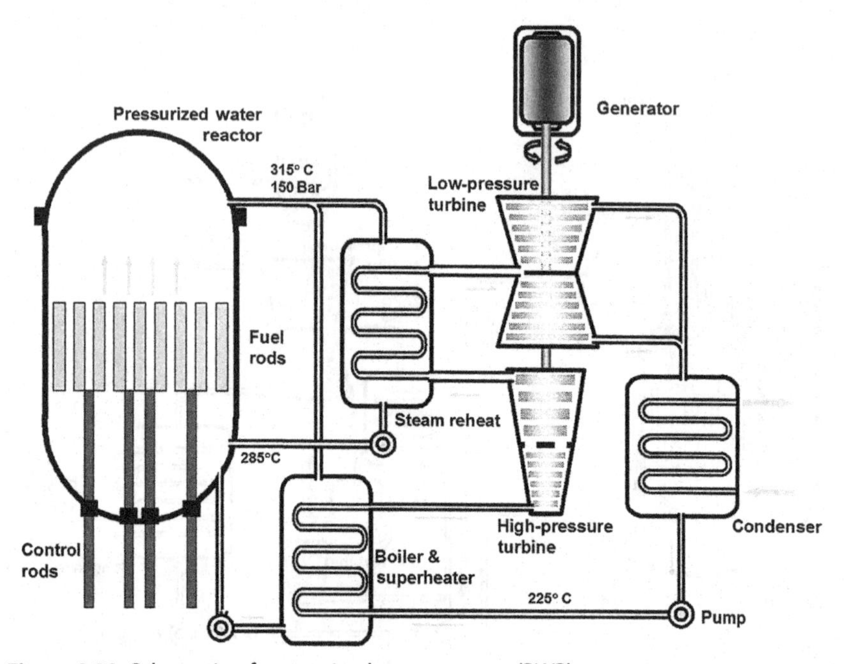

Figure 8.20 Schematic of pressurized water reactor (PWR).

diagram of a PWR. Borate (boric acid) is added to this water to absorb neutrons during the early part of new fuel cycle. As fission products build up in the fuel, they absorb neutrons and the borate concentration is reduced to maintain uniform power production. The water remains liquid under pressure and leaves the reactor at 315°C and 150 bar (the bubble point pressure of water is 105.4 bar).

The closed cycle design of PWR all but eliminates possible radioactive contamination of the power cycle's working fluid (steam/water). If there is a leak in a fuel rod the radioactive "fuel spill" is contained in the reactor cooling water loop. This keeps the radioactive elements for the steam turbine—a "mess" to decontaminate. This is one reason, commercial PWRs outnumber commercial BWRs by about 3:1.

A boiler, super heater, and reheat are used with the BWR similar to a coal-fired facility, but operating at lower temperature and pressure. In principle, the PWR reactor can attain higher efficiencies than the BWR, but the extra water circulation loop limits the upper end of the efficiency at about 33%.

GENERATION IV NUCLEAR POWER PLANTS

The light water reactors (BWR and PWR) are Generation II reactor designs with the BWR and PWR comprising 90% of the nuclear reactors in the United States and 80% of the nuclear reactors in the world. Table 8.12 lists the most promising of the Generation IV reactors along with typical maximum temperatures for the power cycles associated with each design.

This table shows that the anticipated maximum cycle temperatures will be over 500°C; each of these Generation IV systems will attain thermal efficiencies in excess of 40%. Thermal efficiencies up to 50% are possible with the higher temperature systems. This means that a 1 GW power plant becomes a 1.3–1.5 GW power plant using the same amount of fuel.

Generation IV Reactor Systems

The early or prototype nuclear power reactors built in the 1950s and 1960s are classified as Generation I energy systems. This experience provided the technology improvement to the Generation II light water moderated reactors. These were deployed in the 1970s and are most of the commercial reactors in the United States today. The evolution of these

Table 8.12 Summary of nuclear reactor designs and operating temperatures

System	Abbreviation	Typical Tmax[a] (°C)	Fast flux
Generation II			
Boiling water reactor	BWR	288	No
Pressurized water reactor	PWR	300	No
Generation IV			
Gas-cooled fast reactor system	GFR	850	Yes
Lead-cooled fast reactor system	LFR	540/790	Yes
Molten salt reactor system	MSR	680/780	Other, with full actinide recycle
Sodium-cooled fast reactor system	SFR	540	Yes
Supercritical-water-cooled reactor system	SCWR	510/540	Option
Very-high-temperature reactor system	VHTR	990	No

[a]Temperatures are for working fluid in the power cycle. A 10°C minimum approach temperature is assumed for each heat transfer process for the indirect systems.

designs with advances in control, safety, and economics make up the Generation III light water reactors that have been deployed outside the United States in the 1990s. An indication of the success of nuclear reactors deployed in the United States is the numbers from 2002: They produced 790 billion kilowatt-hours of electricity at an average cost less than 1.70 cents per kilowatt-hour. Three billion tons of air emissions would have been released by fossil fuel plants producing this electrical energy in 2002 [11]. This historical record drives the plan that the future of energy sources move toward nuclear energy replacing fossil fuels. These are designated Generation IV nuclear energy systems.

Ten nations have joined to develop technology goals for Generation IV energy systems with sharp focus on four areas: sustainability, economics, safety and reliability, plus proliferation resistance and physical protection [12]. Experts working in teams selected six Generation IV energy systems that should be considered as candidates for long-term (30 years) development and deployment. They are listed alphabetically in Table 8.12.

Electricity is the primary product of the current fleet of commercial nuclear power plants. Some of the Generation IV energy systems will be designed to serve the dual role of providing high-temperature thermal energy for chemical processing as well as commercial electricity. The near term nuclear power system development program for the United States will focus on electric power generation and hydrogen production. Hydrogen will be used as an "environmentally clean" transportation fuel to gradually replace gasoline and diesel fuel, a major source of pollution in high population density regions. The US DOE is pressing research and development for near-term deployment of the very-high-temperature reactor (VHTR) system and the supercritical-water-cooled reactor (SCWR) nuclear energy systems.

Supercritical-Water-Cooled Reactor

The mission of the SCWR is the production of low-cost electricity. There are two proven technologies that support the selection of this energy system: Liquid water-moderated reactors (LWRs) are common and therefore provide operating history for development of the SCWR. Coal-fired super critical water boilers are in operation around the world so the steam end of this energy system has been developed. The SCWR reactor core, based on the US LWR experience, would be contained in a pressure vessel with the high temperature, high-pressure supercritical water expanding directly into the steam turbine. The fuel would be low enriched uranium oxide with no need for new fuel development or new fuel reprocessing technology. The increased temperature and pressure will require additional study of the structural material oxidation, corrosion, stress cracking, embrittlement, and creep (dimensional and microscopic stability) all required to assure the design life of the reactor system. The SCWR design would increase thermal efficiency gained with the higher temperature to the steam turbine, but there will remain the once-through fuel cycle that characterizes the current LWR reactors. Long-term sustainability will require reprocessing additional LWR spent fuel [13]. Figure 8.21 is a schematic diagram of the SCWR. It is much like the BWR Schematic in Figure 8.19.

Very-High-Temperature Reactor

The VHTR will be designed to produce both electricity and hydrogen [14]. Helium will be circulated through the reactor core at high pressure to pick up thermal energy. Some of the hot helium is passed through a

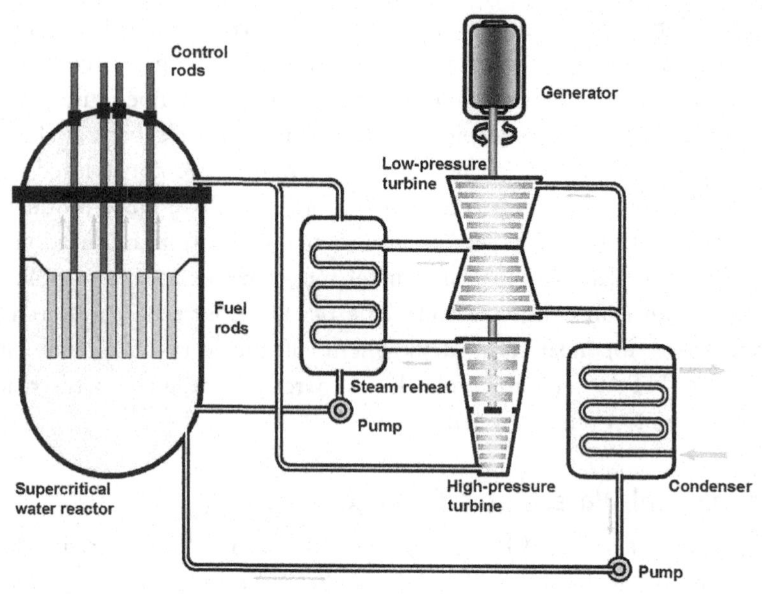

Figure 8.21 Schematic of supercritical-water-cooled reactor (SCWR).

high-temperature heat exchanger to provide process heat. Most of the hot helium will be expanded through a gas turbine to generate electricity and turn the compressors that return the cooled, low-pressure helium to the reactor core pressure. This is an application of the classical Brayton cycle gas-turbine engine for producing work from a hot gas [15]. One proposal uses the electricity to produce hydrogen by electrolysis of high-temperature steam. Figure 8.22 is a schematic of the VHTR system.

The design planned for the VHTR will be a graphite moderated thermal neutron spectrum reactor. The reactor core might be a prismatic graphite block core or it could be a pebble bed reactor. The fuel pebbles could be uranium metal or oxide particles uniformly distributed in porous graphite surrounded by solid graphite, and coated with silicon carbide (tricoated-isotropic (TRISO)-coated gas reactor fuel particles). Each pebble would contain the fission product gases and solids during the irradiation lifetime of the pebble. One proposal is to circulate the pebbles; withdrawing them from the bottom of the reactor vessel and introducing them at the top. The pebbles could then be inspected for physical damage and monitored using the emitted gamma radiation to measure fuel burn-up. Damaged or spent fuel pebbles would be sent to fuel reprocessing and new fuel added to maintain the fuel inventory.

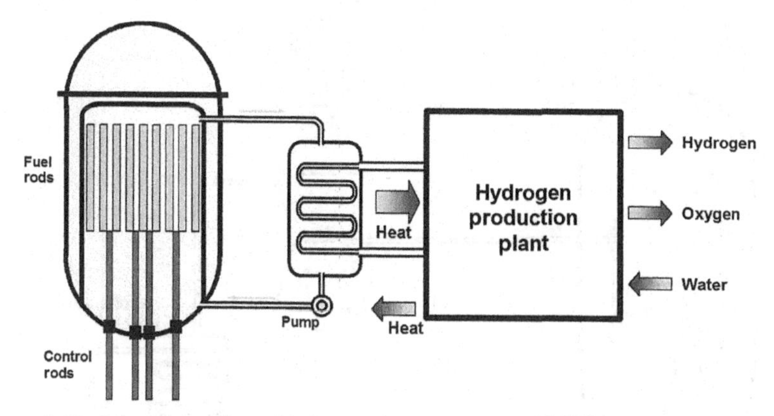

Figure 8.22 Schematic of very-high-temperature reactor (VHTR).

The prismatic graphite block core would be rigid material that would contain the low enriched uranium fuel and provide a thermal neutron spectrum reactor. The size of this reactor would be much smaller than the first graphite piles fueled with natural uranium. These next generation reactors will be designed to use a low-enriched uranium fuel and increase the nuclear fuel burn-up beyond that attained with the LWR reactors.

Both of these reactor systems use thermal spectrum neutrons and therefore cannot efficiently fission the minor actinides present in the spent nuclear fuel. A primary objective of the Generation IV nuclear energy program is to develop fast flux reactors that will fission all of the transuranium elements in recycled spent nuclear fuel. This will reduce the volume and long-term radiotoxicity of the fission product waste stream. Passing from a once through to a closed fuel cycle extends the useful energy yield of the world supply of uranium many fold, a long-term energy sustainability objective. The research and development programs on the fast flux reactor options is designed to select the fast flux energy system(s) for commercial development and deployment by the year 2050.

Gas-Cooled Fast Reactor

The gas-cooled fast reactor (GFR) offers the advantage of building on the high-temperature fuel technology that will be used in the VHTR. The GFR offers the sustainability feature with reduction of the volume and toxicity of its spent fuel and the added potential to use reprocessed LWR spent fuel that continues to accumulate with the once-through fuel cycle [16]

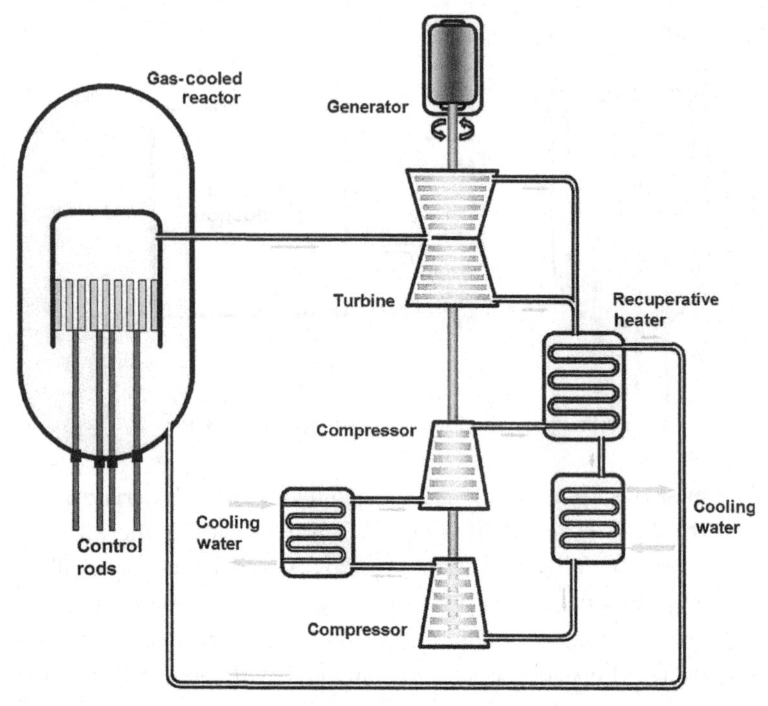

Figure 8.23 Schematic of gas-cooled fast reactor (GFR).

The GFR fuels and in-core structural components must be shown to survive the high temperatures and the fast neutron radiation. Since recycled fuel will contain the minor actinides and some fission products, the serviceable life of the fuel will depend on the integrity of these multicomponent fuel elements. Tests must demonstrate the fuel integrity and performance over the irradiation time between refueling. Figure 8.23 illustrates the GFR.

Successful deployment of the GFR, a new reactor system, will require detailed safety analysis. The development of computational tools is to design the energy system hardware, run simulations of operating transients (e.g., failure of gas coolant flow), and identify data gaps that must be filled with experimental measurements and material qualification data.

Sodium-Cooled Fast Reactor

The sodium-cooled liquid metal energy system features a fast spectrum reactor and a closed fuel cycle [17]. Sodium is the reactor core coolant of choice because liquid sodium has a small collision cross section for

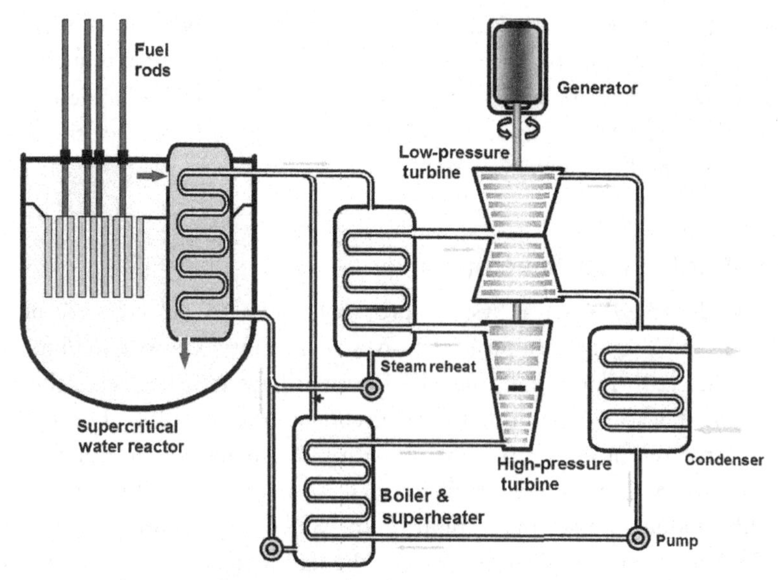

Figure 8.24 Schematic of sodium-cooled fast reactor.

neutrons allowing neutrons to pass without slowing down. There has been significant development of the SFR system; the EBR-II program in the United States [18] is the primary source of fast flux reactor data. This program included on-site reprocessing of the spent fuel to recycle uranium and plutonium. The EBR-II was a pool type reactor with a low pressure, inert gas pad above the sodium pool. Figure 8.24 is a schematic of the sodium-cooled fast reactor.

The French have the most experience with commercial fast flux reactors. A big jump to the Super Phenix, commercial sodium cooled, fast flux power reactor was built in France [19]. It operated from 1985 to 1997 when it was shut down when materials of construction problems and sodium leaks caused a poor electric power production record. The decision to proceed to build this 1200-megawatt (electric) energy system may have been premature, but the commercial failure did produce valuable technical data, operating experience, and identified construction material problems.

The SFR option includes on-site recycling of the spent fuel. This would close the fuel cycle and provide security assurance that weapons grade nuclear material would not be produced. The plutonium would not be separated from the uranium and minor actinides in this process. There would be some fission products in the recycled fuel that would render it radioactive, an additional protection from diversion to weapons. Fission products can be tolerated in fast spectrum fuels and reducing the fuel purity makes spent fuel reprocessing much easier. The design and safety characteristics of these recycled fuels will be the focus of the development of the SFR energy system.

Lead-Cooled Fast Reactor

The lead-cooled reactor system proposal seeks to advance all of the Generation IV goals; nonproliferation, sustainability, safety, reliability, and economics [20]. For some time, the Russians have been studying the substitution of lead for sodium in fast-spectrum reactor [21]. The fuel for this reactor might be a mixed oxide with 80% depleted uranium and 20% plutonium. As the plutonium fission occurs, neutrons captured by U-238 would produce replacement plutonium. Since the accumulating fission products do not significantly change the fast neutron energy spectrum, this fuel might continue in service for 10 or more years with burnup to 15%. This would decrease the number of reprocessing cycles required to use all of the uranium to produce energy. Figure 8.25 illustrates the lead-cooled fast reactor configuration.

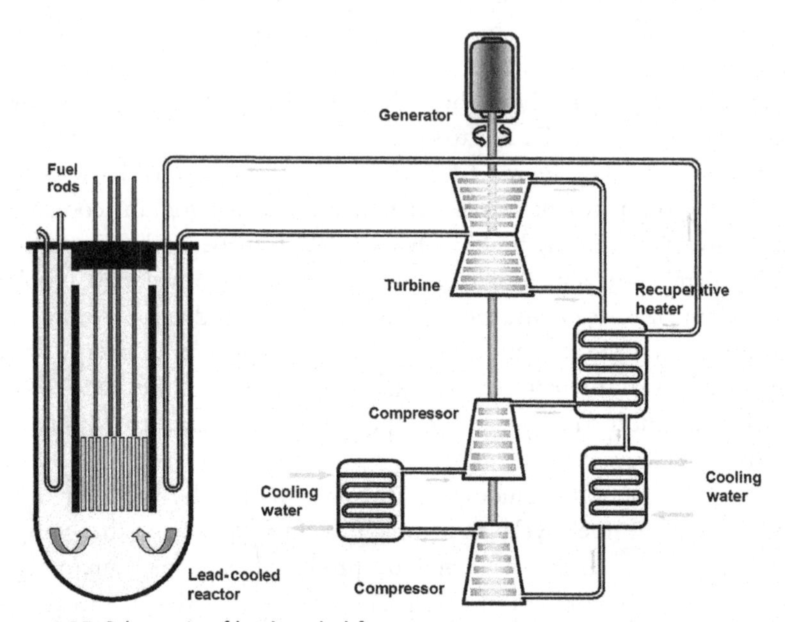

Figure 8.25 Schematic of lead-cooled fast reactor.

The experience with lead-cooled reactors comes from the Russian Navy. They built eight reactors to power submarines that used a lead-bismuth eutectic mixture (to lower the melting point of the liquid metal coolant) and there is about 80 years of reactor operation experience from this program.

The plan for this proposed reactor system includes establishing the necessary features of fuel and core materials that will provide a 20-plus year core life. The benefits of this long core life can be achieved if the construction materials are developed to resist the corrosive effects of hot lead. The reactor core is set in a lead pool and thermal energy removed from the reactor core by natural convection. Heat exchangers in the upper section of the lead pool transfer the heat to high-pressure gas serving a Brayton cycle or to steam and a conventional steam turbine.

Molten Salt Reactor

Two experimental molten salt reactors (MSRs) were built in the United States during the 1950s and 1960s to study the basic technology of this reactor scheme. These results with the ongoing MSR research in Europe provide the basis to develop an Advanced Molten Salt Reactor with an emphasis on fuel cycles. Figure 8.26 illustrates the molten salt reactor.

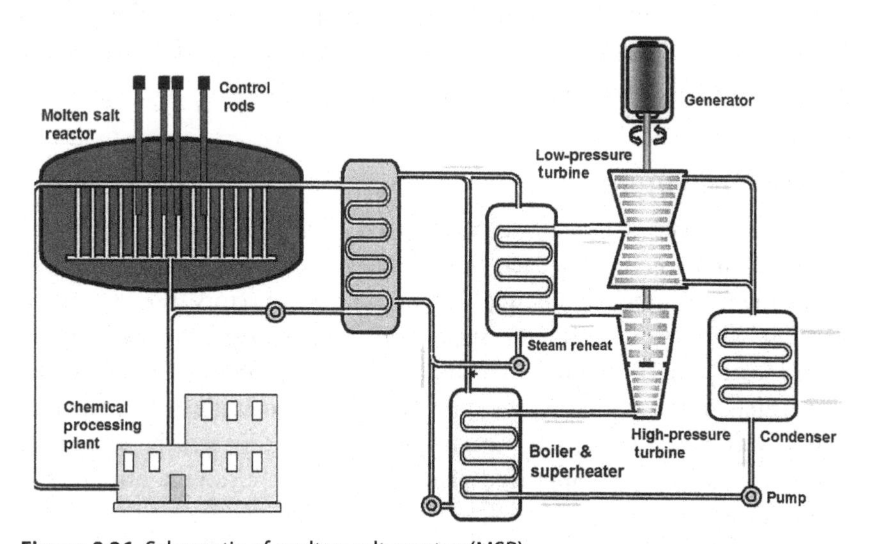

Figure 8.26 Schematic of molten salt reactor (MSR).

The MSR is a liquid fueled reactor that can use actinides as fuel and produce electricity, hydrogen, and fissile fuels. Molten fluoride salt with a 1400°C boiling point is used as a solvent for the nuclear fuel and fission product metals. This primary salt is circulated through a reactor core that contains a graphite moderator. Fissionable metals fission producing excess neutrons that promote fertile metals to fissile metals by neutron capture as the heated salt mixture passes to a heat exchanger. The heat is transferred to a secondary molten salt loop, isolating the radioactivity in the primary salt, and transferring the thermal energy to a second heat exchanger to supply a Brayton cycle (using nitrogen or helium) or a conventional steam cycle turbine-generator unit to produce electricity. The operating temperature of the MSR system can be increased to provide thermally assisted hydrogen production described in the VHTR section.

A portion of the reactor salt is continuously passed to a chemical processing unit. The fission products are removed and nuclear fuel components can be removed or added to maintain the optimal fuel composition. The reactor salt contains radioactive fission products so the chemistry to remove them requires gamma ray shielding and remote handling.

The basic technology of the MSR has been demonstrated, but the concept has a low priority for near-term development. The conceptual design for an (advanced) AMRS will provide an understanding of the economic factors for this reactor system. There is the promise of using all of the actinides as AMRS fuel. Disposal of the minor actinides is an important objective of the Generation IV program. Most of the AMRS activities will be performed under the higher priority studies since the technology of all the proposed energy systems overlap.

Toward the Future

The current fleet of light water moderated nuclear power plants provide technical and economic data that suggest there should be increased deployment of nuclear energy systems. These reactors use a "once-through" fuel cycle producing spent fuel, a very long-term radiological hazardous material. There was a highly contested proposal to place the spent fuel in the Yucca Mountain, Nevada geological repository. This option was closed citing the capacity of this repository will be exceeded. There are currently no approved plans for permanent storage of domestic spent nuclear fuel.

Nuclear power systems are large, expensive, and inherently hazardous. This means the evolution to new nuclear power systems will be slow and must be accomplished with great care. At present, there is no commercial power system based on a fast neutron spectrum reactor that uses plutonium as fuel. The long-term goal of the Generation IV energy systems is to provide high temperature thermal energy for chemical processing (to produce hydrogen) and improve the thermal efficiency of electricity production. The next step in this international effort will be to deploy fast neutron flux breeder reactors making reprocessing the irradiated fuel necessary. The first reprocessing to produce pure plutonium (for World War II weapons) will be greatly simplified since only the fission products need to be removed. The recycled heavy metals (actinide elements) fission are transmuted to fission or can be extracted as fuel in the next reprocessing cycle.

The commitment of the international community to share in the nuclear energy system project should make it happen. Reprocessing spent nuclear fuel is chemistry of heavy metals. The gamma radiation from the fission products and the toxicity of the heavy metals make remote processing necessary. Chemical analysis of the rocks on Mars is being done today from a control room in Houston, Texas. The challenge of doing remote chemistry has been met and the experience with nuclear fuel reprocessing provides the basis for future expansion of nuclear energy in the twenty-first century.

SMALL MODULAR NUCLEAR REACTORS

The US government and industry has initiated a SMR program where with key elements of: (i) certifying a reactor design that can be replicated many times without requiring recertification; and (ii) building these reactors off site from which they will be transported to different locations around the globe for power generation. This approach would eliminate the time needed for certification of just the first couple units and would expedite fabrication through use of a single manufacturing site. The electric companies would realize a single-year turnaround from the time the facility is approved for purchase to ready for installation, testing to certify ready for operation. It is a game-changing technology because it overcomes the two major problems of uncertainty and long construction/certification times characteristic of new nuclear power plants.

SMRs are smaller than 300 MWe that is less than one-fourth the size of the current of generation base load nuclear power plants. The smaller sizes of these facilities could also serve a broader market that includes sites that are too small to accommodate larger current-generation plants. This includes sites with limited cooling water and smaller electrical markets.

An advantage of these facilities is that they could establish a sustainable approach to replacing and expanding the current nuclear power industry. This sustainable demand would also provide the incentive to advance to new generation SMRs that could use reprocessed fuel and produce power at higher efficiencies.

In May 2014, Oregon-based NuScale Power and the US DOE officially signed a contract including $217 million in federal matching funds to support development, licensing, and commercialization of the company's 45 MW SMR design.

Earlier government contracts for two 180-MW SMRs with Babcock & Wilcox's Generation and a 225-MW unit with Westinghouse appear to have been scaled back from initial plans. According to Power Engineering [22] Westinghouse President and CEO Danny Roderick stated, "Simply put, an investment in SMR nuclear technology does not compare favorably right now with other options currently available in the marketplace, and the need to accelerate deployment of the SMR is not as compelling as it was only two years ago." Twelve one GW Westinghouse reactors are under construction in Korea, China, and the United Sates and more planned in Europe.

The nuclear reactor construction industry appears to be healthy. The bad news is that resources are being devoted to traditional rather than game-changing technologies.

LESSONS FROM HISTORY

Nuclear Safety

A nuclear core meltdown is considered the worst-case accident in a nuclear power plant. Both U-235 and Pu-239 are >90% pure for bomb-grade applications compared to 3.4–5% the usual enrichment for nuclear reactor fuel. In the diluted forms (<80% U-235 or Pu-239), the fuel cannot explode in a nuclear chain reaction. In the absence of the right purity or configuration, the initial energy released by a chain reaction will rapidly splat the heavy metals apart, too distant to continue the chain reaction. Worst-case nuclear reactor incidents would potentially release radioactive materials in the form of hot vapors.

This was reported to occur at the Chernobyl nuclear power plant on April 26, 1986. The Chernobyl reactor was a graphite pile (pure solid carbon—think the United States Hanford, WA, plant that produced plutonium in the 1940s) block with horizontal holes uniformly spaced to slow the neutrons. Metal tubes contained slugs of uranium metal were placed in the holes in the graphite block. Water was pumped through the space between the uranium slugs and the graphite holes. During a scheduled power shutdown the operators decided to maintain electric power production by slowing down the flow of water to maintain steam temperature and pressure Fission heat production dropped, some of the tubes "went dry" the metal cans melted and the uranium chemically reacted with the water to form uranium oxide and hydrogen. This produced an initial explosion that exposed lots of nuclear fuel to lots of water and a second explosion destroyed the reactor. The graphite caught fire (graphite is high grade coal) vaporizing some of the fission products. The huge cloud of particles and vapor blew north and west, detected and reported by operators at a Swedish nuclear power plant. The remains of the reactor have been encased in concrete.

All personnel were evacuated from large region surrounding the power plant. This region has become an experiment on the natural recovery of all life forms in a radiation-contaminated area. Reports out of that "experiment" are interesting reading.

Fukushima in Review (3/11/2011)

In March 2011, the Tohoku 9.0 magnitude earthquake caused a 14-m tsunami that overwhelmed the seawall that was designed to turn back a 10-m tsunami. The earthquake automatically shut down the nuclear reactors that were running. All emergency power was cut off by the "wall of water." This resulted in the meltdown of three of the six nuclear reactors at that site. There was a release of radiation estimated to be between 10–30% of that released in the Chernobyl Catastrophe. Much of the radiation contamination went into the sea where it will add less than 0.01% to the background radiation there. No short-term radiation fatalities were reported. The number of deaths attributed directly to the earthquake and Tsunami was 15,884.

The primary problem caused by the incident was the total evacuation of about 300,000 people from the region mandated by the Japanese government. There were some deaths reported, probably due to temporary housing and to hospital closures. In response to questions about the need for evacuation, the World Health Organization reported that populations that would have stayed would have realized a 70% higher risk for thyroid cancer and a 4–7% increase risk for certain other types of cancer such as leukemia and breast cancer.

The primary causes of the meltdowns were the failure of a backup cooling water system. The electric power connection to the local distribution lines were also destroyed. Nothing was designed to withstand such a tsunami event. One must wonder if it would have been possible to have helicopter-mobile cooling water engine/pumps and water purification units that could have been brought in from a location known to be safe from natural disasters that would impact the plant. Certainly, this could have been possible and certainly power plants could be made for easy attachment to such mobile facilities.

While the nuclear contamination is truly unfortunate, the fact that no deaths were directly attributed to radiation poisoning is a result of good reactor design. It should be possible to learn lessons from this incident and use precautions to further reduce the risk of any similar occupancy.

The Chernobyl Catastrophe (1986)

The Chernobyl accident is an example of failure to follow operating procedures that led to the catastrophe. The plant managers were attempting to produce electric power as the reactor was being shutdown for refueling. The reactor was operating at very low power. The control rods were almost fully withdrawn to allow fission to occur even though there was significant decrease in fission due to the fission products in the spent fuel. Operators did not observe safety precautions that led to the dangerous situation. They had not informed the reactor safety group that they were going to run this experiment.

The Chernobyl reactor used graphite to slow down the neutrons to improve the fission probability of the U-235 in the fuel. This graphite contained the natural uranium fuel elements located in the same channels for circulating water to remove the heat produced by the fission reaction. This reactor design made it unstable and susceptible to loss of control when the operators made an error running their experiment. When cooling water is lost, the nuclear chain reaction and the power output increased. During the experiment the operators were running that day, cooling water flow was lost and there was a power surge. Some of the fuel element canisters ruptured and the hot fuel reacted with the water forming hydrogen leading to an explosion. This lifted a 1000-ton cover from the reactor, rupturing most of the remaining tubes causing a second more powerful explosion. The reactor core was now completely exposed to the atmosphere.

The graphite in the core caught fire and this very hot fire vaporized or produced aerosol particles of the reactor core materials. These materials included most of the radioactive fission products and reactor fuel that were not scattered as shrapnel during the explosions. The radioactive cloud from this accident spread for thousands of miles. It is easily the worst industrial nuclear accident in history. There were deaths of those who fought the fire due to radiation poisoning. There were many cases of radiation sickness and increased incidents of cancer in those exposed to the radioactive cloud. The whole region is still a laboratory for the study of radiation poisoning of the land and vegetation. The shadow of this accident will always fall across the best efforts to use nuclear reactors to produce electricity.

The Three Mile Island Accident

The Three Mile Island accident was the worst commercial nuclear disaster in US history. In this incident, the reactor operators did not follow emergency procedures, misread water level indicators on the control room panel and turned off water pumps that would have cooled the reactor core. Some of the upper part of core melted with the melt contained in the lower part of the reactor pressure vessel. The hot fuel formed hydrogen and the high temperatures caused steam to be released into the reactor containment building. There was some radiation release to the environment. The most significant health related effects were *due to the psychological stress on the individuals living in the area* [23]. Scientists still disagree whether the radiation vented during the event was enough to affect the health of those who lived near the plant [24].

The Three Mile Island Accident occurred in March 1979 in a reactor located near Harrisburg, PA. This was a PWR that had been brought to full power late in 1978. The accident began when the feed water pumps to the steam generator stopped. The pressure increased in the vessel containing the reactor core where the heat is generated. This caused a relief valve on the reactor core pressure vessel to open and drop the control rods that stopped the neutron chain reaction. The fission product decay produced about 200 megawatts of heat immediately following reactor shutdown. The pressure in the reactor vessel continued to drop when the pressure relief valve failed to close and the water level in the reactor core continued to drop as water evaporated and vented as steam through the relief valve that was stuck open.

The reactor operators did not know that the vent valve was stuck open since the indicator on the control panel showed it was closed. They did not replace the water in the reactor vessel. There were emergency feed water pumps that should have been running, but the reactor was now being operated in violation of safety rules. The dry fuel rods melted and the hot fuel pellets reacted with the water forming hydrogen. This high-pressure hydrogen bubble prevented water from covering the reactor core for several days. The core melt down did release fission products, but they were held in the reactor containment building. A small amount of the volatile fission products did escape the containment structure to the environment.

This accident did alert the nuclear power industry to the possibility of a core melt down. Operator training and emergency responses have been put in place with emphasis on loss of cooling accidents. Operators receive extensive training to respond properly to any emergency. Engineering changes have been made that assure the control system will respond automatically to emergency shutdown conditions.

Lessons Learned

Valuable lessons were learned from these three nuclear incidents. The Fukushima accident revealed that either more robust backup power systems or mobile systems stored at secure locations are not likely to be impacted by a "flood" that hits the power plant.

The most important lesson from the other two incidents is they were the result of actions of one or a few individuals who "overrode" the safety system. A good reactor design can overcome major operator blunders—not to avoid an incident but to prevent it from becoming a disaster.

Improved reactor design, improved operating procedures, improved operator-override protocols, and location in unpopulated areas provide nuclear power that is safer than coal or natural gas. As recent amendments to NO_x and particulate matter emission standards are not enforced, these emissions continue to lead to environmental and health risks. The safety history of nuclear power generation in the United States is better than natural gas or coal.

RECYCLING AND GREEN CHEMISTRY

Professor Stan Manahan defines green chemistry as "the practice of chemical science and manufacturing that is sustainable, safe, and nonpolluting and that consumes minimum amounts of materials and energy while producing little or no waste material." In many instances, green chemistry can increase process profitability. This is typically possible when a process waste stream has higher concentrations of a metal than the concentration in the natural ore that is mined to recover the metal.

The spent fuel rods of commercial reactors consist of metal tubes held in racks that contain uranium oxide pellets. For each 2000 tons of heavy metals there are 660 tons of cladding and about 269 tons of oxygen. After a 3.4% burn, the uranium content of the spent fuel stream is about 66%, this compares to 0.1–0.5% uranium that is typical in mined uranium ore. The fissile material content (U-235 + Pu-239) is about 1.12% compared to 0.0007–0.0036% in the uranium ore.

The reprocessing of spent fuel rods fits the definition of "green chemistry." Based on the concentration of the metals in the spent fuel rods compared to the natural ore, there is an opportunity to use reprocessing to increase the profitability of nuclear power (without considering the costs of long-term storage of spent fuel rods).

Most of the commercial nuclear power plants in the world use light water moderated reactors and uranium oxide fuel enriched to 2.6–4% U-235. The fuel elements remain in the reactor about 3 years and then are stored in a pool of water on the reactor site as "spent fuel." Figure 8.27 illustrates this process of enriching the natural uranium followed by the nuclear burn.

The spent fuel emits high-energy gamma rays and produces thermal energy as the radioactive fission products decay. The water pool serves as gamma radiation shield and a heat sink for the decay heat. The average decay rate and the energy release decreases with time but for some metals it persists for many years requiring a permanent repository. Such a storage facility was under construction at Yucca Mountain, Nevada, but litigation ended the plan for this long-term storage site. The inventory of spent fuel in the US is now accumulating at the rate of about 2000 metric tons per year and disposal of this material remains an open question in the future of nuclear power.

Current international consensus suggests nuclear energy will be required to assure future energy security. The challenging technology goals to provide long-term sustainable nuclear energy must focus on resource utilization and waste management. Key issues include economics, safety and

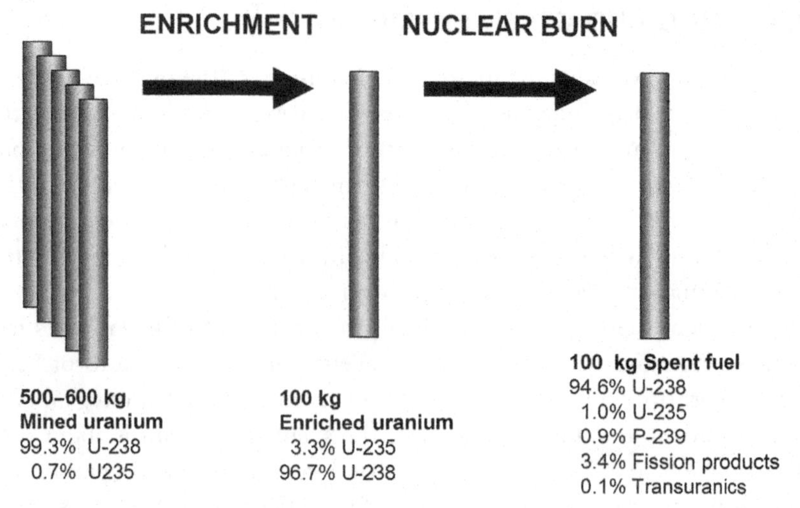

Figure 8.27 Illustration of mass balance for reprocessing. Mass is mass of heavy metal plus fission products of heavy metal.

reliability, weapons proliferation resistance, and physical protection of plant personnel and the public. These factors apply to the power plants now in operation and this experience will guide the development of the new generation of technologies that will become the nuclear energy systems.

In the United States, chemical reprocessing of domestic spent nuclear fuel is currently not as cost-effective as using uranium ore but is an important goal for long-term nuclear energy systems. The benefits include:

- Extending the nuclear fuel supply by recovering fuel values in the current inventory of spent fuel that has produced about 3.4% of the total energy available in the fuel.
- Combining these recycle fuel values with depleted uranium (the U-238 that remains producing the U-235 enriched fuel) extends the fuel supply into future centuries (see Figure 8.28).

Several of the reactors in Europe and Japan use recycled fuel. Reprocessing technology is readily available but the technology could be improved to reduce costs and waste. Reprocessed fuel in France costs 0.90 ¢/kWh (busbar) while new uranium fuel costs 0.68 ¢/kWh.

The current management plan in the United States is to place the spent fuel bundles from the "once-through" fuel cycle in long-term storage. Centuries of fuel values would be buried and labeled as hazardous. Since the radioactive hazard of this waste will last for thousands of years,

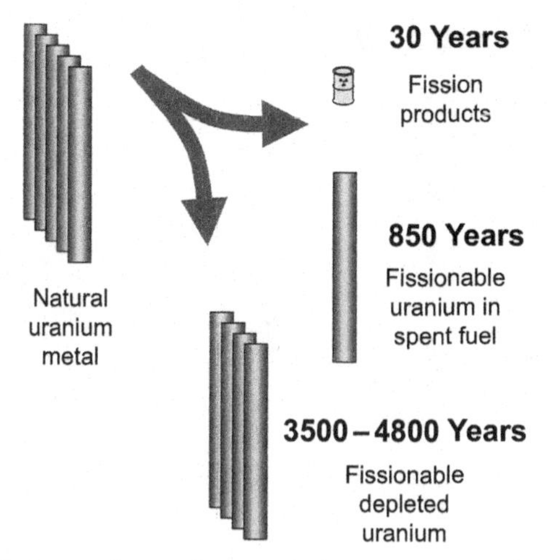

Figure 8.28 Mass balance of once-through fuel use as practiced in the United States. The fission products are the "waste." The years indicate years of available energy if used at the same rate as used in once-through burns. Both France and England have immobilized the concentrated fission products in glass for long-term storage [25].

considerable thought has been given to develop a warning/labeling system that would last as long as the hazard.

Spent fuel reprocessing is an alternative that would recover the fuel values. Reprocessing can be broadly categorized three steps: (i) recovery of unused fuel; (ii) waste minimization; and (iii) full use of uranium/thorium as fuel.

Recovery of Unused Fuel

Reprocessing options include a process of separating the fuel bundles into the fractions illustrated in Figure 8.29. The goal is to recover the heavy metals for use as nuclear fuel—the heavy metals are fissionable or fissile materials including U-235, Pu-239, and U-238. The casings are structural metal/ceramic components that can be washed to a low level radioactive waste. The fission products are the main source of radiation hazard.

Recovery of Unused Fuel

The spent nuclear fuel is about 96.6% heavy metal (uranium, plutonium, and trans-uranium metals). These can be separated from the fission products and processed to become fuel for light water or next generation reactors. Of the remaining 3.4%, about 0.4% has high radioactivity after

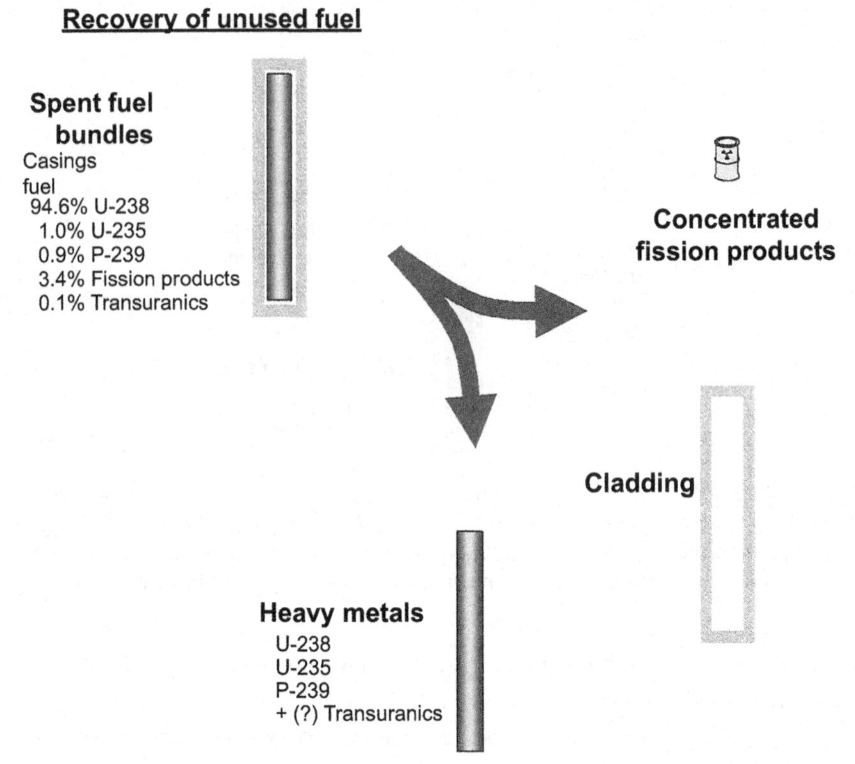

Figure 8.29 Recovery of unused fuel is the first phase of fuel reprocessing. Casings are physically separated as the initial step in future nuclear waste management—separation of structural metals in the bundles and recovery of fissionable fuel.

30 years of storage. Figure 8.30 illustrates the different fractions into which the 3.4% that is fission products could be separated. It is this 0.4% that requires long-term burial or transmutation.

Waste minimization using spent fuel reprocessing extends the life of a burial site by reducing the quantity of wastes and the troublesome radioactive decay heat. The scientific analysis and demonstration of safe repository performance would be simplified by reducing the radioactive lifetimes of the materials going into the geological burial site. The storage time for the hazardous materials is significantly reduced, from about 300,000 to less than 1000 years.

Full Use of Uranium/Thorium

In today's light water nuclear reactors, 33–40% of the energy is provided by the fertile U-238 that is in the reactor fuel. The new fissile material

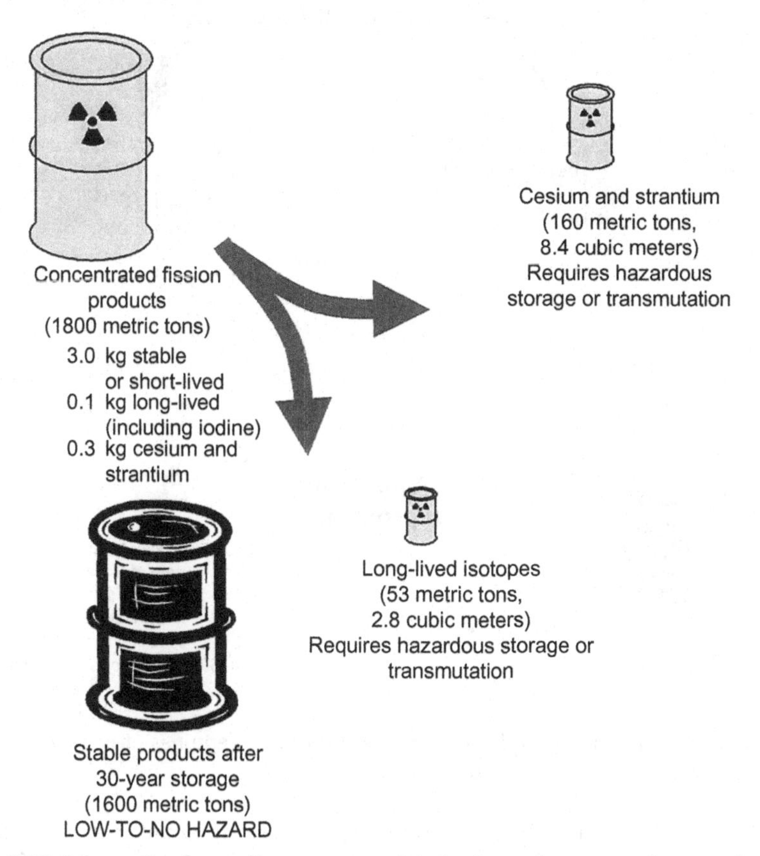

Figure 8.30 Schematic of overall process to minimize hazardous waste from nuclear power. The masses in tons represent total estimated US inventory from commercial reactors in 2014, about 65,000 tons. The volumes are based on uranium density—the actual fission produce volumes would be about twice the values shown.

(Pu-239) is produced during the three year fuel burn cycle The change from new fuel to spent fuel is represented by:

$$3.0 \text{ Parts } U\text{-}235 + 2.1 \text{ parts } U\text{-}238$$
$$\rightarrow 3.4 \text{ Parts burned} + 0.8 \text{ parts } U\text{-}235 + 0.9 \text{ parts Pu-239}$$

For every 3.0 parts of fissile U-235 material that enters the reactor, 1.7 parts of fissile material leaves the reactor (U-235 and Pu-239). Using the "once-through" light water reactor, insufficient fissile material is available to justify recycle. But the U-238 can be recovered yielding centuries worth of energy value available in those U-238 inventories.

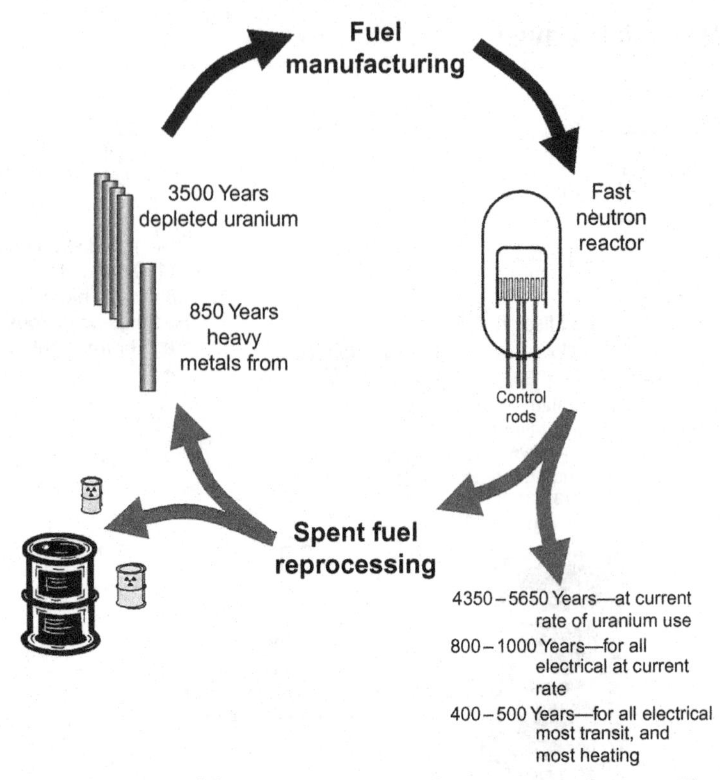

Figure 8.31 Illustration of full-use uranium providing centuries of energy. Thorium is also a fertile fuel that can be used in this closed cycle.

The Generation II light water reactors are designed for fission propagation with thermal neutrons. This leads to a depletion of fissile material during the fuel burn. Most of the Generation IV reactors are designed for fission propagation with fast neutrons. In these reactors much more U-238 is burned than U-235. The breeder reactors are designed to produce more fissile material than is burned.

Reactors based on fast neutron fission are an important part of reaching the potential of nuclear fuel reprocessing. Figure 8.31 illustrates the overall reprocessing cycle including use of fast neutron reactors, properly referred to as fast-spectrum reactors.

DISCOVERY AND RECOVERY

The 1930s was the decade of discovery for the structure of the atom nucleus. The zero-charge neutrons with a mass nearly the same as a proton was found to be tightly packed in the nucleus of atoms. Isotopes of many elements were prepared by placing a sample in a stream of neutrons, the nucleus is changed with the capture of one or more neutrons (forming an isotope) without altering the balanced charge of the protons and electrons (without changing the chemistry of the atom).

It was during these experiments that irradiation of uranium produced barium. It was proposed that the uranium nucleus splits into two nuclei, one of which was the barium found in the experiment. The new atomic nuclei repelled each other with release of about 200 MeV of energy. This event also produced a pulse of fast neutrons. The term *nuclear fission* was coined to describe this process [26].

The huge energy release plus the release of two or three high-energy neutrons suggested a proper amount and configuration of uranium 235 could sustain a chain reaction. The resulting energy release would very fast and produce an explosion much more powerful than any chemical explosive. Lisa Meitner and Otto Hahn (February 1939) authored the paper in Nature describing fission just before the beginning of World War II—shortly thereafter all nuclear research was classified *Top Secret.* The remarkable developments during the 1940s nuclear decade were conducted to produce a nuclear weapon with all of the research and development done under that strict veil of military secrecy.

First Production of Plutonium

Natural uranium is composed of two primary isotopes, 99.27% U-238 and 0.71% U-235, and it was known that the 235 isotope will fission. Isotopic separations are difficult and to produce enough of the fissionable U-235 to make a nuclear weapon was a stiff technological challenge.

In January 1941, Glen Seaborg and coworkers reported the discovery of a new element, plutonium 239, produced by neutron capture of the U-238 isotope [27]. Cunningham and Werner irradiated a sample of natural uranium with neutrons. They found that lanthanum fluoride, LaF_3, precipitate was an efficient carrier of plutonium to make the separation possible [28]. This separation procedure produced the first few micrograms of Pu-239 metal used to determine it had a fission cross section about 50% greater than U-235. This made plutonium an attractive alternative to U-235 as a weapon material.

Continued laboratory preparation of plutonium provided about 20 μg (micrograms) of metal used to establish its radiological properties, chemical oxidation states, and to estimate its physical properties.

The threat that Germany might develop a nuclear weapon convinced President Roosevelt that the Army should be in charge of the new nuclear weapon program. In June 1942, a new unit was formed called the Manhattan District [29]. There were five construction sites under this project. There was little sharing of task information between the groups and security must have been good. One measure of the size of this secret project is the total workforce on the Hanford site grew to over 40,000 in May 1943 [30].

A group under the direction of Enrico Fermi successfully demonstrated that a nuclear chain reaction could be sustained using a matrix of natural uranium placed in a stack of graphite bricks (a nuclear pile) [31]. The pile was initially operated at 0.5 watt and raised to 200 watts thermal energy on December 12, 1942. It was clear that the uranium metal fuel in such a reactor is continuously exposed to the fission neutrons that produce plutonium when a neutron was captured by a U-238 atom.

This successful demonstration of the nuclear pile reactor motivated the decision to fund the secret Manhattan District Project to produce plutonium for a nuclear weapon. Plans proceeded to build a nuclear pile to produce kilogram quantities of plutonium using known data and a 10^9 engineering scale factor. This plutonium production nuclear reactor would generate about 250 megawatts of thermal energy. The remote site near Hanford in eastern Washington with the nearby Columbia River was selected for this plutonium plant. Construction was underway before the chemical steps for recovering the plutonium from the irradiated uranium fuel were known.

It was recognized early that it would be easier to produce an atomic bomb using plutonium than with U-235. The chemical purification of plutonium from uranium was much easier than the physical separation of U-235 from U-238 (natural uranium). In addition, the larger fission cross section relaxed concentration and configuration constraints on making the bomb. For this reason, nonproliferation efforts tend to focus on restricting access to plutonium and attempts to stop countries from operating piles/reactors capable of converting U-238 into plutonium.

The laboratory recovery methods used ether extraction and produced a gelatinous LaF_3 precipitate, clearly not a good choice for large-scale plutonium production. The chemical explosion hazard using ether and the corrosion problems using aqueous fluorides recommended alternatives.

Precipitation remained the method of choice for the separation process because it offered quick development to large scale. Early in 1943, S. G. Thompson showed that $BiPO_4$ precipitate strongly carried Pu^{+4} [32]. The precipitate is crystalline and easily collected by filtration or with a centrifuge. This was the method selected for plutonium production.

Construction of the Hanford plutonium production reactors were underway based on the physics of the nuclear chain reaction but with no plutonium metal to demonstrate it could be made to explode. The scale of the project can be imagined by the Hanford nuclear pile reactors that consisted of 1200 tons of pure graphite containing about 250 tons of uranium slugs, each slug consisting of a few kilograms of uranium sealed in an aluminum can. The fuel slugs were placed in horizontal aluminum tubes passing through the graphite. Cooling water was pumped through the tubes to remove the thermal energy (heat).

A plutonium production run lasted 100 days that converted about 1/4000 of the U-238 atoms to U-239. The uranium slugs were pushed out of the reactor with new fuel to start the next irradiation cycle. The irradiated uranium was stored under water to remove fission product decay heat and provide biological shielding from the gamma radiation the radioactive decay produces.

The steps shown in Figure 8.32 to form plutonium from U-238 is now well established. The neutron capture of a U-238 nucleus results in the immediate release of gamma rays and a total energy corresponding to the

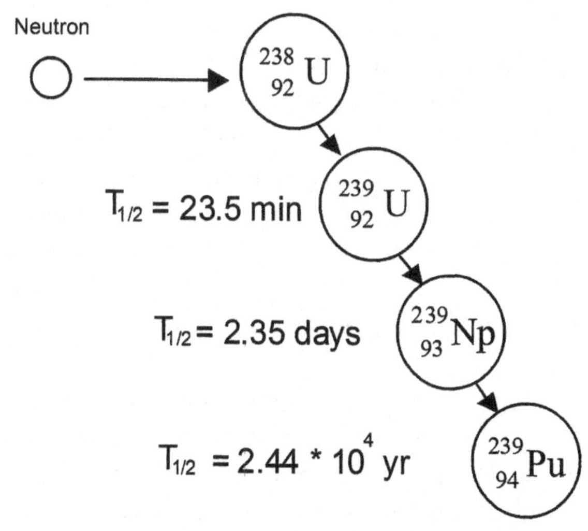

Figure 8.32 Conversion of U-238 to Pu-239.

BE of the neutron. The new U-239 atom has a half-life of 23.5 min and decays releasing a beta particle to form neptunium 239 (Np-239). Np-239 has a half-life of 2.35 days and decays releasing a beta particle to form Pu-239. The plutonium also decays releasing an alpha particle, but the plutonium half-life is 24,400 years making it relatively stable atom [33].

The recovery of the plutonium began with removal of the aluminum cans covering the fuel slugs (either mechanically or chemically dissolving). The fuel elements containing uranium, plutonium, and fission products were dissolved in nitric acid. The plutonium in solution was reduced to the +4 state and precipitated with $BiPO_4$. The uranium was held in solution using sulfate ion, SO_4^{-2} to form a soluble uranium ion. This separation step split the very small amount of plutonium from the uranium and most of the fission products. Some of the fission products stayed with the plutonium. Dissolving the plutonium-$BiPO_4$ precipitate and oxidizing the plutonium to the +6 state soluble in acid solution yields nearly pure plutonium. The impurities remained insoluble and were separated with the precipitate. The plutonium in solution was again reduced to the +4 state and the precipitation cycle repeated two times with the extracted plutonium ending in a final acid solution. The third $BiPO_4$ precipitation cycle was followed by a cycle using LaF_3 to remove the last traces of fission products [34].

It is remarkable that following all of these steps, the recovery of plutonium was greater than 95% and the decontamination factor of the plutonium product exceeded 10^7. This might have been considered good luck, since the decision to use the $BiPO_4$ precipitation process was made well before the chemistry of plutonium was known—"hats off" to the insight of the scientists working on the project.

The "fast track" military schedule ignored some of the disadvantages of the process. A batch process always requires careful operator attention. The quantity of process chemicals produced a large volume of high-level radioactive liquid waste that still remains stored in huge tanks on the Hanford site and represents a World War II legacy waste treatment problem. Nearly all of the uranium remained in the waste stream and additional processing steps are required to separate the uranium from the fission products.

There is an additional difficulty working with irradiated uranium. All of the reactor operations and the chemical treatment must be done behind radiation shielding to protect the plant personnel from the high-energy gamma radiation produced by the decay of the fission products. Mechanical manipulators were designed to provide remote services to operate and maintain the equipment handling irradiate fuel. A 1945 audit indicated the Hanford plutonium production facility cost over $300 million (1945 dollars) [35]. When we include the uranium enrichment plant at Oak Ridge, TN, the weapons development site at Los Alamos. NM, the metallurgical laboratory in Chicago, the Manhattan Project represents the most ambitious, expensive, successful research and development project in history.

PUREX Process—Cold War Plutonium Production

The signing of the German and Japanese surrender documents was completed in August 1945 and the end of World War II brought a national "sigh of relief." Military personnel quickly returned to private life. This was also true for many of the science and technology people working on the Manhattan project. The advantage of possessing the most powerful weapon was obvious, but the authority over the program was transferred from the Army (military) to a new civilian committee, the Atomic Energy Commission. Production of plutonium was continued under Atomic Energy Commission (AEC) control.

The first nuclear explosion in the Soviet Union in 1949 came as no big surprise. Winston Churchill's "Iron Curtain" speech in March 1946 certainly gave warning of the aggressive attitude of the Stalin led Soviet Union toward its World War II allies. This represents the introduction into the decades long Cold War.

The Cold War assured the military would demand additional plutonium production. Countercurrent extraction was a mature technology when the batch process was selected for recovery and purification of plutonium. The operation of a continuous extraction train is more complicated than a batch process and there was little time available for process development. The bismuth phosphate batch process continued in service until 1951.

The source of plutonium would remain natural uranium irradiated in the graphite-moderated reactors at Hanford Washington and newer reactors built at Savannah River, South Carolina. The irradiation times remained short with about 1/4000 for the U-238 atoms converted to plutonium. There were three essential requirements of any proposed separation process:

1. All of the plutonium must be recovered as weapons grade material.
2. The uranium must be recovered essentially free of radioactive fission products.
3. The mass of the fission product waste stream should be greatly reduced (as small as practical.)

The radioactive fission product content of the uranium and the plutonium was set at about the level of natural uranium so these metals could be handled and machined without the cumbersome gamma radiation shielding.

It was known in 1945 that tri-n-butyl phosphate (TBP) could be used as an extraction agent for nuclear fuels [36]. The PUREX (*P*lutonium-*UR*anium-*EX*traction) process was designed to remove the fission products, the source of the essentially all of the gamma radiation in the product uranium and plutonium, requiring separation factors of 10^6-10^7. The toxicity of plutonium, inhaled or ingested, required plutonium separation from uranium to be set at 10^8. A little uranium in the plutonium was not considered a problem. The recovery of both the uranium for recycle and the plutonium product was to be 99^+%. These separation process specifications far exceeded the highest standards in industrial practice at the time.

Since the PUREX process was designed to replace the bismuth phosphate batch process, the feed irradiated uranium fuel was the same. The first step was to chemically remove the aluminum cans on the uranium fuel slugs and then dissolve the fuel in hot nitric acid. The pH and metal concentration of this feed solution was adjusted to maximize the solubility of the plutonium and uranium in the organic extract phase. This feed solution entered the center of the first extraction train.

The first extraction separated the plutonium and uranium from the fission products. The organic extraction solvent was a nominal 30 (vol.)% TBP in a paraffinic hydrocarbon (much like kerosene) that gives good flow characteristics in the liquid extraction contact stages. The organic solvent enters one end of the extraction train with the first contact stage removing all but a trace of the uranium and plutonium from the acid (aqueous) phase containing the fission products. The TBP solution continues to load with U and Pu at each stage until it reaches the feed stage. Then 2–3 molar nitric acid is fed at the other end of the extraction train and scrubs fission product metals from the TBP phase. The contact time between the TBP and acid phase containing the fission products was kept short to minimize the gamma radiation damage to the organic phase. The acid solution goes to a nitric acid recovery unit that recovers the nitric acid values, removes water and concentrates the fission product solution for storage.

A second extraction train receives the TBP solvent stream with one end fed dilute nitric acid containing chemicals that reduce the plutonium to the acid soluble Pu^{+3} state. Fresh TBP is fed to the other end of the train to remove traces of uranium stripped into the acid stream that now contains the plutonium. This extraction step completes the separation of the large fraction that is uranium from the small plutonium fraction.

The uranium is released from the TBP with a scrubbing train using dilute acid. The stripped TBP goes to solvent recovery and is cleaned up for recycle. Water is evaporated from the dilute acid containing the uranium. The pH is adjusted, and the uranium extracted with countercurrent TBP-acid scrub to remove traces of plutonium and fission products. The spent nitric acid goes to nitric acid recovery and the fission product waste, including a trace of plutonium that goes to waste concentration.

The uranium is recovered from the TBP with very dilute nitric acid scrub. The TBP is recycled and the uranium solution evaporated. There may be a final uranium "polishing step" before the steps to produce the final uranium product, a uranium nitrate solution or de-nitration to form UO_3.

The plutonium that was left in the acid solution is oxidized to the Pu^{+4} state and center fed to an extractor which collects the plutonium in the TBP phase and leaves impurities in the acid phase. The TBP phase goes to a dilute acid stripper to recover the product as plutonium nitrate solution. There usually is a product plutonium-polishing step to attain the maximum purity of the product. The stripped organic TBP phase and the aqueous acid phase are recycled to recover nitric acid values, renew the organic phase, and reduce the volume of the waste stream.

There are several variations of the general separation steps described above. The first PUREX process plant to produce weapons grade plutonium was located at the AEC Savannah River Plant and began production in November 1954 [37]. With this successful demonstration, another plant began operation at the Hanford site in January 1956. Improvements in the operation of the PUREX process continued with operating experience.

The demands for high purity plutonium and uranium made the development of the PUREX process a real separation technology challenge. Add to this the demands for personnel safety, protection from the toxicity of the heavy metals, and the continuous gamma radiation from the fission products. The engineering task included developing mechanical manipulaters to perform all of the process operations and equipment maintenance tasks protected by the radiation shielding. This technology was developed for the military and is the basis for modified PUREX to reprocess domestic spent nuclear fuel.

PUREX Process—Domestic Spent Fuel

The first nuclear reactors designed to produce electricity were installed in nuclear submarines. Such a reactor must provide the electric and thermal power required to sustain the crew under water, long term, and then to provide the additional variable power required during battle maneuvers. The total mass and the size of the reactor must fit on the submarine and protect the crew from fission product gamma radiation. These reactor designs, the nuclear fuel composition and configuration were classified "Top Secret" by the military. Civilian contractors built these reactors and their engineers saw an opportunity, and were encouraged, to extend this technology to civilian electric power production under the Atoms for Peace Initiative.

The nuclear reactors for domestic electric power production were designed to provide base line power, continuous operation at power plant design capacity for long periods between refueling and mechanical equipment maintenance shutdowns. The two designs widely adopted and deployed in the United States were the BWR and the PWR reactors. The fuel for these reactors is uranium oxide slightly enriched to between 2.6–4% U-235.

Uranium oxide fuel is commercially manufactured into small, cylindrical pellets about 12–13 millimeters in diameter and the same length. These pellets are loaded into metal tubes (about 1 centimeter OD), originally stainless steel but soon replaced by Zircaloy (mostly pure zirconium alloyed with tin, nickel, chromium, and iron). The end cap on each end of the tube is welded to isolate the uranium fuel and all the fission products (gases and solids) from the water in the reactor. Often, these tubes are pressurized with helium to improve the heat transfer from the fuel pellets to the tube wall. Zircaloy has a low neutron capture cross section, is corrosion resistant, and quickly becomes the material of choice for this application [38].

The fuel tubes for a typical PWR are fixed in a fuel assembly consisting of a 15 × 15 array of fuel tubes fixed in place with space to circulate pressurized water to remove heat and to serve as the neutron moderator (to slow down the neutrons). Such a fuel assembly is about 4 meters long and weighs about 658 kg. It contains about 523 kg of uranium oxide (461 kg of uranium metal). There is 135 kg of Zircaloy and hardware metal in each fuel assembly [39]. These fuel assemblies contain the spent nuclear fuel that is the feed for spent fuel reprocessing.

The composition of the spent fuel is determined by the initial composition of the fuel and the radiation history of the fuel assembly. There are three sources of radioisotopes formed during the power cycle of the fuel:

- Fission products formed by the splitting of the fissile elements, the initial U-235 and the Pu-239 that is formed by neutron capture of the U-238 in the fuel during the fuel cycle)
- The transuranic elements formed by neutron capture (neptunium, plutonium, americium, and curium)
- The activation products formed by exposure of atoms to the high radiation field in the reactor.

Immediately following reactor shutdown, the fuel will contain more than 350 nuclides, many with very short half-lives that decay in seconds or minutes [40]. These radioactive decay processes produce thermal energy (heat) and gamma radiation that must be managed when the spent fuel elements are stored.

For example, consider a reactor that operates with a fuel burn up to 30,000 megawatt days per ton (1000 kg) of uranium metal in the fuel. Immediately after shutdown, these fuel elements will produce nearly 2000 kW of thermal energy and a nuclide radioactivity of about 2×10^8 curies per ton. The fuel assemblies are stored in a deep pool of water (containing soluble boron, a neutron absorber) located at each power plant where the thermal energy is removed and the water serves as radiation shielding [41]. The thermal energy release and the gamma radiation decrease with time as each radioactive isotope decays and after 10 years the thermal energy release is about 1.1 kW and the radioactivity is about 3.9×10^5 curies per tonne. The thermal energy release is still much too high to allow isolated underground storage. The gamma radiation requires bulky biological shielding to protect persons transporting spent fuel to any remote storage or reprocessing site.

REPROCESSING: RECOVERY OF UNUSED FUEL

In the beginning of the twenty-first century, reprocessing technology in the United Sates has not been sufficiently developed for commercial spent nuclear fuel to be economically competitive with new uranium. An agreement signed during President Carter's administration closed out the option of reprocessing domestic spent fuel. Reprocessing would make plutonium more easily attainable to terrorists—if kept in the spent fuel and with the highly radioactive fission products it is not as available.

The fear of nuclear weapons proliferation is a major obstacle to domestic nuclear fuel reprocessing. Security assurances of excess military plutonium and highly enriched uranium will be important in any decision to proceed to commercial spent fuel reprocessing.

The plutonium in the spent nuclear fuel is not suitable for bombs, even if concentrated, since it contains too much of the Pu-240 isotope [42]. The separation of the Pu-240 from the Pu-239 would be more difficult than concentrating bomb-grade U-235 from fresh uranium ore because the atomic weights of Pu-239 and Pu-240 are even more nearly the same than U-235 and U-238 [43]. It is the mass difference between the isotopes that allow the mechanical centrifuge separations. Uranium isotope separation is expensive—plutonium isotope separation would be more expensive.

Since nuclear waste handling is a known significant expense of nuclear power generation, most governments levy a tax on nuclear electrical power to be applied toward disposal. In the United States, this tax is 0.1 cents per kilowatt-hour. This fund has more than $36 billion available (without interest computed) [44]. If existing nuclear plants reprocessed spent fuel rods, they could store the concentrated waste indefinitely. If the revenues collected for handling nuclear waste were used to reprocess nuclear waste, the waste problem would be solved and reprocessing would be viable. One means to supplement fissile fuel content in reprocessed fuel is to use the plutonium in excess nuclear weapon's inventories.

The British and French have over 35 years of experience in reprocessing spent nuclear fuel. The PUREX process is the primary technology in use.

PUREX Process

Figure 8.33 summarizes the PUREX process steps for recovering spent nuclear fuel. The first step for commercial nuclear fuel reprocessing is opening the fuel tubes so the irradiated fuel can be dissolved in nitric acid. Chemical de-cladding used in military plutonium production is replaced by mechanical shearing the commercial reactor fuel assemblies into short lengths. This releases helium (if helium was filled during fuel manufacture) and the fission product gases (isotopes of krypton, xenon, tritium are examples) that must be collected. After a reasonable time these gases convert to stable isotopes and can be released. Radioisotopes, including Kr-84, I-131, and Xe-133, are currently vented to the atmosphere [45]. Other treatment may also be necessary. The long-lived radioactive iodine released during this step is given special attention [46].

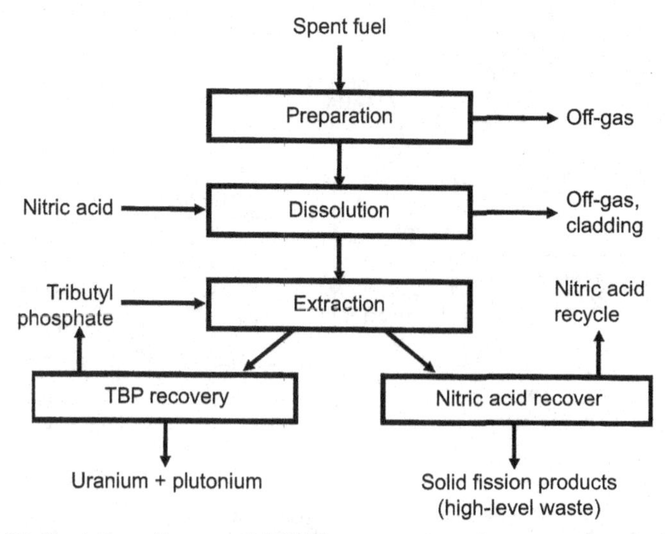

Figure 8.33 Block flow diagram of PUREX reprocessing of spent nuclear fuel.

The next step is to *dissolve the fuel metal oxides* containing fission products, uranium, plutonium, and transuranic metals in nitric acid. The stainless steel and Zircaloy pieces from the fuel assemblies do not dissolve and are separated from the nitric acid solution, washed to remove all of the uranium, fission products, and transuranic elements, dried and packaged as low-level radioactive waste. The nitric acid solution pH is adjusted to assure that uranium and plutonium are in the most favorable oxidation states for extraction.

Some of the fission products are (or form) metal compounds that exceed solubility limits and these are *filtered out* before entering the extraction train. For each metric ton of metal in the spent fuel, there will be about 944–946 kg of U-238, 8–11 kg of U-235, 5–9 kg of Pu-239, and 1–9 kg of heavy metal isotopes with atomic numbers greater than uranium (transuranic) in the periodic table. The total mass of the fission product metals (more than 40 elements) is about 34 kg [47].

A small fraction of the fuel does not dissolve in nitric acid. These residues vary depending on the fuel characteristics, the time the fuel is irradiated, and the procedure used to dissolve the fuel [48]. The acid solution must be clarified before it is fed to the extraction train. These residue solids will be radioactive and a heat source that requires special handling, especially for "young spent fuel" (spent fuel aged less than 10 years).

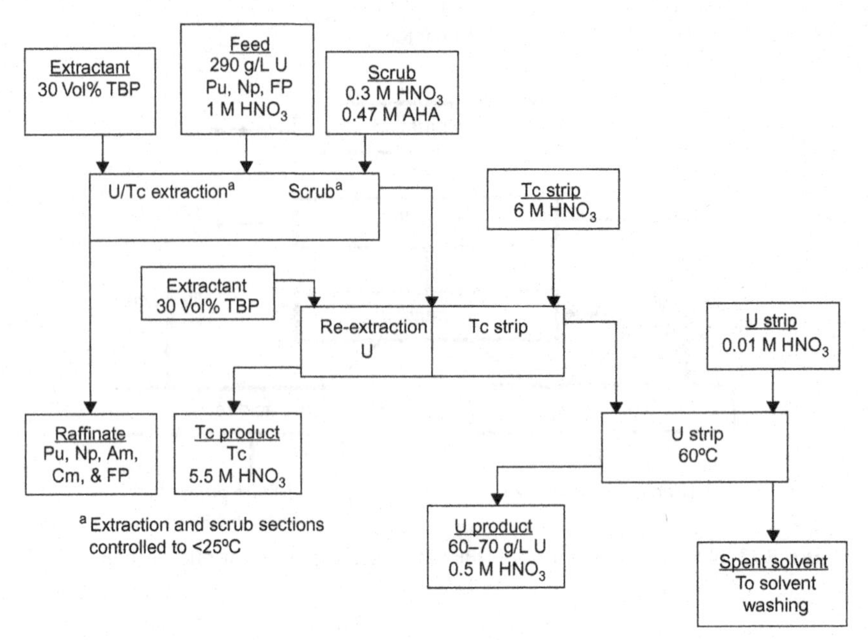

Figure 8.34 UREX block flow diagram [49].

The *first extraction train* removes the uranium and plutonium to the organic TBP phase and leaves the 43 g of minor actinides and fission product metals in the aqueous phase. The strong gamma radiation of the fission products that cause radiological damage to the TBP phase is essentially all removed in this first extraction step. Extraction steps to strip the remaining fission products and to separate the uranium and plutonium follow, with minor modification of the process for the production of plutonium.

This extraction process can be operated to extract only uranium (UREX) (see Figure 8.34) or plutonium and uranium (PUREX). A very pure uranium is required to feed the U-235 enrichment process for new uranium oxide fuel—the fuel for which light water reactors were designed. UREX uranium contains 0.7–1.1% U-235, about the same as natural uranium.

There are other uranium isotopes in this recycle stream that "tag along" and these are neutron absorbers in the recycled fuel for the LWR. These isotopes accumulate each time the fuel is recycled and the different decay routes produce thermal energy and high-energy gamma radiation which may require protective clothing or remote handling of recycled uranium metal [50].

The PUREX process does produce pure plutonium, the source of strong objection from the members of the nuclear weapons nonproliferation people. The proposal to mix plutonium oxide with uranium oxide to form a mixed oxide fuel (MOX) that can be used as fuel for light water reactors has been used in commercial reactors in Europe. MOX fuels have been successfully used on a limited basis and this fuel mixture does not present a weapons threat.

Plutonium represents the key ingredient for closing the uranium fuel cycle and it is an important source of energy for future nuclear power plants. The current fleet of LWRs in the United States produce about 2000 tons of spent fuel per year containing about 0.5–0.9 % Pu, yielding 10–20 tons of plutonium. The age of the spent fuel (how long it has been in storage) changes the ratio of the isotopes of plutonium and the performance of the plutonium fuel in the LWR reactor. The energy producing fissile materials in the light water moderated reactor are U-235 and Pu-239. The other isotopes accumulate with each reprocessing cycle. This disincentive for reprocessing can be overcome using the reprocessed fuel in a new generation of fast flux (fast neutron) reactors discussed below.

Further processing can be performed on the mixture of uranium and plutonium leaving the PUREX process to prepare pure uranium and pure plutonium. The PUREX process produces gaseous effluents and cladding hulls that are not hazardous. The fission products are concentrated into a solid high-level waste. Since the nitric acid is recycled the solvents (primarily tributylphosphate, TBP) is also recycled. In general, the acid and solvents do not add to the volume of waste resulting is a substantial decrease in the total volume of radioactive waste from these new processes.

The PUREX process separates the uranium and plutonium from the spent fuel, but the fission products and minor actinides that remain in the waste stream represent a very long-term waste storage problem. An advanced form of PUREX known as URanium EXtraction (UREX) has environmental and antiproliferation advantages over PUREX as described by the following excerpt from the DOE 2003 Report to Congress [51].

In the UREX process, plutonium, and other transuranics, and fission products are extracted in a single stream from which transuranics could be extracted for reuse in nuclear fuel. The feature of UREX that makes it much more proliferation resistant than PUREX is the continual presence of minor actinides, the high radioactivity and thermal characteristics of which make these materials relatively unattractive to potential proliferaters.

Additionally, because UREX does not place these actinides in the waste stream, there could be a significant reduction in the amount of highlevel waste produced. Short-lived radioactive isotopes are separated and may be stored and allowed to decay to harmless elements over several decades.

Further, experiments completed in 2002 have proven UREX to be capable of removing uranium from waste at such a high level of purity that we expect it to be sufficiently free of high-level radioactive contaminants to allow it to be disposed of as low-level waste or reused as reactor fuel. These laboratory-scale UREX tests have proven uranium separation at purity levels of 99.999 percent. If spent fuel were processed in this manner, the potential exists to reduce significantly the volume of high-level waste. An additional advantage of UREX is the use of acetohydroxamic acid, which enables the use of chemical processes that are far more environmentally friendly than PUREX.

AFCI (Advanced Fuel Cycle Initiative) Series One research would include the continued development of aqueous chemical treatment technologies including the possible demonstration of UREX at a scale relevant to its eventual commercial use.

An advanced development of UREX, referred to as "UREX+," would be a key element of an AFCI program. This additional research would evaluate different aqueous chemical treatment methods to separate selected actinide and fission product isotopes from the UREX stream after the uranium has been extracted in a manner that minimizes waste. For example, UREX+ would provide mixtures of plutonium and selected minor actinides for preparing proliferation-resistant fuels.

Long-lived fission products, iodine-129 and technitium-99, which are major contributors to the long-term radiotoxicity from spent fuel, could be separated for incorporation into targets for destruction in reactors. This work would allow the program to obtain a detailed understanding of all waste streams, the data needed for understanding what would be needed in a commercial scale treatment facility, and provide the basis for estimating the cost to design, build, and operate such a facility.

If implemented successfully, this treatment technology could significantly reduce the volume of high-level waste from commercial nuclear power. This accomplishment would reduce the cost of the first repository and potentially eliminate the technical requirement for a second.

Key programmatic treatment technology elements for AFCI Series One would include: (i) laboratory demonstration of UREX+ using radioactive materials; (ii) engineering scale demonstration of UREX+; (iii) laboratory demonstration of PYROX (pyro-chemical dry treatment) technology using spent LWR fuel; (iv) demonstration of PYROX actinide recovery;

(v) engineering scale demonstration of PYROX using radioactive materials; (vi) demonstration of large-scale metal waste form technology; and (vii) treatment facility requirements, costs, and design studies.

The UREX treatment technology, combined with additional processing steps, provides a way to produce proliferation-resistant transmutation fuels for use in LWRs or gas-cooled reactors.

Successful implementation of this technology would require dealing with several issues such as fabrication and testing of trans-plutonic-bearing fuels, which would require remote fabrication.

To support this effort, research on transmutation fuels would focus on the development of proliferation-resistant fuel forms, preliminary fuel irradiation testing, and analysis of the resulting transmutation system (including waste streams). In the case of LWR transmutation fuels, several technology options would be considered, including the French CORAIL, Advanced Plutonium Assembly systems, advanced assembly designs, and inert matrix/non-fertile fuel concepts.

Gas-cooled reactors use very small spherical fuel particles, which if manufactured with advanced coating technology, are strong enough to permit much higher burnup than are possible with LWRs, and are difficult to reprocess. Very high destruction levels of plutonium (over 90%) have already been demonstrated using pure plutonium fuels; however, the challenge remains to achieve these impressive burnups with proliferation-resistant fuels. Research is needed to address this challenge and will include the development of proliferation-resistant fuel forms, fuel irradiation testing at the High Flux Isotope Reactor at Oak Ridge National Laboratory or the ATR at the Idaho National Engineering and Environmental Laboratory (INEEL), and analysis of the resulting transmutation system performance for gas cooled reactor fuel.

Advanced Aqueous Separation

More recently, Argonne National Laboratory personnel have developed an advanced aqueous process called UREX+ has been demonstrated on a laboratory scale [52]. The UREX+ process consists of five solvent extraction steps that separate the dissolved spent nuclear fuel (the PUREX feed) into seven fractions. In the first stage, the uranium and technetium are recovered in separate streams with high total recovery and purity. Next, the cesium and strontium (heat producers, a problem in

Table 8.13 Summary of separations in PUREX and UREX processes

PUREX	• Extracts of pure U and Pu, free from fission product contamination • Minor actinides go to waste along with fission products
UREX	• Separates pure U • All transuranics are recovered as a group • Cesium and strontium removed to improve effective repository capacity • Lanthanide fission products can be retained with TRUs if needed to provide limited self-protection radiation barrier • Hybrid modification sends the process stream after U, Cs, and Sr removal to a pyrochemical process for separation of TRUs from fission products for fast reactor recycle
UREX+	• Several variants of the UREX process are being studied; all include separation of pure U and removal of Cs and Sr • Each variant provides different options for the recovery of transuranics, either as a group, or as subgroups, for use in different recycle scenarios in thermal and fast spectrum systems • Provides flexibility of response to evolving nuclear systems in the United States.

repository waste storage) are removed. A little feed adjustment allows the plutonium and neptunium to be recovered with impurity levels that allow these actinides to be incorporated into MOX fuel. The fourth step recovers the minor actinides and the rare earth elements. The final step separates the minor actinides from the rare earth elements.

The many metal species in spent fuel dissolved in nitric acid form a complex chemical mixture. The UREX+ process demonstration indicated that with additional work to understand this chemistry and to refine the separation parameters, the product streams can yield recycle nuclear fuel, radioactive isotopes that can be formed into targets for transmutation, and a waste product that can meet the demands for long-term geological storage. This work demonstrates an option for the treatment of the inventory of spent fuel accumulating from the current fleet of light water reactors.

Tables 8.13 and 8.14 provide a summary comparison of these aqueous treatment processes.

Experimental Breeder Reactor II

The nuclear reactors designed to produce military plutonium used uranium metal as fuel. As early as the mid-1940s there were attempts to demonstrate the use of plutonium as a fuel for power production [53].

Table 8.14 Further comparison of aqueous treatment processes

Process	PUREX	Advanced aqueous separations	
		UREX	UREX+
Pure Pu separation	Yes	No	No
Remote fuel fabrication Required	No	Yes	Some do
Technology development completion	Available	2010	2012
Commercial Experience	Large in old plants	None	None

In 1963, Experimental Breeder Reactor II (EBR-II) was the first fast flux (the high-energy neutron spectrum produced at fission) reactor. The fuel core, the circulation pumps, and the primary heat exchanger were submerged in a pool of liquid sodium contained in the reactor vessel.

Sodium has a small neutron cross section that minimizes the neutron slowing down effect. The primary sodium coolant that is exposed to high-energy radiation in the reactor core becomes radioactive (a portion of the sodium becomes Na^{24} with a 15 hr half-life). The secondary liquid sodium loop is circulated through the heat exchanger and transports the thermal energy to a steam generator to supply steam to a turbine. This higher pressure secondary sodium would leak into the reactor pool if there were a leak in the sodium heat exchanger. This would prevent radioactive sodium release from the reactor vessel.

The EBR-II operated until 1994, over 30 years without heat exchange problems. The facility was designed to "breed" plutonium to extend the inventory of fissile uranium. Later, the fuel loading was modified to demonstrate it could be a "plutonium burner" (to reduce the inventory of surplus military plutonium). EBR-II provided the test bed for irradiation studies of many proposed metallic and oxide fuels for military and commercial applications [54].

The EBR-II was designed to be an integral nuclear power plant. This included on site nuclear fuel reprocessing and new fuel production on the reactor site. The fertile atoms (U-238) in the fuel absorb neutrons to form fissile atoms (Pu-239). These would join the fissile metals in the initial fuel to serve as fuel in the fast neutron flux. The demonstration included the production of steam to drive a turbine to produce electric power to complete the simulation of a commercial power plant.

The engineers on the EBR-II program steadily increased the performance of the fuel for the reactor. Initial problems with the fuel showed the uranium metal expanded during irradiation, mechanically stressing the tubes containing the fuel limiting service to about 1% of the U-235 in the fuel. The swelling problem was solved using a uranium–plutonium–zirconium alloy fuel that achieved 10% fission of the fuel (compared to 3.4% for commercial LWR metal oxide fueled reactors). The metal fuel was sealed in stainless steel tubes with sodium metal filling the space between the fuel and the tube wall (sodium bonded fuel). The all-metal fuel elements in contact with the liquid sodium pool provided high-heat transfer rates, smaller in core temperature gradients making possible a smaller reactor core.

The original EBR-II fuel was enriched uranium to serve as a *driver* (neutron source) for the production of plutonium. The demonstration of "breeding" plutonium in uranium 238 rods surrounding the reactor core required the recovery and recycling of the new plutonium as fuel. Collaboration between Argonne National Laboratory in Illinois and the Idaho National Laboratory developed a "dry" (no water, no nitric acid) process they called pyropartitioning [55].

Pyrometallurgical Reprocessing

Unlike the PUREX and UREX processes, pyrometallurgical processing is based on electrochemical separation. Spent fuel metal is dissolved in the salts of the electrorefiner. Oxide fuels must first be reduced to metal prior to electrolysis as illustrated in Figure 8.35. Electrochemical separation can be compact and relatively simple; however, the basic process mixes the cladding with fission products, forming more high-level waste than some alternatives.

In the pyropartitioning process, the spent fuel elements are chopped into short pieces and placed in metal basket in a pool of molten salt. A minimum melting (eutectic) mixture of potassium chloride and lithium chloride (KCl–LiCl) was used as the solvent. An electric current passed from the chopped fuel basket serving as an anode where all the heavy metals and fission products in the spent fuel oxidize to form metal chlorides dissolved in the molten salt. An inert metal cathode serves to collect the selectively reduced heavy metals from the salt. Uranium is the first metal collected and since it is the most abundant metal in the spent fuel the deposition voltage separating them as they selectively deposit on the

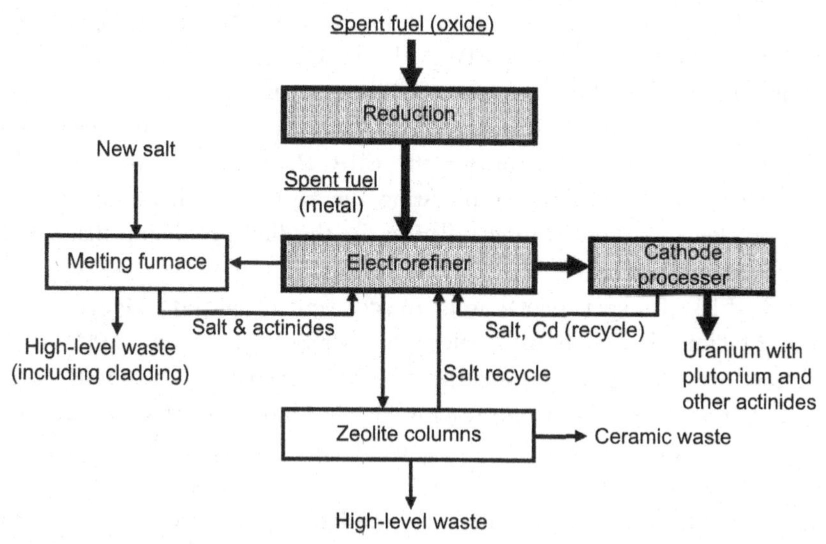

Figure 8.35 Pyropartitioning process to recover heavy metals.

cathode based on the electronegative potential for each metal chloride in the salt (the voltage for deposition of each metal is like the voltage of the lead-acid battery used in an automobile). This electrolytic process separates the uranium and plutonium from the fission products and the minor actinides producing uranium–plutonium metal. The metal deposited on the cathode is harvested, separated from adhering salt, and cast into new metallic fuel pins for fuel or sent to storage.

The removal of the uranium and plutonium (about 94–95% of the mass) leaves the fission products and the transuranic elements in the salt. The metallic sodium bonding agent in the fuel elements forms sodium chloride. The melting temperature of the KCl–LiCl eutectic salt increases as metal chlorides build up. The saturated solvent salt is removed, most of the KCl–LiCl recovered, and the remainder formed into ceramic waste form suitable for long-term storage.

There are noble metal fission products that remain solid during the electrodeposition and these are combined with the fuel cladding pieces that remain in the anode basket. These solids are stabilized into a metallic waste form and sent to storage.

All of the processing steps must be done in an inert atmosphere (argon) that is essentially free of water, hydrogen, nitrogen, and oxygen. The actinide and rare earth metals in the spent fuel are chemically active and readily form oxides, nitrides, and hydrides that are insoluble in the molten salt. They collect as insoluble solids in the processing equipment. The radioactive fission products in the spent fuel produce high-energy gamma radiation making radiation shielding necessary.

Remote handling is required for all of the steps in the fuel reprocessing and new fuel fabrication for the EBR-II fuel. Protection from the health hazards of inhaling or ingesting the radioactive heavy metals are minimal because there is total isolation for the fuel processing that serves to protect the workers.

The EBR-II experimental program achievements include: Generation of over 2 billion kilowatt-hours of electricity; irradiation of over 30,000 specimens of fuel, structural, and neutron absorber materials; advance instrumentation testing; a test of inherent reactor safety with demonstration of total loss of coolant flow; and advanced computer technology applied to diagnostics and control [56]. In December 1995, James Toscas (Executive Director of the American Nuclear Society) stated, "EBR-II is arguably the most successful test reactor ever." The technology informing the next generation of fast flux reactors depends on data collected during the EBR-II experimental run.

The pyropartitioning process can be used to separate the nuclear fuel components from fission products to make fuel for the next generation of fast flux reactors. The LWR spent fuel in the United States is estimated to reach 70,000 metric tons in 2015. The current fleet of reactors in the United States produce about 2000 metric tons of spent fuel per year. Anticipating commercial scale processes will be required to recover fuel values and reduce the mass of radioactive waste going to a repository, the Chemical Engineering Division at Argonne National Laboratories has completed a demonstration of an electrochemical process that reduces the oxide fuel to metal [57]. The reduced metal would be fed to the pyropartitioning process to recover the fuel values from the fission products.

The promise and future of nuclear power systems depends on the success of research and development programs. There must be investments to develop and deploy new energy systems for nuclear energy to play its role in the future of electric power generation.

Mining and Processing

Uranium ore is mined and processed to uranium oxide concentrate (U_3O_8) that is sold as a feedstock for further processing into reactor fuel. For use as reactor fuel, the uranium must be enriched in the U-235 isotope. This is performed in a gas-phase process by forming uranium hexafluoride (UF_6).

Enrichment is used to increase the 0.7% natural U-235 concentration to 2.6–5%. The enriched uranium is processed to uranium dioxide (UO_2) and formed into fuel pellets that are placed in tubes as reactor fuel rods.

Today's world nuclear power production includes about 437 nuclear reactors with a capacity of 381,000 megawatts. Every year each 1000 megawatts of power production converts 750 kg of U-235 (and some U-238/Pu-239) into 750 kg of waste fission material mixed with about 30,000 kg of unused fissionable elements (U-235, U-238, and Pu-239).

Proven world uranium reserves are 3.3 million tons [58] with vast deposits in Australia and Canada [59]. Estimated reserves in addition to the proven reserves include another 10.7 million tons. Recently discovered uranium deposits in Canada are so rich in uranium that they must be mined with robots to avoid exposing miners to their natural radiation. In the United States, 56,000 tons of spent fuel plus 224,000 tons in depleted uranium represent 400 years of fuel for essentially all energy needs (electricity, transportation, and heating). The world estimated reserves are more than 50 times this 400-year supply—at least 20,000 years. Thorium can also be used to fuel nuclear reactors. Of course, these projections are rather superfluous since man can hardly predict how technology will create new ways to meet energy needs 30 years in advance; let along 500 or 5000 years in advance.

In the 2005 AFCI Report to Congress [60] an attempt was made to estimate unconventional uranium reserves. These reserves include 180 (sandstone), 4300 (seawater), and 800,000 (phosphate) million metric tons. If 16 million metric tons correspond to 20,000 years of uranium, 804,500 million metric tons correspond to 1 billion years of energy from uranium.

The Report to Congress classifies these massive phosphate reserves as "unconventional," which is an ambiguous term. If not commercially viable, these reserves are at least on the edge of commercial viability.

> *Uranium from phos-acid is extremely (technically) viable and phosphates represent a major source of U_3O_8. Plants were operated in the 50's in Florida but with the discovery of uranium in the western U.S. the facilities were shut down due to unfavorable economics [61].*

In the early 1970s, interest was revived with the growth of the nuclear power industry and a significant development to improve earlier processes. Much of this work was done at Oak Ridge National Laboratory. Private companies also developed processes: United Nuclear, Freeport Chemical and Westinghouse.

International Minerals and Chemical (IMC) initially worked with United Nuclear to develop a process to recover the uranium. Their process (after installation by WR Grace) had some problems so they began their own development and eventually worked with Oak Ridge National Laboratory.

Several commercial installations resulted from these developments. Freeport installed a facility at their plant in LA. IMC installed three facilities to extract uranium. Their approach was to install primary extraction facilities at the individual phos-acid plants, then produce a concentrated uranium solution (in phos-acid). This material was then trucked to their main refinery for U_3O_8 recovery.

A plant was also installed at the Gardinier facility in Tampa. There were also commercial facilities in Europe. In the late 1970s, there was an excess of 3 million pounds/year of U_3O_8 recovered from phosphoric acid capacity in operation. (As a side note, IMC had about 2.2 MM lbs/yr of capacity and was at that time the fourth largest U_3O_8 producer in the United States.)

Unfortunately, Three Mile Island occurred and nuclear plant activity came to a virtual halt. The long-term U_3O_8 outlook diminished and existing supply contracts were not renewed as the price of U_3O_8 plummeted. The result was that by the early mid-1990s, all of the phos-acid based production ended due to low prices for the product.

Based on the earlier economics (adjusted for inflation), U_3O_8 pricing in the \$35+ range in the mid-1970s, \$25/lb was the price where interest was generated. Uranium from phos-acid was a commercial industry from the mid-1970s to early 1990s, and the technology is well established to restart this industry.

What is important is that nuclear fuel can be used to produce all the power needs, and the lessons of US commercial utilization demonstrate that it is safer and has a lower environmental impact than alternatives. Furthermore, since about 40 years of spent fuel have already been stockpiled, nuclear fission can provide abundant energy without additional mining and actually using more waste than it generates.

WASTE GENERATION FROM REPROCESSING

In the rush to develop the nuclear bomb and the subsequent arms race, vast amounts of nuclear waste were generated. The volume of the high-level

Table 8.15 Estimated quantities in metric tons of chemicals, glass, and salt used in reprocessing waste by different methods

Process	Net chemical consumption	Net glass/ salt frit consumption	High-level waste	Reduction in waste
Once-through			2000	0%
PUREX	4.2	420	490	76%
UREX+	7	124	232	88%
UREX/PYRO	5.6	322	280	86%
PYROX	80	500	490	76%
Advanced aqueous process	0.8	124	232	88%

Assume 3.4% of spent fuel is fission products and ends up in waste and that all the glass, salts, and chemicals end up in waste.
Data from the US DOE Office of Nuclear Energy, Science and Technology Advanced Fuel Cycle Initiative (AFCI) Comparison Report.[205]

waste was compounded by landfilling and containerizing chemicals dilute in the actual radioactive materials—concentrated wastes would be much less voluminous. The liquids can leach through soils and contaminate the waters around storage areas. It is this history that leads many to believe that reprocessing will generate more waste than treated. These perceptions are inaccurate.

Table 8.15 summarizes the five reprocessing options put forward by the 2004 US DOE Office of Nuclear Energy, Science and Technology AFCI Comparison Report [62]. These methods are based primarily on the recovery of uranium, plutonium, and other actinides (in some cases). These technologies reduce the volume of high-level waste by 76–88%.

Table 8.16 summarizes the fuel and nonhazardous product from the reprocessing options. The components marked for recycle for future reactors includes minor actinide transuranics. The cladding would be washed to low- or zero-level waste. The secondary waste is broken contaminated equipment and materials contaminated in fuel transport. The cladding could be recycled as cladding on recycled fuel.

The fate of the uranium product stream from reprocessing is not obvious. The problem is that the amount of U-235 (about 1%) in the uranium is about the same as natural uranium, and this means it has little or no fuel value. The motivation for use of the uranium reprocessing stream is to avoid it becoming a low-level radioactive waste.

Table 8.16 Fuel and low-level wastes from reprocessing options

	Fissile and fertile materials			Low-level waste	
Process	**Uranium**	**Recycled Pu, Np, Am, Cm**	**Recycle for future reactors**	**Cladding**	**Secondary waste**
PUREX	1892	17.0	0.0	660	2.1
UREX+	1892	18.0	3.2	660	3.5
UREX/PYRO	1892	21.2	21.2	660	4.2
PYROX	1892	21.2	21.2	660	2.1
Advanced aqueous process	1892	18.0	3.2	660	1.4

Data from the US DOE Office of Nuclear Energy, Science and Technology Advanced Fuel Cycle Initiative (AFCI) Comparison Report.

Waste Minimization

The reprocessing technologies in Table 8.16 are based on recovering uranium and plutonium. This is only one of the following three phases of spent fuel reprocessing:

Phase 1—recover unspent fuel

Phase 2—minimize waste by separating stable fission products from high level fission products

Phase 3—transmute the high level fission products into non-hazardous materials.

Phases 2 and 3 bring additional costs and require technology developments. Instead of depositing the glass-stabilized wastes, the fission products could be placed in temporary storage (30–60 years) with the objective of processing in the future when larger volumes of these materials have accumulated. These wastes could be stored at the reprocessing facility, and the temporary storage could reduce the consumption of glass.

Waste minimization provides as opportunity to reduce 3.4 parts of high-level fission product waste to 0.4 part of high-level waste. Due to the different properties of the elements (nuclides) in fission waste, a variety of chemical processes would provide the desired waste minimization. The stable isotopes would not be high-level waste. In this processing, the 0.4 part of unstable products could again be placed in temporary storage.

The volume of the unstable products would be low. Technology is available to transmute many of the unstable isotopes—using fast-spectrum reactors and/or accelerators. Storage for centuries would be unnecessary. The waste would be managed with a policy of continuous reduction of the waste that requires long-term storage.

REPORT TO CONGRESS

In January 2003, the US DOE prepared the Report to Congress on the AFCI: The Future Path for Advanced Spent Fuel Treatment and Transmutation Research [63]. This official document confirms the potential of nuclear technology to meet our energy needs without continued growth of spent nuclear fuel as illustrated by the maximum in civilian spent fuel storage at about 2040 according to Figure 8.36 prepared from this DOE report.

The mass of radioactive waste would actually reduce if the stable elements were removed and used. For every 3.4 parts of fission products, 3 parts are stable after moderate storage times. The volume of remaining materials having high radioactive hazards is quite small. If the entire inventory of these materials (for 40 years) were melted into a metal cube, that cube would be less than the size of a small house. It is an option to store this concentrated material for several decades or even a century.

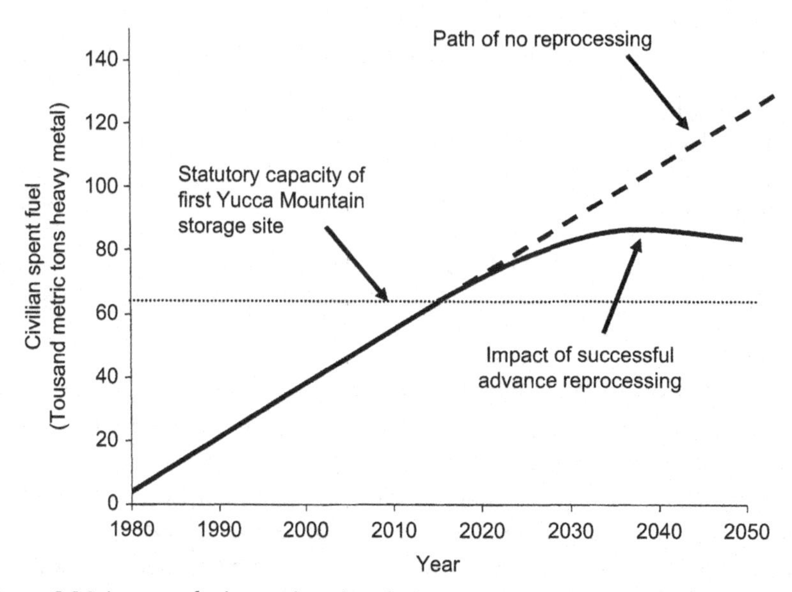

Figure 8.36 Impact of advanced nuclear fuel reprocessing.

When the time is right (due to accumulation of volume or availability of technology), transmutation technology could be used to transform even that small fraction into benign, stable waste.

EXAMPLE CALCULATIONS

Example calculations are from DOE Fundamentals Handbook: Nuclear Physics and Reactor Theory. Volume 1 of 2.

Calculation of Mass Defect—The mass defect can be calculated using Equation 8.5. In calculating the mass defect it is important to use the full accuracy of mass measurements because the difference in mass is small compared to the mass of the atom. Rounding off the masses of atoms and particles to three or four significant digits prior to the calculation will result in a calculated mass defect of zero.

$$\Delta m = [Z(m_p + m_e) + (A - Z)m_n] - m_{atom} \tag{8.5}$$

where

Δm = mass defect (amu)
m_p = mass of a proton (1.007277 amu)
m_n = mass of a neutron (1.008665 amu)
m_e = mass of an electron (0.000548597 amu)
m_{atom} = mass of nuclide ${}^{A}_{Z}X$ (amu)
Z = atomic number (number of protons)
A = mass number (number of nucleons)

Example:

Calculate the mass defect for lithium-7. The mass of lithium-7 is 7.016003 amu.

Solution:

$$\Delta m = [Z(m_p + m_e) + (A - Z)m_n] - m_{atom}$$
$$\Delta m = [3(1.007826 \text{ amu}) + (7 - 3)1.008665 \text{ amu}] - 7.016003 \text{ amu}$$
$$\Delta m = 0.0421335 \text{ amu}$$

Calculation of Binding Energy—Since the mass defect was converted to BE (binding energy) when the nucleus was formed, it is possible to calculate the BE using a conversion factor derived by the mass–energy relationship from Einstein's Theory of Relativity. Einstein's famous equation relating mass and energy is $E = mc^2$ where c is the velocity of light (c = 2.998 × 108 m/sec). The energy equivalent of 1 amu can be determined by inserting this quantity of mass into Einstein's equation and applying conversion factors.

$$
\begin{aligned}
E &= m\,c^2 \\
&= 1 \text{ amu} \\
&= 1 \text{ amu}\left(\frac{1.6606 \times 10^{-27}\text{ kg}}{1 \text{ amu}}\right)\left(2.998 \times 10^{8}\frac{\text{m}}{\text{sec}}\right)^2 \times \left(\frac{1\text{ N}}{1\,\frac{\text{kg} \times m}{\text{sec}^2}}\right)\left(\frac{1\text{ J}}{1\text{ N}\cdot\text{m}}\right) \\
&= 1.4924 \times 10^{-10}\text{ J}\left(\frac{1\text{ MeV}}{1.6022 \times 1^{0-013}\text{ J}}\right)\left(2.998 \times 10^{8}\frac{\text{m}}{\text{sec}}\right) \\
&= 931.5 \text{ MeV}
\end{aligned}
$$

Conversion Factors:

1 amu = 1.6606×10^{-27} kg

1 N = 1 kg m/sec^2

1 J = 1 Nm

1 MeV = 1.6022×10^{-13} J

Since 1 amu is equivalent to 931.5 MeV of energy, the BE can be calculated using Equation 8.6.

$$
\text{BE} = \Delta m\left(\frac{931.5\text{ MeV}}{1\text{ amu}}\right) \tag{8.6}
$$

Example:

Calculate the mass defect and BE for U-235. One U-235 atom has a mass of 235.043924 amu.

Solution:

Step 1: Calculate the mass defect using Equation 8.5.

$$
\begin{aligned}
\Delta m &= [Z(m_{\text{p}} + m_{\text{e}}) + (A - Z)m_{\text{n}}] - m_{\text{atom}} \\
\Delta m &= [92(1.007826\text{ amu}) + (235 - 92)1.008665\text{ amu}] \\
&\quad - 235.043924\text{ amu} \\
\Delta m &= 1.91517\text{ amu}
\end{aligned}
$$

Step 2: Use the mass defect and Equation 8.6 to calculate the BE.

$$
\begin{aligned}
\text{BE} &= \Delta m\left(\frac{931.5\text{ MeV}}{1\text{ amu}}\right) \\
\text{BE} &= 1.91517\text{ amu}\left(\frac{931.5\text{ MeV}}{1\text{ amu}}\right) \\
&= 1784\text{ MeV}
\end{aligned}
$$

REFERENCES

[1] DOE Fundamentals Handbook, Nuclear Physics and Reactor Theory, vols, 1 of 2, U.S. Department of Energy, Washington, D.C., DOE-HDBK-1019/1-93, 1993.

[2] Kadi Y, CERN. From, Garwin RL, Charpak G. Megawatss and megatons. New York: Alfred A. Knopf; 2001.

[3] Report to Congress Advanced Fuel Cycle Initiative: Objectives, Approaches, and Technology Summary. Prepared by U.S. DOE, Office of Nuclear Energy, Science, and Technology, May, 2005.

[4] Report to Congress on Advanced Fuel Cycle Initiative: The future path for advanced spent fuel treatment and transmutation research, Prepared by the U.S. DOE, January, 2003.

[5] Morgan D. Magnetic Fusion: The DOE Fusion Energy Sciences Program, CRS Issue Brief for Congress, Order Code IB91029, January 15, 2002.

[6] Understanding Radiation. See <http://www.epa.gov/radiation/understand/>.

[7] Fact Sheet on Biological Effects of Radiation National Council on Radiation Protection (NCRP) Report 93, 1987. See <http://www.nrc.gov/reading-rm/doc-collections/fact-sheets/bio-effects-radiation.html>.

[8] A Technology Roadmap for Generation IV Nuclear energy Systems, Issued by the U. S. DOE Nuclear Energy Research Advisory Committee and Generation IV International Forum, December 2002, p. 1.

[9] World Nuclear Association, World Nuclear Power Reactors & Uranium Requirements, April 2015. See <http://www.world-nuclear.org/info/Facts-and-Figures/World-Nuclear-Power-Reactors-and-Uranium-Requirements/>.

[10] World Nuclear Association, International Framework for Nuclear Energy Cooperation, March 2015. See <http://www.world-nuclear.org/info/Facts-and-Figures/World-Nuclear-Power-Reactors-Archive/Reactor-Archive-February-2015/>.

[11] Generation IV Nuclear Energy Systems Ten-Year Program Plan, Fiscal Year 2005 Volume 1, Office of Advanced Nuclear Research, DOE Office of Nuclear Energy, Science and Technology, Released March 2005, p. 1.

[12] A Technology Roadmap for Generation IV Nuclear Energy Systems. Issued by the U. S. DOE Nuclear Energy Research Advisory Committee and the Generation IV International Forum, December 2002, p. 11.

[13] Generation IV Nuclear Energy Systems Ten-Year Program Plan, Fiscal Year 2005 vol. 1, Office of Advanced Nuclear Research, DOE Office of Nuclear Energy, Science and Technology, Released March 2005, p. 28–31.

[14] Generation IV Nuclear Energy Systems Ten-Year Program Plan, Fiscal Year 2005 vol. 1, Office of Advanced Nuclear Research, DOE Office of Nuclear Energy, Science and Technology, Released March 2005, p. 23–28.

[15] Howell JR, Buckius RO. Fundamentals of engineering themodynamics. 2nd ed. New York, NY: McGraw-Hill, Inc.; 1992. p. 425–436.

[16] Generation IV Nuclear Energy Systems Ten-Year Program Plan, Fiscal Year 2005 vol. 1, Office of Advanced Nuclear Research, DOE Office of Nuclear Energy, Science and Technology, Released March 2005, p. 31–34.

[17] Generation IV Nuclear Energy Systems Ten-Year Program Plan, Fiscal Year 2005 vol. 1, Office of Advanced Nuclear Research, DOE Office of Nuclear Energy, Science and Technology, Released March 2005, p. 39–41.

[18] ANL-W History – Reactors (EBR-II). See <http://www.anlw.anl.gov/anlw_history/reactors/ebr_ii.html>.

[19] Garwin RL, Charpak G. Megawatts and megatons. New York, NY: Alfred A. Knopf; 2001. p. 131–135.
[20] Generation IV Nuclear Energy Systems Ten-Year Program Plan, Fiscal Year 2005 vol. 1, Office of Advanced Nuclear Research, DOE Office of Nuclear Energy, Science and Technology, Released March 2005, p. 35–38.
[21] Garwin RL, Charpak G. Megawatts and megatons. New York, NY: Alfred A. Knopf; 2001. p. 164.
[22] S. Dotson. The Progress of U.S. SMR Developments, August 1, 2014. See <http://www.power-eng.com/articles/npi/print/volume-7/issue-4/nucleus/the-progress-of-u-s-smr-developments.html>.
[23] See <http://www.libraries.psu.edu/crsweb/tmi/accidnt.htm>.
[24] See <http://www.washingtonpost.com/wp-srv/national/longterm/tmi/tmi.htm>.
[25] Adams ML. vol 32, Number 4 Sustainable energy from nuclear fission power. National Academy of Engineering Publications; Winter, 2002.
[26] Adloff JP, Guillaumont R. Fundamentals of radiochemistry. Ann Arbor: CRC Press; 1993. p. 15–20.
[27] Roland LK, Jerry BG, Gary T, editors. The plutonium story: journals of Professor glenn T. Seaborg, 1939–1946. Columbus, OH: Benefiel, Battelle Press; 1994.
[28] Cunningham BB, Werner LB. The first separation of the synthetic element 94PU-239. In: Seaborg GT, Katz JJ, Manningam WM, editors. The transuranium elements, research papers. New York: McGraw-Hill Book Company; 1949. p. 51–78.
[29] Groueff S. Manhatten Project: the untold story of the making of the atomic bomb. Boston, MA: Little, Brown and Company; 1967. p. 9.
[30] Thayer H. Management of the hanford engineer works in world war II. New York, NY: ASCE Press; 1996. p. 94.
[31] Glasstone S. Sourcebook on atomic energy. 3rd ed. Princeton, NJ: D. Van Nostrand Company; 1967. p. 519–523.
[32] Thompson SG, Seaborg GT. Progr. Nuclear Energy Ser. III, 1, 1956. p. 163.
[33] Adloff JP, Guillaumont R. Fundamentals of radiochemistry. Ann Arbor: CRC Press; 1993. p. 20.
[34] Benedict M, Pigford TH, Levi HW. Nuclear chemical engineering. 2nd ed. New York, NY: McGraw-Hill Book Company; 1981. p. 458–459.
[35] Thayer H. Management of the hanford engineer works in World War II. New York, NY: American Society of Civil Engineers; 1996. p. 17.
[36] Siddall III TH. Solvent extraction processes based on Tri-n-butyl phosphate. In: Flagg JF, editor. Chemical processing of reactor fuels. New York, NY: Academic Press; 1961. p. 199–207.
[37] Benedict M, Pigford TH, Levi HW. Nuclear chemical engineering. 2nd ed. New York, NY: McGraw-Hill Book Company; 1981. p. 461.
[38] Wymer RG, Vondra Jr BL, editors. Light water reactor nuclear fuel cycle. Boca Raton, FL: CRC Press; 1981. p. 39–41.
[39] Wymer RG, Vondra Jr BL, editors. Light water reactor nuclear fuel cycle. Boca Raton, FL: CRC Press; 1981. p. 65.
[40] Cochran RG, Tsoulfanidis N. The nuclear fuel cycle: analysis and management. 2nd ed. LaGrange Park, IL: American Nuclear Society; 1990. p. 270.
[41] Wymer RG, Vondra Jr BL, editors. Light water reactor nuclear fuel cycle. Boca Raton, FL: CRC Press; 1981. p. 69–72.
[42] See <http://www.uic.com.au/nip09.htm>.
[43] See <http://www.uic.com.au/nfc.htm>.

[44] See <http://www.world-nuclear.org/info/Nuclear-Fuel-Cycle/Nuclear-Wastes/Appendices/Radioactive-Waste-Management-Appendix-4--National-Funding/>.
[45] Hinrichs RA, Kleinbach M. Energy its use and the environment. 3rd ed. New York: Brooks/Cole; 2002. p. 475.
[46] Benedict M, Pigford TH, Levi HW. Nuclear chemical engineering. 2nd ed. New York, NY: McGraw-Hill Book Company; 1981. p. 475.
[47] Wymer RG, Vondra Jr BL, editors. Light water reactor nuclear fuel cycle. Boca Raton, FL: CRC Press; 1981. p. 67–68.
[48] Wymer RG, Vondra Jr BL, editors. Light water reactor nuclear fuel cycle. Boca Raton, FL: CRC Press; 1981. p. 92.
[49] Thomas MS, Norato MA, Kessinger GF, Pierce RA, Rudisill TS, Johnson JD. Demonstration of the UREX solvent extraction process with Dresden reactor fuel solution. Report WSRC-TR-2002-00444. Springfield, VA: U.S. Department of Commerce, NTIS; 2002.
[50] Cochran RG, Tsoulfanidis N. The nuclear fuel cycle: analysis and management. 2nd ed. LaGrange Park, IL: American Nuclear Society; 1990. p. 270, 215–237.
[51] Report to Congress on Advanced Fuel Cycle Initiative: The future path for advanced spent fuel treatment and transmutation research, Prepared by the U.S. DOE, January, 2003.
[52] G.F. Vandegrift, M.C. Regalbuto, S.B. Asse, H.A. Arafat, A.L. Bekel, D.L. Bowers, et al., Lab-Scale Demonstration of the UREX+ Process, WM'04 Conference, February 29 – March 4, 2004, Tucson, AZ.
[53] Waltar AE, Reynolds AB. Fast breeder reactors. New York, NY: Pergamon Press; 1981. p. 24–26.
[54] History of Nuclear Energy, ANL-W, 1998. See <http://www.anlw.anl.gov/anlw_history/general_history/gen_hist.html>.
[55] A Report to Congress on Electrometallurgical Treatment of Waste Forms, March 2001, p. 4–6.
[56] ANL-W History – Reactors (EBR-II). See <http://www.anlw.anl.gov/anlw_history/reactors/ebr_ii.html>.
[57] Electrochemical process for spent fuel treatment. See <http://www.cmt.anl.gov/science-technology/nuclear/>.
[58] Survey of Energy Resources: Part 1 Uranium. World Energy Council. See <http://www.worldenergy.org/wec-geis/publications/reports/ser/uranium/uranium.asp>.
[59] See <http://www.uic.com.au/uran.htm>.
[60] Report to Congress Advanced Fuel Cycle Initiative: Objectives, Approach, and Technology Summary. U.S. DOE Office of Nuclear Energy, Science, and Technology, May 2005.
[61] Personal conversation with Wes Berry.
[62] U.S. DOE Office of Nuclear Energy, Science and Technology Advanced Fuel Cycle Initiative (AFCI) Comparison Report, FY 2004. Published by the U.S. DOE, September, 2004.
[63] Report to Congress on Advanced Fuel Cycle Initiative: The Future Path for Advanced Spent Fuel Treatment and Transmutation Research. Prepared by the U.S. Department of Energy, Office of Nuclear Energy, Science, and Technology, January 2003. See <http://www.ne.doe.gov/reports/AFCI_CongRpt2003.pdf>.

RECOMMENDED READING

Advanced Fuel Cycle Initiative (AFCI) Comparison Report, FY 2003, October 2003. Available at: <http://www.ne.doe.gov/reports/reports.html>.

Design Features and Technology Uncertainties for the Next Generation. Nuclear Plant, June 30 2004. Available at: <http://www.ne.doe.gov/reports/reports.html>.
DOE Fundamentals Handbook, Nuclear Physics and Reactor Theory, vol. 1 of 2, U.S. Department of Energy, Washington, D.C., DOE-HDBK-1019/1-93, 1993. Available at: <http://www.eh.doe.gov/techstds/standard/hdbk1019/h1019v1.pdf>.
Report to Congress Advanced Fuel Cycle Initiative: Objectives, Approach, and Technology Summary. U.S. DOE Office of Nuclear Energy, Science, and Technology, May 2005. Available at: <http://www.ne.doe.gov/reports/reports.html>.
Report to Congress on Advanced Fuel Cycle Initiative: The Future Path for Advanced Spent Fuel Treatment and Transmutation Research. Prepared by the U.S. Department of Energy, Office of Nuclear Energy, Science, and Technology, January 2003. Available at: <http://www.ne.doe.gov/reports/AFCI_CongRpt2003.pdf>.

CHAPTER 9

Options for Remote Locations

Contents

If the United States had all of today's technology options and an absence of the current electrical power grid, it is likely that a different infrastructure would emerge. Such a mental exercise has important implications on how the infrastructure in the United States should change as money is spent replacing failing old infrastructure and on how third-world countries might do things better for their new infrastructure.

RESURGENCE OF FARMHOUSE AND SMALL-TOWN POWER NETWORKS

The remote farmhouse is a good example. A good alternative to running miles of electrical grid transmission lines to a remote farm house would be the combination of a wind turbine, some on-site batteries, and a plug-in hybrid electric vehicle (PHEV) that could either be charged by wind power or be used to provide power to the farmhouse. In such a system, the PHEV would typically be at the farmhouse when the owner is home and in the evening which are the times when backup power would be most needed. The maintenance and upkeep of the "engine backup" would also be synonymous with maintenance and upkeep of a primary or secondary PHEV needed for transportation. Such a system would have minimal maintenance above and beyond that maintenance already expended to keep the PHEV operational. Also the system would have up to triple redundancy.

Sustainable Power Technologies and Infrastructure.
DOI: http://dx.doi.org/10.1016/B978-0-12-803909-0.00009-9

The concept of using PHEVs to provide power upon demand to the grid during peak time has been suggested by many. However, the possibilities may be greater for off-grid communities and farms. A key factor in this approach is the availability of PHEVs capable of providing 120 VAC electrical power which is technically relatively easy.

Another incentive that local networks offer is the full realization of the rewards and benefits associated with peak load shifting and energy efficiency. While it is clearly beneficial to the power grid for consumers to insulate houses that reduce peak demand air conditioning and to run clothes dryers at night instead of during times of peak demand, the consumer does not receive any benefit from actions that benefit the grid. The result is that investments into efficiency and decisions as to when clothes dryers and water heaters operate tend to be based on convenience and short-term plans. In off-grid networks, the full benefits of peak load shifting and energy-efficient designs are realized, and so, these actions are more likely to be performed in a manner that realizes real benefits.

Other advantages also exist for such off-grid networks, including:

- Creation of jobs locally
- Reduced losses (e.g., 10%) associated with electrical power transmission
- Increased robustness against failure of power distribution due to weather or other phenomena.

Table 9.1 summarizes fuel costs for various options of fuels for use in gasoline engines (30% efficiency) with two benchmark comparisons to power plants. A problem with using PHEVs to provide electricity for

Table 9.1 Example comparative fuel prices for remote electrical power

Fuel	Price ($/MMBtu)	AVG ($/kWh)	Conversion efficiency (%)	Electricity cost ($/kWh)
Remote natural gas	$0.50	$0.002	30%	$0.006
US uranium (powerplant)	$0.62	$0.002	33%	$0.006
Coal (powerplant)	$1.30	$0.004	40%	$0.011
Natural gas	$3.00	$0.010	30%	$0.034
Natural gas	$6.00	$0.020	30%	$0.068
Gasoline ($2.00/gal)	$16.81	$0.057	30%	$0.191
Gasoline ($2.50/gal)	$21.01	$0.072	30%	$0.239

residences is the high cost of gasoline fuel for the electricity at 19–24 ¢/kWh. For many locations, an electricity cost of 10 ¢/kWh is a high price.

Alternatively, if the PHEV was powered by natural gas, the option of using a PHEV to back up wind power and battery packs at a remote residence can be attractive. This depends on the prices of natural gas for vehicle use, which can vary considerably.

COST OF WIND TURBINE SYSTEMS

For large commercial wind turbine systems (1500 kW) the National Renewable Energy Laboratory's 2013 Cost of Wind Energy Review [1] identified installed capital costs from \$1450 to \$3000 per kW (\$1730 best estimate) and capacity factors of 25–50% (38% best estimate) for wind turbine systems. For the installed costs, average breakdown is 32% for the turbine, 18% for support structure, and 10% for electrical infrastructure. The levelized cost of electricity based on 200 year operational life was about 6.6 ¢/kWh.

Comparative data on small scale wind turbine systems is not readily available. A survey of prices versus capacity for advertized systems is summarized in Figure 9.1 and is compared to the commercial benchmark prices of \$1730 per kW installed and 32% of that for just the wind

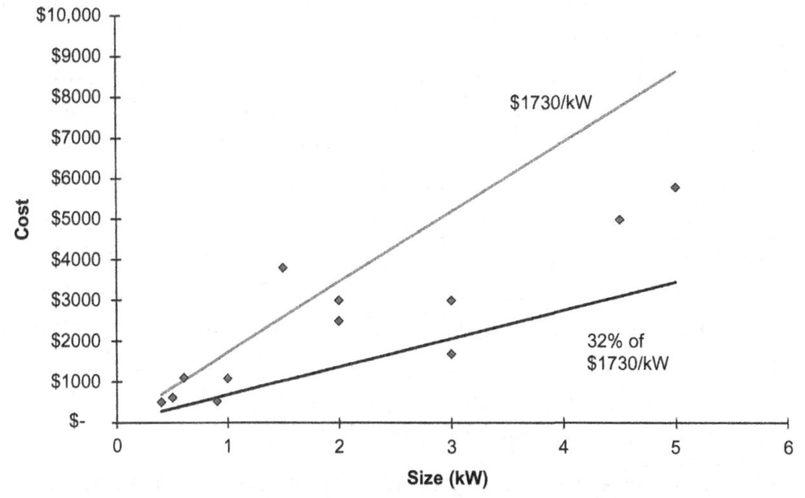

Figure 9.1 Trends in cost of wind turbine power systems as available from internet marketing. Costs generally do not include installation and conversion from DC to AC.

turbine. The conclusion is that commercial price statistics for 1500 kW systems have applicability to residential systems of 0.5–6 kW. This is reflective of the impact of two separate economies of scale.

Commercial wind farms have economies of scale related to the large size of the units which is common in engineering. The residential system has economies of scale related to large-scale production of identical units. In addition, smaller residential systems can benefit for the lower degree of specialization both in skilled labor and equipment needed for installation. None of the systems in Figure 9.1 included shipping and installation, and most did not include a DC to AC converter. The conclusion is that on the topic of residential and small community wind turbines, installed cost of $1530 per kW is a good benchmark to which costs of such systems can be compared.

ECONOMICS OF RURAL AND SMALL-TOWN WIND TURBINE SYSTEMS

According to the US Energy Information Agency, the average US residential utility customer consumed 10,837 kWh per year. This is 29.7 kWh per day or 2.5 kW expended during a 12-h day (50% capacity factor) which is a best-case scenario of little consumption at night and reliable winds during the day. This ranged from 1.5 kW in Maine to 3.4 kW in Louisiana. The numbers reflect a high use of air conditioning during the summer in warmer climates and do not necessarily account of heating during the winter.

These numbers are consistent with recommendations of the American Wind Energy Association [2] which recommends that "on average, a typical American home would require a small turbine with a 5-kilowatt generating capacity to meet all its electricity needs..., however, can range from 2 kW to 10 kW (12–25 ft. diameter)."

Based on these numbers, Table 9.2 presents example costs for the recommended 5 kW wind turbine system with 2 days of battery backup. The levelized cost is 88 ¢/kWh. This example illustrates how battery backup costs can dominate the economics of remote power systems.

The example of Table 9.1 illustrates the primary and major fundamental flaw of wind power as a form of energy that flaw is:

> ***The advertized economics only provide for the production of electrical power by wind and do not account for the cost of having backup energy ready to cover the consumer when the wind is not blowing.***

Table 9.2 Cost of 5 kW wind turbine system with 2-day battery backup

5 kW Wind turbine with 2-day bttery backup	
Capital costs	
Wind turbine installed ($1730/kW)	$8650
2 days battery storage ($600/kWh, 29.7 kWh per day)	$35,640
Total	$44,290
Yearly costs	
Annualized capital cost over 10 year	$4429
Annual Mmintenance (10% of capital)	$4429
Cost over 10,837 ($/kWh)	$0.88

Table 9.3 Cost of 5 kW wind turbine system with gasoline engine backup

5 kW Wind turbine, 5 kWh backup, 5 kW gasoline generator	
Capital costs	
Wind turbine installed ($1730/kW)	$8650
5 kW Gasoline engine generator	$1000
5 kWh Battery storage ($600/kWh)	$3000
Total	$12,650
Yearly costs	
Annualized capital cost over 10 yr	$1265
Annual maintenance (10% of capital)	$1265
Per kWh gasoline cost (20% kWh, 20 ¢/kWh)	$0.04
Cost over 10,837 ($/kWh)	$0.29

Without the backup power, the wind energy system in Table 9.2 costs 14.6 ¢/kWh, which is still expensive, but considerably less than the cost with battery backup. The cost of 14.6 ¢/kWh can be further decreased to numbers approaching 10 ¢/kWh if a 30-year life is used rather than 10 years and if the maintenance cost is put at less than 10%. While the purchase and installed prices of 2–8 kW system may have comparable per kWh prices as large-scale systems, the lifetime and relative costs of repairs will tend to be more and less, respectively. The National Renewable Energy Laboratory report does not include the cost for backup power options. Other factors also impact the costs as summarized by the following examples.

An alternative is to have a gasoline engine generator as backup as summarized in Table 9.3. Use of the gasoline engine would be a last resort option since fuel costs for operation would be around 20 ¢/kWh as

Table 9.4 Summary of economics for small town wind turbine systems

Wind turbine (kW)	Battery backup (kWh)	Engine backup type	Engine cost	% Engine online	Fuel cost (kWh)	Free PHEV	Credit	Electricity cost ($/kWh)
5	59.4	–	–	–	–	–	–	0.88
5	5	Gasoline	$1000	20%	0.2	–	–	0.29
2.5	2	Gasoline	$1000	35%	0.2	–	–	0.20
2.5	2	Nat. gas	$1,000	35%	0.05	–	–	0.15
2.5	2	Nat. gas	$0	35%	0.05	Yes	–	0.13
2.5	2	Nat. gas	$0	35%	0.05	Yes	$7500	0.00

indicated by Table 9.1. Batteries are still needed to reduce gasoline engine operation and to avoid having to frequently start and stop the engine. The cost of this system is still high at 29 ¢/kWh.

These calculations can be performed under various scenarios as summarized in Table 9.4. Downsizing of the wind turbine to 2.5 kW and batteries to 2 h, the estimated cost of electricity for the remote site drops to 20 ¢/kWh. If the generator were operated from a natural gas engine, the price reduces to 15 ¢/kWh. If a natural gas PHEV were used for backup with the driver of the vehicle was home, the cost would drop to 13 ¢/kWh with elimination of additional refueling or maintenance for a separate engine generator.

The greatest breakthroughs in both risk and electricity costs are realized if the costs of powerlines going to the site are avoided.

Example Florida Power and light costs for new subdivisions is *between $1223 and $2025 per lot to install our standard overhead service* [3]. Based on lot distances of 100 to 200 feet, this translates to costs of $10–12 per foot. This can be compared to another independent estimate of $2.50 per foot. These estimates place the costs of overhead powerlines at $11,000 per mile ±50% for lowest cost service to an isolated farmhouse. If the $7500 in costs is realized in savings by the consumer, the cost of electricity is essentially free. Certainly, avoiding those costs through use of wind power with backup is a better option at distances greater than about 1 mile.

The most important aspect of Table 9.4 is how it summarizes the scenarios that work for remote electricity. Due to advanced rural electrification in the United States, the applications are not widespread; however, there is an opportunity for a sustainable industry to emerge around this market as battery costs decrease.

In third-world countries the combination of: (i) undersized wind turbines; (ii) extreme conservation in electricity utilization, and (iii) avoiding rural powerline infrastructure in reality. Lower cost batteries will be instrumental in the transformation from extreme conservation of electricity giving way to improved standards of living. While a natural gas PHEV may not be widely adopted in the United States in third-world countries it could the foundation of both the electrical and transportation infrastructures.

In certain locations, the remote natural gas is not able to reach pipelines. In those locations, that remote gas could be used to refuel vehicles which would further reduce costs as compared to Table 9.4 scenarios.

The discussion in this section may have appeared to be negative and even biased against wind power; however, the lesson to be learned from Table 9.2 example is valid. The lesson is that wind power is very expensive and has a major hidden cost related to providing power when the wind is not blowing.

The latter examples in this section illustrate how there are wonderful opportunities for wind power when augmented with other technologies and applied in the right markets.

The true strength of wind power as an energy source may not have been realized (or at least conveyed) if it was not identified that the initial applications were not effective and that improved options needed to be pursued. For a societal perspective, poor investments (government incentives resulting from single-issue lobbyists) not only take away other technologies more appropriate for an application, the poor investments can also take away critical paths of technologies that will ultimately lead to the greatest impact and sustainability.

SOLAR APPLICATIONS

The previous section on wind power illustrated how combinations of renewable energy, energy storage, low-cost engines, and avoidance of electrical infrastructure costs can have optimal combinations.

An additional benefit of these remote systems is the full accountability of actions. This translates to directly realizing the full benefits of keeping the house a little warmer during the summer where this additional peak demand would require that the entire system be scaled to a large size.

By comparison, for residences connected to a grid, the increased use of the air conditioner typically only translates to paying for more kWh of electricity. The real cost of that air conditioning is both more kWh of electricity and electricity that costs more per kWh because of the additional facilities needed to produce peak-demand electrical power.

The full accountability for actions also translates to full benefit for good choices. Those good choices could include solar thermal for hot water heating and solar photovoltaic to create a more reliable supply of energy. In this approach, the merits of different options will emerge and result in sustainable industries that can evolve constantly and increase in impact.

As with wind power, solar power is available at the discretion of nature; a cloud can shut down the energy collection of a solar photovoltaic cell. Backup power is a major cost and is needed to provide power during night and when the skies are cloudy. And as with wind power, niche markets exist where solar power is cost-effective. Improved technology, both on photovoltaic cells and backup power, will expand the applications where solar power can be cost-effectively applied.

COGENERATION

Communities can realize advantages comparable to the remote farmhouse examples of the previous section. In these communities, there would be a need for distribution of power to houses, but the costs of powerlines to the community may be avoided. In the United States, communities already have powerlines to the national grid, and so, this is not likely to occur. However, as incremental additions are needed, the full benefit of wind power scenarios such as those summarized in Table 9.4 could be realized.

In these communities that are serviced by natural gas pipelines, there can be opportunities for cogeneration. In cogeneration, the heat that goes out of the radiator and the exhaust of an engine are used for heating and cooling. The heat can be directly used for space heating and hot water heating. The heat can be used to generate air conditioning for cooling cycles like the ammonia cycle.

Hot water heaters rated at recovering greater than 90% of the heat of natural gas combustion are readily available on the market. If the natural gas is burned in an engine, 30% would go to the mechanical power, 63% would be converted to heat, and 7% would be lost. Here 63% of the heating value is realized for heating as compared to 90%. This translates to the price of 5 ¢/kWh natural gas being reduced to 1.5 ¢/kWh for electricity. Here the generator not only provides backup electrical power, it also provides low cost electrical power. Useful applications would be limited to where the heat can be easily distributed such as in a downtown network of public buildings or a downtown hotel.

Such cogeneration systems could be the foundations for many useful applications in small communities, including: (i) optimal combinations with renewable energy sources, (ii) backup power key public buildings such as schools or hospitals, and (iii) peak demand power generation that could be used to avoid purchasing of costly peak demand power from the grid (as can be an option for municipalities). Communities could have multiple of these units in place; it is less expensive to tap into an existing natural gas line infrastructure than it is for a similar distribution of hot water.

These cogeneration systems are being considered and prototyped for natural gas fuel cells. Since the waste heat can be so effectively used for heating, the economics are just as good for natural gas engines.

THIRD-WORLD ELECTRIFICATION

Figure 9.2 summarizes electricity usage in Africa [4]. While the US consumption is over 13,000 kWh per year, half of Africa consumes less than

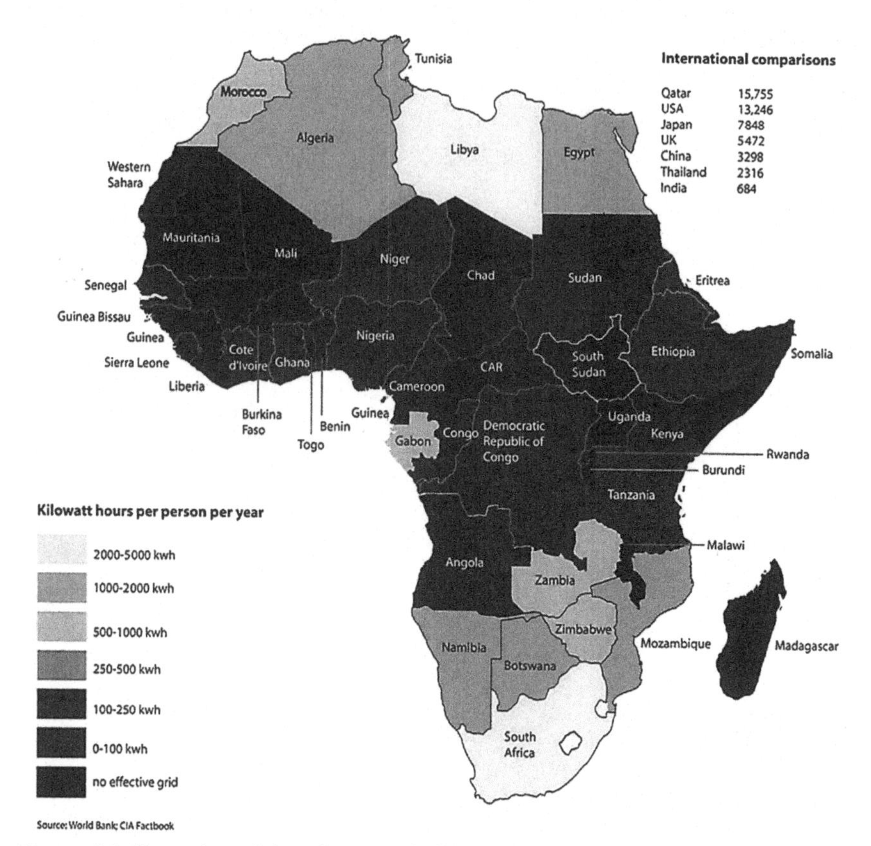

Figure 9.2 Illustration of electrification of Africa in 2013. Available from CIA Factbook.

250 kWh per year. At places of low consumption, bare-essential home facilities like lighting are provided by batteries that are hand carried to charging stations.

It is in this environment where the building of an electrical grid is not the best solution. It is in this environment where lower cost batteries will find ready and eager markets and small-scale wind turbines can be mass produced to achieve even lower costs. Many of the African countries have remote natural gas, and it is here where PHEV vehicles will transform both transportation and the electrical infrastructure.

REFERENCES

[1] Moné C, Smith A, Maples B, Hand M. Cost of wind energy review. National Renewable Energy Laboratory, Technical Report NREL/TP-5000-63267, February 2015. See <http://www.nrel.gov/docs/fy15osti/63267.pdf>; 2013.
[2] See <http://www.awea.org/>.
[3] See <http://www.fpl.com/faqs/underground.shtml>.
[4] See <http://www.globalconreview.com/markets/electrification83737367363636363636363636-africa/>.

CHAPTER 10

Strategies and Critical Paths to Sustainability

Contents

TAXES AND SOCIAL COST

Adam Smith (1723–90) is cited as the "father of modern economics," and his approaches to national economic strategies were and remain a strength that helped mold the United States into the great nation it is today. The foundation of that philosophy is free trade. However, as with many good philosophies that were the right approach at the right time, for some entities this concept of free trade has evolved to the single issue of tariff-free trade. And, Adam Smith's free trade is not and never should be considered in the simplified version of tariff-free trade.

Sustainable Power Technologies and Infrastructure.
DOI: http://dx.doi.org/10.1016/B978-0-12-803909-0.00010-5

Adam Smith identified two legitimate exceptions to tariff-free trade that would advance economies: (i) when industry was necessary to the defense of the country, and (ii) when tax was imposed on domestic production [1]. Corporate income taxes are an obvious example of a tax imposed on domestic production. Personal income taxes may be a less obvious example, but, from the perspective of foreign competition, the personal income tax has the same impact as the corporate income tax—so do property taxes, unemployment taxes, dividend taxes, and Federal Insurance Contributions Act (FICA).

A case is not presented that there should be no taxes, but rather, that the last thing any tax structure should do is place domestic manufacturers at a disadvantage to foreign manufacturers for sales in their home country. Alternative to this undesirable impact of income taxes, sales, and consumption taxes are applied without bias to the origin of the item sold, and so, do not place a burden selective to domestic producers.

Today, the US government goes past these biased corporate income taxes, personal income taxes, property taxes, dividend taxes, and FICA. After imposing all these economic barriers, the government then uses incentives specific to certain industries such as corn-based ethanol production for fuel. At the time of Adam Smith, the concept of governments taking in taxes with one hand and either giving out money (or providing tax breaks) to select businesses with the other hand was unheard of. Clearly, this approach is an additional violation of the philosophy of Adam Smith.

The end result of current taxes and selective incentives is a system where lobbyists set the agendas of politicians which in turn set up favorite industries to succeed at the expense of others.

A particularly problematic aspect of the current system of domestic tax and social costs as complemented by incentives for certain industries is that that the default laws place advantage for foreign producers and to those industries receiving incentives.

CURRENCY AS A METRIC FOR COMPARING ALTERNATIVES

Energy policies in Germany provide an extreme example of how incentives impact energy policy. Germany has had a particularly strong antinuclear movement in recent decades. The use of nuclear power decreased from 22.4% in 2010 to 17.7% in 2011. In fact, *Within days of the March 2011 Fukushima Daiichi nuclear disaster, large antinuclear protests occurred in Germany. Protests continued and, on 29 May 2011, Merkel's government announced that it would close all of its nuclear power plants by 2022. Eight of the seventeen operating reactors in Germany were permanently shut down following Fukushima* [2].

A cited advantage of phasing out nuclear and phasing in renewable energy was that it would position Germany for the important and growing renewable energy sector. Here again, a fundamental flaw of supply–demand economics occurred when politics was used to determine a technology.

A hidden cost in the wind power option that is favored in Germany is the cost of backup power when the wind is not blowing. The price for making such single-issue decisions by a company can be high.

Cost alone is the best method to select technologies. The reason is that cost is the opposite of "single issue" decisions. Essentially, all tangible products and activities trade in terms of the dollar (in the United States); and so, the price of an item is impacted by all tangible alternatives on the market. For all practical purposes, costs that are not impacted by incentives and tax policies are the most comprehensive method to select a technology that is opposite the use of a single-issue method.

For example, if a society decides to spend $2 billion on wind power when $1 billion in nuclear power could have provided the power, the society gives up what could have been purchased with the $1 billion of additional expenditures. That $1 billion could have gone to cancer research, providing more youth with college educations, alleviating restaurant taxes, or any of thousands of options. This may be a simplified version of immediate alternatives that can be purchased; however, if the market is given time to adjust, such items as the work force will adjust making options essentially as universal as the currency.

Also, when making decisions on where to invest in technology infrastructure, the up-front cost of that technology is only one factor to consider. Operating costs must also be considered. In the case of wind power, the price of supplying energy when the wind is not blowing needs to be considered. The following describes the levelized cost method used by companies to take all these factors into account.

LEVELIZED COST APPROACH

To allow cost comparisons for capital, fuel, and operation/maintenance, a levelized cost formula provides a basis for comparing annual contributions of each components with Equation 10.1.

$$LC_{total} = LC_{capital} + LC_{fuel} + LC_{O\&M} \tag{10.1}$$

where LC is the levelized cost in \$/kWh of electricity produced, commonly referred to as the busbar cost.

In a levelized cost comparison, the cost of providing electrical power to the grid allows the capital costs paid prior to startup to be directly compared to the fuel, operating, and maintenance costs paid during power production.

The recent comparison of electrical costs by Ilten [3] provides a basis for discussion and sensitivity analyses. Table 10.1 compares Ilten's nuclear base case to his base cases for coal and natural gas. The fuel costs are further clarified by Table 10.2.

Table 10.1 concisely and accurately conveys those factors that, on the average, give nuclear, natural gas, and coal their unique competitive advantages in the United States.

Natural gas has an advantage due to reduced capital costs. When natural gas prices are about \$4/MBTU, coal is the better option for base load power generation. For seasonal power plants, natural gas can be a better option than coal at gas prices higher than \$4/MBTU due to the low-capacity factors of these facilities.

Table 10.1 Summary of US busbar costs

Case	Plant type	Capacity factor[a]	Capital	O&M	Fuel[b]	Total
Nuclear	Open cycle	80%	3.55	1.03	0.68	5.26
Natural gas	Combined cycle	80%	0.93	0.32	3.94	5.19
Coal	Pulverized	80%	2.53	0.60	1.05	4.18

All costs are in ¢/kWh for base load power generation. Values are from the report Ilten except the price of natural gas has been adjusted to reflect trends since 2003. A 25-year plant life and 10% discount factor are assumed.

[a]A capacity factor of 80% was used rather than 70% to be more internally consistent overnight capital costs estimated later in this chapter.

[b]For natural gas, a combined cycle plant efficiency of 50% is assumed. For case 1, this translates to 3.94 ¢/kWh on a natural gas (not busbar) basis. Here, \$0.0394/kWh-e × 0.53 kWh-e/kWh-f × 2.778 kWh/0.009486 MBTU results in a fuel cost of \$6.12/MBTU for natural gas.

Table 10.2 Fuel cost basis for Table 10.1

Case	Fuel cost ($/MBTU)	Efficiency (fuel to busbar)	Busbar fuel cost (¢/kWh)
Nuclear	0.62	31%	0.68
Natural gas	6.11	53%	3.94
Coal	1.29	42%	1.05

Coal has lower fuel costs than natural gas and lower capital costs than nuclear. This combination makes coal a better long-term investment than natural gas and it has a lower capital risk than nuclear. In recent years, the ability of coal to compete with natural gas for new power plant construction has changed with changing prices of natural gas. Natural gas prices tend to fluctuate wildly; coal prices tend to be rather stable near $1.29/MBTU. In 2006, prices of natural gas of $6, and even $12, per MBTU gave favor to coal. With new gas from shale fracking, prices of $3–6 (2013) gave natural gas favor and resulted in losses of coal-based to gas-based electrical power. Expansion of wind power with natural gas backup has become particularly popular since 2010.

Nuclear power's advantage is the abundance of uranium at low cost. A primary premise of this book is available; inexpensive and abundant nuclear fuel offers sustainable power generation. The advantage of near-zero greenhouse gas emissions strengthens the case for nuclear. To realize its potential, the levelized capital costs of nuclear will need to decrease from the $5.26 per kWh to values comparable to coal at $4.18 per kWh. Additional cost reductions can be realized by increasing power cycle efficiency, which can reduce both fuel and capital costs.

CAPITAL COSTS

Base Case Assumptions

The levelized capital costs of Equation 10.1 and Table 10.1 take into account the actual dollars spent to build the facilities (K in dollars, over-night capital), the discount factor (r in percent, used in place of the borrowing) used to represent costs in terms of a reference year, and the construction time (c in years). Consequently, reductions in each of these factors can reduce the capital costs.

By selecting the reference year as the first year of electrical power production (year one), the levelized capital cost is calculated by first converting the overnight capital to a capital cost discounted to year one (I in dollars) and then dividing this by the lifetime electrical power generation of the plant discounted to year one.

Equation 10.2 provides the method of calculating I including an approximation that assumes an even expenditure of overnight capital for each year of construction.

$$I=\sum_{t=1} S_t K(1+r)^{T-t+1} \approx K\sum_{t=1}(1+r)^{T-t+1}/T \qquad (10.2)$$

where the summation is from 1 to T the construction period in years and S_t is the percent of overnight capital spent each year. This approximation assumes that the overnight capital expenditure is the same each year.

Equation 10.3 relates the levelized capital cost ($LC_{capital}$ in \$/kWh) to the total overnight capital costs (K, in dollars).

$$LC_{capital}=I\bigg/\left[FE\sum_{t=1}(1+r)^{t-1}\right] \qquad (10.3)$$

where the summation is over the number of years of electrical power production, F is the capacity factor, and E is the theoretical maximum amount of power that can be produced the power plant. For a 1 GW by facility, E is 8.76E9 kWh. ($E = 1$ Gw × 365 day × 24 h).

Ilten's estimate of $LC_{capital}$ for the once-through nuclear facility is 3.55 ¢/kWh (see Table 10.1). This is consistent with F[1] = 80% (capacity factor), $r = 10\%$, 25-year production life, and a construction time of 7 years. In equation form, this is \$0.0355/kWh = 1.491 K/(9.89 × 0.80 × 8.76E9) or K = \$1.65 billion for a 1 GW facility. With these assumptions the $LC_{capital}$ of 3.55 ¢/kWh corresponds to an overnight capital cost of \$1650/kW.

Table 10.3 provides the overnight capital costs and assumptions for the levelized capital costs provided in Table 10.1. Values of the total overnight capital costs reported from other sources are also included and substantiate that the values are reasonable and consistent. The assumptions and values listed in Table 10.3 provide the base case and the basis for discussions of factors that impact the $LC_{capital}$ portion of projected nuclear busbar costs.

[1] The calculations by Ilten could not be replicated, namely with the use of a 70% capacity factor did not result in an overnight capital cost of \$1853/kW. As a compromise, an 80% capacity factor was used to achieve an overnight capital cost of \$1830.

Table 10.3 Summary of overnight capital costs (K) with assumptions that link them to levelized capital costs ($LC_{capital}$). OC is Open Cycle, CC is Combined Cycle, P is Pulverized

Case	$LC_{capital}$ (/kWh)	T (yr)	F[a]	K (/kW)	Comparative K (/kW)[22]
Nuclear OC	3.55 ¢	7	80%	$1650	$1600
Natural gas CC	0.93 ¢	3	80%	$531	$590
Coal P	2.53 ¢	4	80%	$1374	$1350

Assumes a 25-year production life and 10% discount factor.
[a]A capacity factor of 80% was used rather than 70% to be more internally consistent overnight capital costs estimated later in this chapter.

Table 10.4 Summary of parameters to be varied in sensitivity study

Parameter	Base case	Parametric study
Reductions in construction time (years)	7	4
Decreases in overnight capital costs ($/kW)	$1650	$1365
Selective loan guarantees (% interest)	10%	5%
Increases in fuel efficiency	31%	47%
Improved reprocessing technology ($/kWh)	0.68	0.40
State subsidies ($/kWh)	0	Tax neutral

Parameters Impacting Capital Cost

Clearly, the clean burning nature of natural gas gives natural gas an advantage with low capital costs and low risk if natural gas prices are reasonable. Figure 3.4 (Chapter 3) shows how natural gas and wind power have gained importance due to recent lower prices in natural gas and the synergy of wind and natural gas power. Provided the wind power can be maintained without excessive maintenance, the grid is more diverse and stronger as a result of this. The relation between coal and nuclear is more complex.

Comparing coal to nuclear, variations in years of production, discount rate, and capacity factors would arguably be applicable for both coal and nuclear options. Applying the same changes to assumptions would change the levelized cost of coal and nuclear about equally. Therefore, nuclear power would gain only incremental advantages over coal—incremental advantages that are within the accuracy of the calculation. The parameters of greatest impact in a sensitivity analysis are those that uniquely impact nuclear power.

Table 10.4 summarizes factors that uniquely impact nuclear over coal with reductions in construction time being particularly important.

Standardized Designs

Standardized and preapproved nuclear power plant designs could reduce the construction times from 7 to 4 years for new nuclear power plants. The value of 4 years compares with 3.4 years, the minimum construction time for existing nuclear power facilities in the United States, and 5.3 years for the average construction times of facilities since 1993. The summation term of the approximation of Equation 10.2 gives the effect of this change in time. For a 10% discount factor (interest rate), this term changes from 1.491 to 1.276 for a 14.4% decrease in the levelized capital costs, or a 0.51 ¢/kWh decrease.

Use of a common site for construction of small modular reactors (SMRs) could reduce the time from decision to install to operation down to 1 year. This is a game-changing technology.

Improved Reactor Designs

Ilten summarizes capital cost estimates for a number of alternative nuclear reactor designs as estimated by SAIC (Science Applications International Corporation), Scully, and Energy Information Agency (EIA). In each case at least one design option (pebble bed, advanced technology, modular helium reactor) gave a cost at or below $1365/kW while maintaining a similar nuclear fuel cost. A decrease from $1650 to $1365 in the overnight price of a new reactor design translates to a 17.3% decrease in the levelized capital cost, or a 0.614 ¢/kWh decrease.

Guaranteed Loans

Providing guaranteed loans can be a good option for a federal government to stimulate the development of a first-of-a-kind facility. If successful, investments in this first-of-a-kind nuclear facility can offer the public a greater benefit than a single corporation. It is the goal of the federal government to promote actions that benefit of the public. For the corporation building the facility the loan guarantees would offset the risk associated with new technology and the increased construction time building a new design. A loan guarantee would correspond to the corporation qualifying for a low interest rate. Decreasing interest from 10% to 5% during construction, Equation 10.2 summation changes from 1.491 to 1.221 for a 18.1% decrease in the levelized capital costs, or a 0.64 ¢/kWh decrease.

Efficiency

The low efficiency of nuclear power generation is just above peaking gas turbines in the mix of electrical power technologies. The inefficiency source is the low maximum temperatures of pressurized water reactors (about 340°C) and the limits imposed by the Rankine cycle. A 50% increase in this efficiency is possible with a corresponding 33.3% reduction in fuel consumption. This would reduce the fuel cost from 0.68 ¢/kWh to 0.46 ¢/kWh or a decrease of 0.22 ¢/kWh.

Improved reactor designs with higher operating temperature can more efficiently produce power. This translates to a smaller reactor to provide the same power output. Most of the 0.614 ¢/kWh decrease in cost associated with improved design is due to the smaller reactor core made possible by increased efficiency.

Reprocessing

Reprocessing spent nuclear fuel is desirable because it can reduce the mass of radioactive materials to store/landfill by about 96% with the assumption that all fission products must be handled as radioactive waste and other "contaminated" materials can either be reused or disposed of as low level waste with lower disposal cost. Reprocessing consists of several physical and chemical processing steps that could be low in cost. France does reprocess fuel in its closed cycle nuclear processes where fuel cost is 0.90 ¢/kWh compared to once-through technology in the United States at 0.68 ¢/kWh. The once-through technology does not include the undetermined cost of spent fuel disposal.

Despite the years of experience France has with reprocessing, the cost is not near the bottom of this cost curve. One option that has not been explored is the relaxing of the metal purity of the reprocessed fuel. Rather than removing essentially all fission products and actinides, the fuel could be processed to remove only fission products and concentrate the fissile materials. Reactors could then be designed to use reprocessed fuel that contains some of the more difficult to remove fission products. This approach would reduce fuel costs about 0.28 ¢/kWh.

If the reprocessing is performed at the nuclear power plant site, the states realize an elimination of the 0.68 ¢/kWh that would otherwise go to importing of uranium fuel to provide power in the state.

State Incentives

It is within the prerogative of states to provide incentives for electrical power generation technologies that generate state tax revenue and provide quality jobs. A 1 GW power plant produces about 7.9E9 kWh of electrical power per year (90% capacity factor). At a natural gas busbar fuel cost of $0.0394/kWh this translates to $310 million per year in dollar flow from the state for just one natural gas combined cycle power plant that produced base load electrical power.

A nuclear power plant with the same capacity would pay $54 million per year for uranium fuel. The $256 million reduction in fuel cost represents maybe 10,000 jobs in a state with a nuclear power plant rather than a natural gas plant.

Capital investments in a state also benefit the state. The investment in the construction of a new power plant comes from corporate funds—funds that otherwise are invested anywhere. When used for construction in state, about one-third of the capital costs are for site labor and site materials [4]. Adding to this engineering services, secretarial support, and cash flow from the out-of-state workers, the total capital expenditures contribute to local state economies can be 40% or more.

The cash flow from new construction and the money paid for fuel can have a significant impact on the state economy. In a global economy where more and more manufacturing is performed overseas, use of electrical power that results in local job creation is a good option.

Table 10.5 compares the different electrical power generation options from the perspective of the cents per kilowatt-hour of electrical power produced that leave the state.

Table 10.5 Impacts of power plant options on state cash flow from a state that imports natural gas, coal, and uranium

Case	Total & (imported) capital	Total & (imported) O&M	Total & (imported) fuel[a]	Total state import in ¢/kWh
Nuclear	3.55 (−2.13)	1.03 (−0.206)	0.68 (−0.68)	−3.02
Natural gas	0.93 (−0.558)	0.32 (−0.064)	3.94 (−3.94)	−4.56
Coal	2.53 (−1.518)	0.60 (−0.12)	1.05 (−1.05)	−2.69
Reprocessed nuclear	4.05 (−2.43)	1.43 (−0.286)	0 (0)	−2.72

All are busbar in ¢/kWh. The state impact numbers assume 40% of the capital expenditures and 80% of the O&M expenditures stay in the state where construction occurs.

[a]For natural gas, a combined cycle plant efficiency of 50% is assumed. For case 1, this translates to 3.94 ¢/kWh on a natural gas (not busbar) basis. Here, $0.0394/kWh-e × 0.53 kWh-e/kWh-f × 2.778 kWh/0.009486 MBTU results in a fuel cost of $6.12/MBTU for natural gas.

Considering the total impact, a 1 GW coal-fired power plant places about 1.87 ¢/kWh more money into the state economy than a 1 GW natural gas combined cycle facility (natural gas assumed to be $6.11/MBTU. This translates to about $148 million dollars per year in avoided trade deficit. If the 2005 natural gas prices of $12.22/MBTU were used, this differential flow of cash from the state rises to an astronomical $458 million dollars per year.

These calculations indicate that the building of a power plant that has a lower fuel cost component in the total per kWh busbar cost is about the equivalent of attracting a large manufacturer to the state. A power plant would have higher paying jobs than an assembly line. Furthermore, a state has more control over the fuels used to produce electrical power than it has over selecting the next manufacturing plant to build.

The data in Table 10.5 includes the option of a nuclear facility with on-site fuel reprocessing. The costs for this option are based on the 0.9 ¢/kWh costs for reprocessed fuel as a capital and maintenance/operating expenditure. On-site reprocessing reduces state cash export by about 0.3 ¢/kWh representing a $24 million dollar impact from a 1 GW facility.

Because the busbar costs of nuclear power are more in Table 10.5 than for coal, nuclear tends to export more dollars than coal even though the fuel costs for nuclear are less. Table 10.6 shows a similar analysis for nuclear technologies proposed having busbar cost parity with coal. In this comparison, nuclear reduces the export of cash flow relative to coal—a direct annual impact of $25–59 million dollars on the state cash flow.

Table 10.6 Impacts of power plant options on state cash flow for targeted nuclear processes

Targeted process	Total & (imported) capital	Total & (imported) O&M	Total & (imported) fuel[a]	Total state import in ¢/kWh
Targeted nuclear	2.47 (−1.482)	1.03 (−0.206)	0.68 (−0.68)	−2.37
Targeted reprocessed nuclear	2.75 (−1.65)	1.43 (−0.286)	0 (0)	−1.94

All are busbar in ¢/kWh. The state impact numbers assume 40% of the capital expenditures and 80% of the O&M expenditures stay in the state of construction.

[a]For natural gas, a combined cycle plant efficiency of 50% is assumed. For case 1, this translates to 3.94 ¢/kWh on a natural gas (not busbar) basis. Here, $0.0394/kWh-e × 0.53 kWh-e/kWh-f × 2.778 kWh/0.009486 MBTU results in a fuel cost of $6.12/MBTU for natural gas.

If components of the nuclear power plant are produced in a state it would reduce the amount of "imported capital," and nuclear would become the winner in the race to reduce the import costs of electrical power. Here, SMRs can make a big impact in addition to the impact on reduced construction times.

Multiplying factors from 4 to 6 are often used to represent the true economic impact of avoided trade deficits. Multiplying factors and reduced trade deficits typically do not translate to corporate incentives to make investments that are good for the state. One approach that improves corporate incentives is for states to pass incremental tax revenues to the corporations in the form of subsidies. An estimate of the incremental difference in state personal income tax revenues resulting from the different power plant options is 10% of the differential flows of cash from the states. It is both reasonable and within the power of states to pass on these tax revenues to corporations. For once-through nuclear versus coal, these translate to about 0.025– 0.059 ¢/kWh.

For all of the estimates of per kWh cash flow from states, it is assumed that 60% of the capital costs go out of state. These include an average of 37.3% factory equipment costs. Most of the remaining 22.7% is for the administration by the primary construction contractor and out-of-state workers. For those states with commercial power plant construction, nuclear fuel reprocessing, or power plant equipment, the 60% of capital costs assumed to leave the state can be reduced. An additional 1 ¢/kWh or more could remain in the state for fuel reprocessing or power plant construction if firms were located in the state.

Sensitivity Analysis

While coal has cost advantages relative to traditional nuclear power, there is potential for technology and regulatory changes to reduce the costs for nuclear power plants. Table 10.7 summarizes the various parameters on the levelized cost of production for nuclear power. The parameters apply to nuclear power over coal.

The greatest reduction in busbar costs would come from new technology that reduces the cost of a nuclear power plant by increasing the energy efficiency of the facility. Increased efficiency can decrease the capital costs (at constant electrical power generation) and decrease the levelized fuel costs. Combined, these can reduce the cost of nuclear power by about 0.83 ¢/kWh. These savings are cumulative and add to the savings from reducing construction times.

Table 10.7 Summary of parameters to be varied in sensitivity study of nuclear versus coal

Parameter	**Decrease in LC_{total} (¢/kWh)**
Improved thermal efficiency with improved design that also decreases capital costs	0.61 + 0.22
Loan guarantees (reduce construction times)	0.64 (0.51)
Decreases in overnight capital costs alone	0.61
Increases in fuel efficiency (impact on fuel costs)	0.22
Improved reprocessing technology	0.28
State subsidies (upper value)	0.10

The second greatest reduction can be achieved by reducing construction costs with pre-certified designs (e.g., SMRs). For first-of-a-kind facility, the longer construction times could be covered by guaranteed loans. These translate to 0.51–0.64 ¢/kWh. These construction costs are reduced as more pre-certified are built.

Additional incremental gains are for nuclear power from 0.10 ¢/kWh to 0.28 ¢/kWh through states returning incremental tax revenues to corporations and with improved reprocessing technology.

CASE STUDIES

Scenario 1: 40-Year Production Life with 4 Year Construction Times

A shift from a 25-year plant life to a 40-year plant life is reasonable based on operating experience by nuclear power plants currently in use. With this basis and a 4-year construction time for both nuclear and coal facilities, Table 10.8 was prepared to compare the busbar costs. For nuclear, the $LC_{capital} = 1.276 \times \$1.65E9/(10.67 \times 0.80 \times 8.76E9)$ or $LC_{capital} = 2.82$ ¢/kWh for a 1 GW facility. For coal, $LC_{capital} = 1.276 \times \$1.65E9/(10.67 \times 0.80 \times 8.76E9)$ or $LC_{capital} = 2.35$ ¢/kWh. Reducing construction times to 4 years and assuming a 40-year plant life reduced the busbar price difference between nuclear and coal by about half, but coal still has the advantage.

Scenario 2: A $1365/kW Nuclear Power Plant at 47% Thermal Efficiency

New generation nuclear power plants are projected to have overnight capital costs similar to coal and thermal efficiencies of about 47% which

Table 10.8 Scenario 1 busbar costs for 40-year plant lives, 10% discount factor, and 4-year construction times for both nuclear and coal

Case	Plant type	Capacity factor	Capital	O&M	Fuel	Total
Nuclear	Open cycle	80%	2.82	1.03	0.68	4.53
Coal	Pulverized	80%	2.35	0.60	1.05	4.00

Values are in ¢/kWh.

Table 10.9 Scenario 2 busbar costs for 40-year plant lives, 10% discount factor, and 4-year construction times for both nuclear and coal

Case	Plant type	Capacity factor	Capital	O&M	Fuel	Total
New nuclear	Open cycle	80%	2.35	1.03	0.46	3.84
State Import			*− 1.41*	*− 0.206*	*− 0.46*	*− 2.08*
Coal	Pulverized	80%	2.35	0.60	1.05	4.00
State Import			*− 1.41*	*− 0.12*	*− 1.05*	*− 2.58*
New nuclear	Closed cycle	80%	2.75	1.23	0	3.98
State Import			*− 1.65*	*− 0.246*	*0*	*− 1.90*

New generation nuclear power plant has overnight capital cost similar to coal with 47% thermal efficiency. Values are in ¢/kWh.

is slightly better than coal. Table 10.9 summarizes the levelized costs for this scenario. Here, nuclear power has a cost advantage that is independent of plant life and discount factor.

The busbar cost of 3.84 ¢/kWh is consistent with a series of nuclear options reported by Ilten [5] with costs between 3.6 and 4.1 ¢/kWh. Ilten's lowest projected cost for coal facilities is 3.7 ¢/kWh.

The results in Table 10.9 indicate that nuclear can be less costly than coal, but, up to 0.68 ¢/kWh in cash flow leaving the state can be eliminated with nuclear power versus coal. This translates to a direct $54 million per year per 1 GW facility and indirect impact (economic multiplying factor) in excess of $200 million per year to the state economy.

COSTS OF REPROCESSING

A relatively firm estimate of the cost difference between reprocessed nuclear fuel and new fuel is 0.22 ¢/kWh. This is the difference between the cost of once-through US fuel at 68 ¢/kWh and closed cycle French fuel at 90 ¢/kWh.

Table 10.10 Estimates of capital costs for fuel reprocessing and fuel fabrication

Process	Reprocessing facility	Fuel fab. facility
PUREX	8.0	2.0
UREX+	6.0	2.0
UREX/PYRO	6.0	3.0
PYROX	7.0	3.0
Advanced aqueous process	4.0	2.0

Numbers are billions of dollars and are for 2000 metric tons per year (spent fuel generation rate in 2005).
Data from US DOE Office of Nuclear Energy, Science and Technology Advanced Fuel Cycle Initiative (AFCI) Comparison Report [6].

As shown in Table 10.10 reprocessing facility and fuel fabrication capital costs are available from the US DOE to supplement this estimate.

The combined reprocessing and capital cost estimates range from $6 to $10 billion for capacities to handle all US commercial waste. If policy is to store the waste for 30 years before reprocessing, the 2000 metric tons per year capacity is enough to handle US commercial waste for the next 30 years.

In 2002, 103 operating nuclear power plants produced 790 billion kWh of electrical power.

Assuming overnight construction of the reprocessing and fuel fabrication facilities, the capital costs produce levelized busbar costs of 0.76–1.270, and 0.03–0.05 ¢/kWh based on the power produced generating the electricity. Since the U-235 content of the reprocessed fuel is about the same as natural uranium, it offers no premium fuel value. The premium fuel value for the plutonium presents at about 0.9% in the reprocessed fuel. The adjusted levelized cost of the plutonium from reprocessing is 0.092–0.154 ¢/kWh. This is greater than the 0.9 ¢/kWh reported by the French for closed-cycle fuel reprocessing.

The operating costs for 25 years of operation are summarized in Table 10.11. Assuming a 0% discount factor, these costs translate to 0.17–0.072 ¢/kWh on the basis of electricity produced generating the waste or 0.21–0.49 on the use of reprocessed Pu-239 to enrich uranium fuel for power production. These costs are less than the price of new fuel at 0.68 ¢/kWh. However, The US Generation IV Implementation Strategy, FY2003 only provides a $12 billion fuel sale credit versus the $15.2–23.0 billion for Operating and D&D.

Table 10.11 Estimated costs for operating reprocessing and fuel fabrication facilities for 25 years

Process	D&D	Operating
PUREX	3.0	20.0
UREX+	2.4	14.0
UREX/PYRO	2.7	12.5
PYROX	3.0	14.0
Advanced aqueous process	1.8	12.5

Data from the US Generation IV Implementation Strategy, FY2003.

The reactor fuels fabricated from these facilities would be mixed oxide fuels. The fissile material content of the U-235 + Pu-239 from the reprocessing could be topped off with weapon's grade plutonium or U-235.

The highly enriched uranium extracted from nuclear weapons (HEU) agreement signed between the US and Russian Federation (1993) provides for the purchase of 500 metric tons of highly enriched uranium between 1993 and 2013 [7]. If this uranium (>80% U-235) were available and used to enrich the 0.9% Pu-239 and 0.8–1.s1% U-235 in the heavy metals of reprocessed fuel (to attain 3.3% fissile material) it would be sufficient to enrich 29,000–36,000 metric tons of recycled fuel. At 2000 tons per year, this translates to up to 18 years. No new uranium would need to be mined or enriched to prepare this fuel.

In 2007, approximately 30 years of spent fuel will have been accumulated. During this time, 0.1 ¢/kWh has been collected for the purpose of waste disposal. Thirty years time 790 billion kWh/yr times 0.1 ¢/kWh translates to $23.7 billion. While much of this has already gone into the pockets of attorneys, ample is still available for the $10 billion investment into reprocessing and fuel fabrication.

If these funds were used to build these facilities, the cost of reprocessed fuel would need to cover the operating and distribution costs about 0.2 ¢/kWh—reprocessed fuel would be sustainable cost lower than new fuel. Strategies exist to permanently avoid long-term storage by separating the unstable fission products from the unstable isotope and eventually transmuting these materials. The objective of waste management would be met.

In fact, there should be sufficient funds available from the $23.7 billion to cover the incremental costs associated with the construction of first and second of their kind fast-spectrum reactors.

The present course based on once-through fuel would fill the Yucca Mountain repository with waste generated by 2015. The estimated life cycle cost for a second repository similar to Yucca Mountain would be $50 billion. The 2003 US Generation IV Implementation Strategy places the cost of the once-through repository approach to be $33–43.2 billion more than initiating reprocessing.

BARRIER TO THIRD-WORLD SOCIETIES

For the nuclear engineering examples, two barriers to the evolution of the nuclear industry are lack of demand for SMRs due to the lack of certainty in their availability and the unwillingness of the government to use moneys set aside for disposal of spent nuclear fuel to reduce "waste" by reprocessing of that spent nuclear fuel. A path forward based on these two technologies would ultimately lead to:

1. Much improved economics for nuclear power
2. A solution to the spent nuclear fuel problem
3. The tapping into a fuel source that would last multiple centuries
4. A sustainable path where advances in technology can be commercialized.

For the wind power example of Chapter 9, it was clearly identified that published costs for wind power are much lower than the actual cost of wind power because the published costs do not include the cost of backup power when the wind is not blowing. As a result of these flawed economics, wind turbines are commercialized in a manner that deters better options for grid power while attention is diverted from applications that can lead to sustainable paths of high impact in remote locations. More specifically, if all the lobbyist efforts devoted to federal incentives for large-scale wind farms for grids were diverted to smarter application at remote locations and for incrementally increasing needs for small (remote) communities, society would receive much greater benefit.

For the Hyperloop example, the goals (LA to San Francisco in 35 minutes) of the well-publicized system are well attainable, but the Hyperloop suffers critical technical flaws and does not have a reasonable critical path to attain the goals. Alternatively, the Terreplane system (see TerreplaneSystem.com) has the upside potential of the Hyperloop, has further applications, and does have a reasonable critical path to sustainability.

As fourth example, consider how the United States is investing significant research funds on topics of algal biodiesel, photocatalytic hydrogen, and carbon dioxide sequestration. Each of these has fatal flaws that will likely result in the technologies never becoming commercially sustainable.

A conclusion from these four examples is that corruption and ignorance (specifically, not considering the whole picture) are prevalent in the making of key decisions. Decisions are being made based on half truths, information from entities with conflicts of interest, and in the interests of those who pay lobbyists rather than the interests of society. This is the easy answer, but what it really is, is a "fifth" example of a very real problem.

The real problem in each of these, now five, examples is that in each case decisions are made without all the information. It is human nature to pursue answers to challenging questions in a procedure that typically consists of: (i) placement of effort past a self-defined critical threshold of effort; (ii) pursue the answer until an final answer is attained which usually substantiated preconceived ideas on what the answer should be; and (iii) pursue a validation that the final answer is correct. In the case of the NREL estimates for the cost of wind turbine systems, all of these steps were attained and the cost estimates were not only reasonable, but of rather good quality. But that good and reliable method for estimating the price of wind turbine systems is only part of what is needed to make meaningful conclusions about the best choices for power generation; the rest of the information that is needed includes the identifying that wind systems need backup power for when the wind is not blowing and the cost of the backup power substantially changes the economic viability relative to alternatives.

For each of these five examples, the selected technology delivers to the analyst what the analyst wants to hear. The conclusions are correct in the limited domain within which it was evaluated, but the conclusions are faulty from the perspective that real-life application is much more complicated in what impacts performance, in what is needed for achieving the desired performance, and in what competitive technologies have to offer.

To overcome the barriers to achieving sustainable solutions for each of these as well as a host of issues impacting third-world societies, improved and more rigorous economic analyses are needed. This includes demands, by all impacted, that these improved analyses be performed. This includes a systematic advancement of both the science of economic analyses and the standards of analyses that are required before strategic decisions are made.

In the political arena, it is true that politicians frequently yield to the request of lobbyists. It is also true that these lobbyists are well funded and are able to use that funding to provide convincing arguments in favor of their clients. If in every one of these instances of lobbying, there was a more rigorous and more respected/accepted economic analysis that pointed to a different solution, the frequency of politicians yielding to lobbyist would reduce. This is especially the case when such "more rigorous" analysis identified that the lobbyists were advocating regulations that were ultimately bad for the constituencies.

To some extent, the US EIA and the International Energy Agency (IEA) provide raw information and analyses. They have made major strides in the way of providing readily available data. What is missing is the systematic critical and comprehensive comparisons as related to how alternatives impact specific country's economies.

Engineering design regularly uses profitability analysis similar to the levelized cost analysis previously presented. The most important part of that profitability analysis is cost estimation. A majority of cost estimation methods rely on the publication of one paper in 1969 by Guthrie [8]. This author decided to use his experience from years of service in industry to advance this area of engineering design. It was not a trivial step. The point is that key aspects of profitability analysis and resulting comparison of alternatives is not a science, it is a constant work in progress that needs to be constantly updated. For all but the few working under the umbrella of a major corporation, it can be very difficult to get updated costs on items such as the price of lithium metal. The NREL cost summaries on wind turbine prices are a rarity to be found in open literature.

As a starting point, methods of estimating costs for capital equipment and for obtaining costs of materials need to be openly standardized in the open literature. Governments need to take stewardship in keeping the cost estimate methods up to date. And finally, the scientific community needs to keep the methods honest through constant peer review. But this is only the beginning.

The science of economic analysis needs to be advanced. Specifically, it needs to be advanced to the point where it is a "simulation" art that evaluates multiple scenarios such as those summarized in the tables of Chapter 9. The simulations need to be comprehensive in the alternatives considered as well as in providing full performance (e.g., backup power for wind turbine systems).

These analyses need to go even further; the analyses need to be temporal. The analyses need to include how the available solutions will be accepted/used as a function of time and how the solutions will evolve as a function of time.

Consider the HYPERLOOP system. The final system is extremely energy efficient and has high performance. However, in order for that system to become reality, it needs a viable path to commercial operation including suitable returns on investment. A simulation of the temporal alternatives suggests that systems which can begin sustainable operation with smaller investment, while being able to evolve to the indicated performance, are needed. This is where the TERREPLANE system prevails. The same principles apply to evaluating the path forward for nuclear energy.

Hence, the next evolution that is most needed is not new technology, it is for superior economic analyses. Application of today's best economic tools can be challenging to the dedicated scientist. The needed advances will result in greater challenges to effective utilization of the methods and tools, but it is a needed advance. Ultimately, the time and effort wasted pursuing approaches substantiated by "half truths" is much greater than the time and effort needed to advance and apply the needed economic analyses.

DOMESTIC MANUFACTURING INFRASTRUCTURE

The last few decades in the United States have realized a steady decline in manufacturing jobs in the United States. The primary driver for this decrease is that the manufactured items could be produced at lower costs outside the US tax and wage structures. The decisions were made at corporate levels with little regard to how Adam Smith's advocacy of how free markets can be distorted by such things as domestic taxes and the need for national security. In the words as printed in the Harvard Business Review:

> *Too many American companies base decisions about how to source manufacturing largely on narrow financial criteria, never taking into account the potential strategic value of domestic locations [9].*

The authors proceed to identify that outsourcing or offshoring can impact a company's capacity to innovate. Indeed, domestic manufacturing directly and indirectly impacts the ability for "new innovations" to be manufactured. These could be inventions by employees of the company performing the manufacturing or the local inventors seeking a manufacturer. Crossing international borders to locate a manufacturer can be perilous especially if the foreign country where the manufacturing is located is not covered by patents on the invention.

The industrialization of the United States in the nineteenth and twentieth centuries included the appearance of infrastructure simultaneous to thriving innovation. The hands-on exposure to equipment by workers, the witnessing of equipment by visitors to the factory, and even the superficial observation of the manufacturing in the community can all lead to innovation. Manufacturing brings with it expertise that enters the community through word of mouth, participation in education of youth, and the retiring (or simply leaving) of workers. All of these contribute to innovation.

Certainly, manufacturing infrastructure is not the only factor that creates a good environment for innovation. In the United States, there is good basic education system and a sound foundation in patent law. The US patent now has over 200 years of experience with the primary purpose of "promote(ing) the progress of science and useful arts" by rewarding inventors both with acknowledgment of their intellectual contribution and by securing legal rights to prevent others from practicing a patented invention.

The United States simultaneously built infrastructure and created many of machines that were responsible for creating high standards of living. For third-world countries to follow suit, one would think that they would simply have to adopt the machines and methods already invented, and so, with a 20–30 year delay in time their standards of living should follow the course of the United States. Sadly, this is not the case.

This circumstantial evidence goes beyond hypothesis; it validates that manufacturing structure is important to advance innovation. Indeed, most third-world countries resort to importing of manufactured items rather than building a manufacturing infrastructure. The ability to acquire machines and sophisticated electronics is simply not the same as manufacturing those items. Acquiring lacks the components of innovation, education, electrical grid infrastructure, and general dissemination of capabilities and standards that a manufacturing infrastructure brings to a community.

A hypothesis is that for a third-world country to sustainably advance in quality of life, that country must have a healthy manufacturing infrastructure. This is particularly problematic due to the complexity of manufacturing those items that dominate purchases. The automobile emerges at the top of the list of items where purchasing and maintenance dollars are spent.

OVERCOMING COMMERCIALIZATION BARRIERS

For many families, the automobile is the largest purchase made and represents the greatest experienced depreciation. However, the scale of effort needed for competitive automobile manufacturing is beyond what most is reasonably possible for most third-world countries.

A solution is to modularize electric automobile into four or five sections such as: (i) power supply; (ii) chassis; (iii) interior; (iv) body; and

(v) control system. The establishment of standards for plugging the power source into the chassis and control system and for achieving cumulative safety standards is based on safety standards for each component. Multiple manufactures would be possible for each component and would allow assembly of custom combinations at the site of sale. This approach would allow for increased diversification and increased innovation which would be tied to the multiple manufacturing locations.

Here, the use of electrical power as opposed to mechanical "transmissions" make modularization much easier.

It is unlikely that a major automobile manufacturer would open the door to the type of widespread competition that a standardized modularized approach to automobile manufacturing would bring. Such an action would have to be set in motion by an entity like the United Nations with the specific purpose of reducing the cost of third-world countries to enter into the manufacturing arena for many of the high-cost items that are currently imported.

This approach is the identification of a critical path to commercialization.

STRATEGIES AND CRITICAL PATHS

In the discussion of the IEA's summary of the impact of shale oil fracking, it was identified that:

> *The United States is in a strong position to deliver a reliable, affordable and environmentally sustainable energy system, the International Energy Agency (IEA) said today as it released a review of U.S. energy policy. To do so, however, the country must establish a more stable and co-ordinated strategic approach for the energy sector than has been the case in the past.*

A strategy is simply not an emotional or single-issue decision like a knee-jerk decision to phase out nuclear and position for renewables as a reaction to Fukushima. A strategy includes both the consideration of many options and economic analyses that consider all factors. In many cases the technology is already available, but the availability and application of the rigorous economic analyses are lacking.

The results of the strategy are commercialization options that have critical paths to sustainability and reduced cost barriers. Examples of these strategies include remote wind power, the terreplane transit system, SMNRs, and a standardized modular approach to the electric automobile. They take vision, intelligence, and perseverance. The results can transform our world.

REFERENCES

[1] Smith A. Wealth of nations- facsimile of 1776 original. Prometheus Books; 1991; p. 361.
[2] See <http://en.wikipedia.org/wiki/Energy_in_Germany>.
[3] Ilten N. International comparisons of Electricity generation by types & costs. See <http://people.cs.uchicago.edu/~nilten/docs/final.pdf#search='Busbar%20Cost%20of%20Nuclear%20Power'>; August 28, 2003.
[4] The economic future of nuclear power. Table 3.2. A study conducted by The University of Chicago; August, 2004.
[5] The economic future of nuclear power. Table 1.1. A study conducted by The University of Chicago; August, 2004.
[6] U.S. DOE Office of nuclear energy, science and technology advanced fuel cycle initiative (AFCI) comparison report. Published by U.S. DOE; October, 2003.
[7] Report on the Effect the Low-Enriched Uranium Delivery Under the HEU Agreement Between the Government of the United States and the Government of the Russian Federation Has on the Domestic Uranium Mining, Conversion, and Enrichment Industries, and the Operation of the Gaseous Diffusion Plant; December, 31, 2003.
[8] Guthrie KM. Data and techniques for preliminary capital cost estimating. Chem Eng, March 24, 1969;76:114–42.
[9] Pisano GP, Shih WC. Does America really need manufacturing? Harv Bus Rev, March 2012;90(3):94–102.

INDEX

Note: Page numbers followed by "*f*," "*t*," and "*b*" refer to figures, tables, and boxes, respectively.

C

D

E

T

Printed in the United States
By Bookmasters